Student Solutions Manual
for
McKeague's Intermediate Algebra
Seventh Edition

Ross Rueger
College of the Sequoias

THOMSON
BROOKS/COLE

Australia • Canada • Mexico • Singapore • Spain • United Kingdom • United States

COPYRIGHT © 2004 Brooks/Cole, a division of Thomson Learning, Inc. Thomson Learning™ is a trademark used herein under license.

ALL RIGHTS RESERVED. No part of this work covered by the copyright hereon may be reproduced or used in any form or by any means—graphic, electronic, or mechanical, including but not limited to photocopying, recording, taping, Web distribution, information networks, or information storage and retrieval systems—without the written permission of the publisher.

Printed in the United States of America
4 5 6 7 07 06 05

Printer: Globus Printing

ISBN: 0-534-42291-8

For more information about our products, contact us at:
Thomson Learning Academic Resource Center
1-800-423-0563

For permission to use material from this text, contact us by:
Phone: 1-800-730-2214
Fax: 1-800-730-2215
Web: http://www.thomsonrights.com

Brooks/Cole—Thomson Learning
10 Davis Drive
Belmont, CA 94002-3098
USA

Asia
Thomson Learning
5 Shenton Way #01-01
UIC Building
Singapore 068808

Australia/New Zealand
Thomson Learning
102 Dodds Street
Southbank, Victoria 3006
Australia

Canada
Nelson
1120 Birchmount Road
Toronto, Ontario M1K 5G4
Canada

Europe/Middle East/South Africa
Thomson Learning
High Holborn House
50/51 Bedford Row
London WC1R 4LR
United Kingdom

Latin America
Thomson Learning
Seneca, 53
Colonia Polanco
11560 Mexico D.F.
Mexico

Spain/Portugal
Paraninfo
Calle/Magallanes, 25
28015 Madrid, Spain

Contents

Preface

This *Student Solutions Manual* contains complete solutions to all odd-numbered exercises (and all chapter test exercises) of *Intermediate Algebra* by Charles P. McKeague. I have attempted to format solutions for readability and accuracy, and apologize to you for any errors that you may encounter. If you have any comments, suggestions, error corrections, or alternative solutions please feel free to drop me a note or send an email (address below).

Please use this manual with some degree of caution. Be sure that you have attempted a solution, and re-attempted it, before you look it up in this manual. Mathematics can only be learned by *doing*, and not by observing! As you use this manual, do not just read the solution but work it along with the manual, using my solution to check your work. If you use this manual in that fashion then it should be helpful to you in your studying.

I would like to thank a number of people for their assistance in preparing this manual. Thanks go to Rachael Sturgeon and Rebecca Subity at Brooks/Cole Thomson Learning for their valuable assistance and support.

I wish to express my appreciation to Pat McKeague for asking me to be involved with this textbook. His books continue to redefine the subject of intermediate algebra, and you will find the text very easy to read and understand. Good luck!

Ross Rueger
College of the Sequoias
matmanross@aol.com

March, 2003

Chapter 1
Basic Properties and Definitions

1.1 Fundamental Definitions and Notation

1. Written in symbols, the expression is: $x + 5 = 2$

3. Written in symbols, the expression is: $6 - x = y$

5. Written in symbols, the expression is: $2t < y$

7. Written in symbols, the expression is: $x + y < x - y$

9. Written in symbols, the expression is: $3(x - 5) > y$

11. Multiplying: $6^2 = 6 \cdot 6 = 36$

13. Multiplying: $10^2 = 10 \cdot 10 = 100$

15. Multiplying: $2^3 = 2 \cdot 2 \cdot 2 = 8$

17. Multiplying: $2^4 = 2 \cdot 2 \cdot 2 \cdot 2 = 16$

19. Multiplying: $10^4 = 10 \cdot 10 \cdot 10 \cdot 10 = 10,000$

21. Multiplying: $11^2 = 11 \cdot 11 = 121$

23. Using the rule for order of operations: $3 \cdot 5 + 4 = 15 + 4 = 19$

25. Using the rule for order of operations: $3(5 + 4) = 3(9) = 27$

27. Using the rule for order of operations: $6 + 3 \cdot 4 - 2 = 6 + 12 - 2 = 16$

29. Using the rule for order of operations: $6 + 3(4 - 2) = 6 + 3 \cdot 2 = 6 + 6 = 12$

31. Using the rule for order of operations: $(6 + 3)(4 - 2) = 9 \cdot 2 = 18$

33. Using the rule for order of operations: $(7 - 4)(7 + 4) = 3 \cdot 11 = 33$

35. Using the rule for order of operations: $7^2 - 4^2 = 49 - 16 = 33$

37. Using the rule for order of operations: $2 + 3 \cdot 2^2 + 3^2 = 2 + 3 \cdot 4 + 9 = 2 + 12 + 9 = 23$

39. Using the rule for order of operations: $2 + 3\left(2^2 + 3^2\right) = 2 + 3(4 + 9) = 2 + 3 \cdot 13 = 2 + 39 = 41$

41. Using the rule for order of operations: $(2 + 3)\left(2^2 + 3^2\right) = (5)(4 + 9) = 5 \cdot 13 = 65$

43. Using the rule for order of operations: $40 - 10 \div 5 + 1 = 40 - 2 + 1 = 38 + 1 = 39$

45. Using the rule for order of operations: $(40 - 10) \div 5 + 1 = 30 \div 5 + 1 = 6 + 1 = 7$

47. Using the rule for order of operations: $(40 - 10) \div (5 + 1) = 30 \div 6 = 5$

49. Using the rule for order of operations: $24 \div 4 + 8 \div 2 = 6 + 4 = 10$

51. Using the rule for order of operations: $5 \cdot 10^3 + 4 \cdot 10^2 + 3 \cdot 10 + 1 = 5,431$

53. Using the rule for order of operations: $40 - \left[10 - (4 - 2)\right] = 40 - (10 - 2) = 40 - 8 = 32$

55. Using the rule for order of operations: $40 - 10 - 4 - 2 = 30 - 4 - 2 = 26 - 2 = 24$

57. Using the rule for order of operations: $3 + 2\left(2 \cdot 3^2 + 1\right) = 3 + 2(2 \cdot 9 + 1) = 3 + 2(18 + 1) = 3 + 2 \cdot 19 = 3 + 38 = 41$

59. Using the rule for order of operations: $(3 + 2)\left(2 \cdot 3^2 + 1\right) = 5(2 \cdot 9 + 1) = 5(18 + 1) = 5 \cdot 19 = 95$

61. Using the rule for order of operations: $3[2+4(5+2\cdot 3)] = 3[2+4(5+6)] = 3(2+4\cdot 11) = 3(2+44) = 3\cdot 46 = 138$

63. Using the rule for order of operations: $6[3+2(5\cdot 3-10)] = 6[3+2(15-10)] = 6(3+2\cdot 5) = 6(3+10) = 6\cdot 13 = 78$

65. Using the rule for order of operations:
$$5(7\cdot 4-3\cdot 4)+8(5\cdot 9-4\cdot 9) = 5(28-12)+8(45-36) = 5\cdot 16+8\cdot 9 = 80+72 = 152$$

67.   **a.**   Evaluating when $x=5$: $x+2 = 5+2 = 7$ **b.**   Evaluating when $x=5$: $2x = 2(5) = 10$

      c.   Evaluating when $x=5$: $x^2 = 5^2 = 5\cdot 5 = 25$ **d.**   Evaluating when $x=5$: $2^x = 2^5 = 2\cdot 2\cdot 2\cdot 2\cdot 2 = 32$

69.   **a.**   Evaluating when $x=10$: $x^2+2x+1 = (10)^2+2(10)+1 = 100+20+1 = 121$

      b.   Evaluating when $x=10$: $(x+1)^2 = (10+1)^2 = 11^2 = 121$

      c.   Evaluating when $x=10$: $x^2+1 = 10^2+1 = 100+1 = 101$

      d.   Evaluating when $x=10$: $(x-1)^2 = (10-1)^2 = 9^2 = 81$

71. Evaluating when $x=3$ and $y=2$: $6x+5y+4 = 6(3)+5(2)+4 = 18+10+4 = 32$

73. The set is $A\cup B = \{0,1,2,3,4,5,6\}$.   **75.** The set is $A\cap B = \{2,4\}$.

77. The set is $B\cap C = \{1,3,5\}$.   **79.** The set is $A\cup(B\cap C) = \{0,1,2,3,4,5,6\}$.

81. The set is $\{0,2\}$.   **83.** The set is $\{0,6\}$.

85. The set is $\{0,1,2,3,4,5,6,7\}$.   **87.** The set is $\{1,2,4,5\}$.

89. Computing the expression: $12+\frac{1}{4}\cdot 12 = 15$

91.   **a.**   The Venn diagram is:

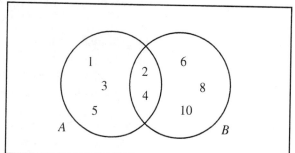

    b.   The Venn diagram is:

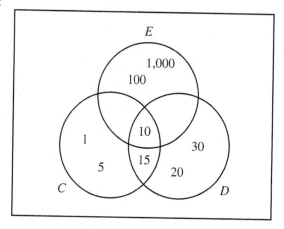

1.2   The Real Numbers

1. Graphing the inequality:

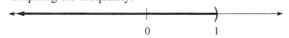

3. Graphing the inequality:

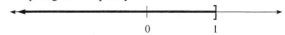

5. Graphing the inequality:

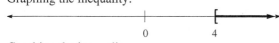

7. Graphing the inequality:

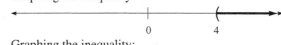

9. Graphing the inequality:

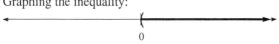

11. Graphing the inequality:

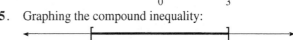

13. Graphing the compound inequality:

15. Graphing the compound inequality:

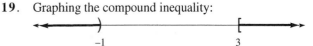

17. Graphing the compound inequality:

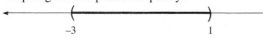

19. Graphing the compound inequality:

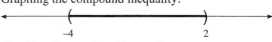

21. Graphing the compound inequality:

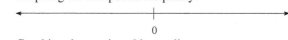

23. Graphing the compound inequality:

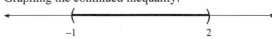

25. Graphing the continued inequality:

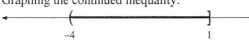

27. Graphing the continued inequality:

29. Graphing the continued inequality:

31. Graphing the continued inequality:

33. Graphing the compound inequality:

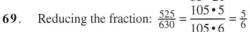

35. Graphing the compound inequality:

37. The inequality is: $x \geq 5$
39. The inequality is: $x \leq -3$
41. The inequality is: $x \leq 4$
43. The inequality is: $-4 < x < 4$
45. The inequality is: $-4 \leq x \leq 4$
47. The counting numbers are: 1,2
49. The rational numbers are: $-6, -5.2, 0, 1, 2, 2.3, \frac{9}{2}$
51. The irrational numbers are: $-\sqrt{7}, -\pi, \sqrt{17}$

53. The nonnegative integers are: 0,1,2
55. Factoring: $60 = 4 \cdot 15 = 2^2 \cdot 3 \cdot 5$
57. Factoring: $266 = 14 \cdot 19 = 2 \cdot 7 \cdot 19$
59. Factoring: $111 = 3 \cdot 37$

61. Factoring: $369 = 9 \cdot 41 = 3^2 \cdot 41$
63. Reducing the fraction: $\frac{165}{385} = \frac{3 \cdot 55}{7 \cdot 55} = \frac{3}{7}$

65. Reducing the fraction: $\frac{385}{735} = \frac{35 \cdot 11}{35 \cdot 21} = \frac{11}{21}$
67. Reducing the fraction: $\frac{111}{185} = \frac{37 \cdot 3}{37 \cdot 5} = \frac{3}{5}$

69. Reducing the fraction: $\frac{525}{630} = \frac{105 \cdot 5}{105 \cdot 6} = \frac{5}{6}$
71. The numbers are either $2 - 5 = -3$ or $2 + 5 = 7$.

73. The numbers are between 3 and 13, which is the inequality $3 < x < 13$.
75. The numbers are either less than 3 or more than 13, which is the inequality $x < 3$ or $x > 13$.
77. **a.** $10 = 3 + 7$ or $10 = 5 + 5$ **b.** $16 = 11 + 5$ or $16 = 13 + 3$
     c. $24 = 19 + 5$ or $24 = 17 + 7$ or $24 = 13 + 11$
     d. $36 = 19 + 17$ or $36 = 23 + 13$ or $36 = 29 + 7$ or $36 = 31 + 5$
79. The inequality is $50 \leq F \leq 270$.   **81.** The inequality is $20 \leq t < 40$.
83. Calculating the value: $5! = 5 \cdot 4 \cdot 3 \cdot 2 \cdot 1 = 120$
85. Calculating the value: $6! = 6 \cdot 5 \cdot 4 \cdot 3 \cdot 2 \cdot 1 = 6 \cdot 5!$. So the statement is true.

1.3   Properties of Real Numbers

1-11. Completing the table:

	Number	Opposite	Reciprocal
1.	4	-4	$\frac{1}{4}$
3.	$-\frac{1}{2}$	$\frac{1}{2}$	-2
5.	5	-5	$\frac{1}{5}$
7.	$\frac{3}{8}$	$-\frac{3}{8}$	$\frac{8}{3}$
9.	$-\frac{1}{6}$	$\frac{1}{6}$	-6
11.	3	-3	$\frac{1}{3}$

13. -1 and 1 are their own reciprocals.

15. 0 is its own opposite.

17. Writing without absolute values: $|-2| = 2$

19. Writing without absolute values: $\left|-\frac{3}{4}\right| = \frac{3}{4}$

21. Writing without absolute values: $|\pi| = \pi$

23. Writing without absolute values: $-|4| = -4$

25. Writing without absolute values: $-|-2| = -2$

27. Writing without absolute values: $-\left|-\frac{3}{4}\right| = -\frac{3}{4}$

29. Multiplying the fractions: $\frac{3}{5} \cdot \frac{7}{8} = \frac{21}{40}$

31. Multiplying the fractions: $\frac{1}{3} \cdot 6 = \frac{6}{3} = 2$

33. Multiplying the fractions: $\left(\frac{2}{3}\right)^3 = \frac{2}{3} \cdot \frac{2}{3} \cdot \frac{2}{3} = \frac{8}{27}$

35. Multiplying the fractions: $\left(\frac{1}{10}\right)^4 = \frac{1}{10} \cdot \frac{1}{10} \cdot \frac{1}{10} \cdot \frac{1}{10} = \frac{1}{10,000}$

37. Multiplying the fractions: $\frac{3}{5} \cdot \frac{4}{7} \cdot \frac{6}{11} = \frac{72}{385}$

39. Multiplying the fractions: $\frac{4}{3} \cdot \frac{3}{4} = \frac{12}{12} = 1$

41. Using the associative property: $4 + (2 + x) = (4 + 2) + x = 6 + x$

43. Using the associative property: $(a + 3) + 5 = a + (3 + 5) = a + 8$

45. Using the associative property: $5(3y) = (5 \cdot 3)y = 15y$

47. Using the associative property: $\frac{1}{3}(3x) = \left(\frac{1}{3} \cdot 3\right)x = x$

49. Using the associative property: $4\left(\frac{1}{4}a\right) = \left(4 \cdot \frac{1}{4}\right)a = a$

51. Using the associative property: $\frac{2}{3}\left(\frac{3}{2}x\right) = \left(\frac{2}{3} \cdot \frac{3}{2}\right)x = x$

53. Applying the distributive property: $3(x + 6) = 3 \cdot x + 3 \cdot 6 = 3x + 18$

55. Applying the distributive property: $2(6x + 4) = 2 \cdot 6x + 2 \cdot 4 = 12x + 8$

57. Applying the distributive property: $5(3a + 2b) = 5 \cdot 3a + 5 \cdot 2b = 15a + 10b$

59. Applying the distributive property: $\frac{1}{3}(4x + 6) = \frac{1}{3} \cdot 4x + \frac{1}{3} \cdot 6 = \frac{4}{3}x + 2$

61. Applying the distributive property: $\frac{1}{5}(10 + 5y) = \frac{1}{5} \cdot 10 + \frac{1}{5} \cdot 5y = 2 + y$

63. Applying the distributive property: $(5t + 1)8 = 5t \cdot 8 + 1 \cdot 8 = 40t + 8$

65. Applying the distributive property: $3(5x + 2) + 4 = 15x + 6 + 4 = 15x + 10$

67. Applying the distributive property: $4(2y + 6) + 8 = 8y + 24 + 8 = 8y + 32$

69. Applying the distributive property: $5(1 + 3t) + 4 = 5 + 15t + 4 = 15t + 9$

71. Applying the distributive property: $3 + (2 + 7x)4 = 3 + 8 + 28x = 28x + 11$

73. Adding the fractions: $\frac{2}{5} + \frac{1}{15} = \frac{2}{5} \cdot \frac{3}{3} + \frac{1}{15} = \frac{6}{15} + \frac{1}{15} = \frac{7}{15}$

75. Adding the fractions: $\frac{17}{30} + \frac{11}{42} = \frac{17}{30} \cdot \frac{7}{7} + \frac{11}{42} \cdot \frac{5}{5} = \frac{119}{210} + \frac{55}{210} = \frac{174}{210} = \frac{29}{35}$

77. Adding the fractions: $\frac{9}{48} + \frac{3}{54} = \frac{9}{48} \cdot \frac{9}{9} + \frac{3}{54} \cdot \frac{8}{8} = \frac{81}{432} + \frac{24}{432} = \frac{105}{432} = \frac{35}{144}$

79. Adding the fractions: $\frac{25}{84} + \frac{41}{90} = \frac{25}{84} \cdot \frac{15}{15} + \frac{41}{90} \cdot \frac{14}{14} = \frac{375}{1260} + \frac{574}{1260} = \frac{949}{1260}$

81. Simplifying the expression: $5a + 7 + 8a + a = (5a + 8a + a) + 7 = 14a + 7$

83. Simplifying the expression: $3y + y + 5 + 2y + 1 = (3y + y + 2y) + (5 + 1) = 6y + 6$

85. Simplifying the expression: $2(5x + 1) + 2x = 10x + 2 + 2x = 12x + 2$

87. Simplifying the expression: $7 + 2(4y + 2) = 7 + 8y + 4 = 8y + 11$

89. Simplifying the expression: $3 + 4(5a + 3) + 4a = 3 + 20a + 12 + 4a = 24a + 15$

91. Simplifying the expression: $5x + 2(3x + 8) + 4 = 5x + 6x + 16 + 4 = 11x + 20$

93. commutative property of addition   **95.** commutative property of multiplication

97. additive inverse property   **99.** commutative property of addition

101. associative and commutative properties of multiplication

103. commutative and associative properties of addition   **105.** distributive property

107. The outside rectangle area is $7(x + 2) = 7x + 14$, while the sum of the two areas is $7x + 7(2) = 7x + 14$. Note that the two areas are the same.

109. The outside rectangle area is $x(y + 4) = xy + 4x$, while the sum of the two areas is $xy + 4x$. Note that the two areas are the same.

111. a. Using clock arithmetic: $10 + 5 = 15 \bmod 12 = 15 - 12 = 3$
   b. Using clock arithmetic: $10 + 6 = 16 \bmod 12 = 16 - 12 = 4$
   c. Using clock arithmetic: $10 + 1 = 11$
   d. Using clock arithmetic: $10 + 12 = 22 \bmod 12 = 22 - 12 = 10$
   e. Using clock arithmetic: $x + 12 = (x + 12) \bmod 12 = x + 12 - 12 = x$

1.4   Arithmetic with Real Numbers

1. Finding the sum: $6 + (-2) = 4$ **3.** Finding the sum: $-6 + 2 = -4$

5. Finding the difference: $-7 - 3 = -7 + (-3) = -10$ **7.** Finding the difference: $-7 - (-3) = -7 + 3 = -4$

9. Finding the difference: $\frac{3}{4} - \left(-\frac{5}{6}\right) = \frac{3}{4} + \frac{5}{6} = \frac{3}{4} \cdot \frac{3}{3} + \frac{5}{6} \cdot \frac{2}{2} = \frac{9}{12} + \frac{10}{12} = \frac{19}{12}$

11. Finding the difference: $\frac{11}{42} - \frac{17}{30} = \frac{11}{42} \cdot \frac{5}{5} - \frac{17}{30} \cdot \frac{7}{7} = \frac{55}{210} - \frac{119}{210} = -\frac{64}{210} = -\frac{32}{105}$

13. Subtracting: $-3 - 5 = -8$ **15.** Subtracting: $-4 - 8 = -12$

17. Subtracting: $-3x - 4x = -7x$ **19.** The number is 13, since $5 - 13 = -8$.

21. Computing the value: $-7 + (2 - 9) = -7 + (-7) = -14$ **23.** Simplifying the value: $(8a + a) - 3a = 9a - 3a = 6a$

25. Finding the product: $3(-5) = -15$ **27.** Finding the product: $-3(-5) = 15$

29. Finding the product: $2(-3)(4) = -6(4) = -24$ **31.** Finding the product: $-2(5x) = -10x$

33. Finding the product: $-\frac{1}{3}(-3x) = \frac{3}{3}x = x$ **35.** Finding the product: $-\frac{2}{3}\left(-\frac{3}{2}y\right) = \frac{6}{6}y = y$

37. Finding the product: $-2(4x - 3) = -8x + 6$ **39.** Finding the product: $-\frac{1}{2}(6a - 8) = -\frac{6}{2}a + \frac{8}{2} = -3a + 4$

41. Simplifying: $3(-4) - 2 = -12 - 2 = -14$ **43.** Simplifying: $4(-3) - 6(-5) = -12 + 30 = 18$

45. Simplifying: $2 - 5(-4) - 6 = 2 + 20 - 6 = 22 - 6 = 16$

47. Simplifying: $4 - 3(7 - 1) - 5 = 4 - 3(6) - 5 = 4 - 18 - 5 = -19$

49. Simplifying: $2(-3)^2 - 4(-2)^3 = 2(9) - 4(-8) = 18 + 32 = 50$

51. Simplifying: $(2 - 8)^2 - (3 - 7)^2 = (-6)^2 - (-4)^2 = 36 - 16 = 20$

53. Simplifying: $7(3 - 5)^3 - 2(4 - 7)^3 = 7(-2)^3 - 2(-3)^3 = 7(-8) - 2(-27) = -56 + 54 = -2$

55. Simplifying: $-3(2 - 9) - 4(6 - 1) = -3(-7) - 4(5) = 21 - 20 = 1$

57. Simplifying: $2 - 4[3 - 5(-1)] = 2 - 4(3 + 5) = 2 - 4(8) = 2 - 32 = -30$

59. Simplifying: $(8 - 7)[4 - 7(-2)] = (1)(4 + 14) = (1)(18) = 18$

61. Simplifying: $-3 + 4[6 - 8(-3 - 5)] = -3 + 4[6 - 8(-8)] = -3 + 4(6 + 64) = -3 + 4(70) = -3 + 280 = 277$

63. Simplifying: $5 - 6[-3(2 - 9) - 4(8 - 6)] = 5 - 6[-3(-7) - 4(2)] = 5 - 6(21 - 8) = 5 - 6(13) = 5 - 78 = -73$

65. Simplifying: $3(5x + 4) - x = 15x + 12 - x = 14x + 12$

67. Simplifying: $6 - 7(3 - m) = 6 - 21 + 7m = 7m - 15$

69. Simplifying: $7 - 2(3x - 1) + 4x = 7 - 6x + 2 + 4x = -2x + 9$

71. Simplifying: $5(3y + 1) - (8y - 5) = 15y + 5 - 8y + 5 = 7y + 10$

73. Simplifying: $4(2 - 6x) - (3 - 4x) = 8 - 24x - 3 + 4x = -20x + 5$

75. Simplifying: $10 - 4(2x + 1) - (3x - 4) = 10 - 8x - 4 - 3x + 4 = -11x + 10$

77. Dividing: $\dfrac{8}{-4} = -2$, since $-4(-2) = 8$

79. $\dfrac{4}{0}$ is undefined

81. Dividing: $\dfrac{0}{-3} = 0$

83. Dividing: $-\dfrac{3}{4} \div \dfrac{9}{8} = -\dfrac{3}{4} \cdot \dfrac{8}{9} = -\dfrac{2}{3}$

85. Dividing: $-8 \div \left(-\dfrac{1}{4}\right) = -8 \cdot \left(-\dfrac{4}{1}\right) = 32$

87. Dividing: $-40 \div \left(-\dfrac{5}{8}\right) = -40 \cdot \left(-\dfrac{8}{5}\right) = 64$

89. Dividing: $\dfrac{4}{9} \div (-8) = \dfrac{4}{9} \cdot \left(-\dfrac{1}{8}\right) = -\dfrac{1}{18}$

91. Simplifying: $\dfrac{3(-1) - 4(-2)}{8 - 5} = \dfrac{-3 + 8}{3} = \dfrac{5}{3}$

93. Simplifying: $8 - (-6)\left[\dfrac{2(-3) - 5(4)}{-8(6) - 4}\right] = 8 - (-6)\left(\dfrac{-6 - 20}{-48 - 4}\right) = 8 - (-6)\left(\dfrac{-26}{-52}\right) = 8 - (-6)\left(\dfrac{1}{2}\right) = 8 + 3 = 11$

95. Simplifying: $6 - (-3)\left[\dfrac{2 - 4(3 - 8)}{1 - 5(1 - 3)}\right] = 6 - (-3)\left[\dfrac{2 - 4(-5)}{1 - 5(-2)}\right] = 6 - (-3)\left(\dfrac{2 + 20}{1 + 10}\right) = 6 - (-3)\left(\dfrac{22}{11}\right) = 6 + 6 = 12$

97. Completing the table:

a	b	SUM $a + b$	DIFFERENCE $a - b$	PRODUCT ab	QUOTIENT a/b
3	12	15	−9	36	$\frac{1}{4}$
−3	12	9	−15	−36	$-\frac{1}{4}$
3	−12	−9	15	−36	$-\frac{1}{4}$
−3	−12	−15	9	36	$\frac{1}{4}$

99. Completing the table:

x	$3(5x - 2)$	$15x - 6$	$15x - 2$
−2	−36	−36	−32
−1	−21	−21	−17
0	−6	−6	−2
1	9	9	13
2	24	24	28

101. For Santa Fe, the final time would be: 6:55 + 2:15 + 1 = 9:70 = 10:10 P.M.
For Detroit, the final time would be: 10:10 + 3:20 + 2 = 15:30 = 3:30 A.M.

1.5   Recognizing Patterns

1. The pattern is to add 1, so the next term is 5.

3. The pattern is to add 2, so the next term is 10.

5. These numbers are squares, so the next term is $5^2 = 25$.

7. The pattern is to add 7, so the next term is 29.

9. The pattern is to add 7, then add 6, then add 5, so the next term is $19 + 4 = 23$.

11. One possibility is: Δ

13. One possibility is: $\odot$

15. The pattern is to add 4, so the next two numbers are 17 and 21.

17. The pattern is to add −1, so the next two numbers are −2 and −3.

19. The pattern is to add −3, so the next two numbers are −4 and −7.

21. The pattern is to add $-\dfrac{1}{4}$, so the next two numbers are $-\dfrac{1}{2}$ and $-\dfrac{3}{4}$.

23. The pattern is to add $\dfrac{1}{2}$, so the next two numbers are $\dfrac{5}{2}$ and 3.

25. The pattern is to multiply by 3, so the next number is 27.

27. The pattern is to multiply by −3, so the next number is −270.

29. The pattern is to multiply by $\dfrac{1}{2}$, so the next number is $\dfrac{1}{8}$.

31. The pattern is to multiply by $\dfrac{1}{2}$, so the next number is $\dfrac{5}{2}$.

33. The pattern is to multiply by −5, so the next number is −625.

35. The pattern is to multiply by $-\dfrac{1}{5}$, so the next number is $-\dfrac{1}{125}$.

37. **a.** The pattern is to add 4, so the next number is 12.   **b.** The pattern is to multiply by 2, so the next number is 16.

39. The sequence is 1,1,2,3,5,8,13,21,34,55,89,144,233,..., so the 12th term is 144.

41. Three Fibonacci numbers that are prime numbers are 2, 3, and 5 (among others).

43. The even numbers are 2, 8, and 34.

45. The air temperature will be 23.5°F – 3.5°F = 20°F at an altitude of 16,000 feet.

47. The sequence of air temperatures is: 41°,37.5°,34°,30.5°,27°,23.5°. This is an arithmetic sequence.

49. The sequence of air temperatures is: 41°,45.5°,50°,54.5°,59°,63.5°. This is an arithmetic sequence.

51. The patient on antidepressant 1 will have less medication in the body, approximately half as much. The patient on antidepressant 2 will still have more of the medication in the body, since a full half-life has not passed by the second day.

53. Completing the table:

Hours Since Discontinuing	Concentration (ng / ml)
0	60
4	30
8	15
12	7.5
16	3.75

Sketching the graph:

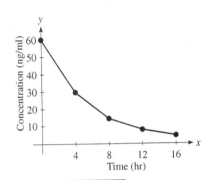

55. Completing the table:

Elevation (ft)	Boiling Point (°F)
–2,000	215.6
–1,000	213.8
0	212
1,000	210.2
2,000	208.4
3,000	206.6

57. **a.** The steep straight line segments show when the patient takes his medication.
b. The graph is falling off because the patient stops taking his medication.
c. The maximum medication is approximately 50 ng/ml.
d. Since the patient takes his medication every 4 hours, the values are A = 4 hours, B = 8 hours, and C = 12 hours.

59. There are 16 branches in the fourth row.   **61.** After five tosses, there are 32 branches.

Chapter 1 Review

1. Written in symbols, the expression is: $x + 2$

3. Written in symbols, the expression is: $\dfrac{x}{2}$

5. Written in symbols, the expression is: $2(x + y)$

7. Multiplying: $3^3 = 3 \cdot 3 \cdot 3 = 27$

9. Multiplying: $8^2 = 8 \cdot 8 = 64$

11. Multiplying: $2^5 = 2 \cdot 2 \cdot 2 \cdot 2 \cdot 2 = 32$

13. Using the rule for order of operations: $2 + 3 \cdot 5 = 2 + 15 = 17$

15. Using the rule for order of operations: $20 \div 2 + 3 = 10 + 3 = 13$

17. Using the rule for order of operations: $3 + 2(5 - 2) = 3 + 2(3) = 3 + 6 = 9$

19. Using the rule for order of operations: $3 \cdot 4^2 - 2 \cdot 3^2 = 3 \cdot 16 - 2 \cdot 9 = 48 - 18 = 30$

21. The set is $A \cup B = \{1, 2, 3, 4, 5, 6\}$.

23. The set is $\{5\}$.

25. Sketching a number line:

27. The opposite is -2 and the reciprocal is $\frac{1}{2}$.

29. Writing without absolute values: $|-3| = 3$

31. Writing without absolute values: $|-4| = 4$

33. The integers are: $-7, 0, 5$

35. The irrational numbers are: $-\sqrt{3}, \pi$

37. Reducing the fraction: $\dfrac{4356}{5148} = \dfrac{396 \cdot 11}{396 \cdot 13} = \dfrac{11}{13}$

39. Multiplying the fractions: $\left(\frac{3}{4}\right)^3 = \frac{3}{4} \cdot \frac{3}{4} \cdot \frac{3}{4} = \frac{27}{64}$

41. Graphing the inequality:

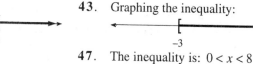

43. Graphing the inequality:

45. The inequality is: $x \geq 4$

47. The inequality is: $0 < x < 8$

49. Simplifying: $-2y + 4y = 2y$

51. Simplifying: $3x - 2 + 5x + 7 = (3x + 5x) + (-2 + 7) = 8x + 5$

53. Commutative property of addition (a)

55. Commutative property of addition (a)

57. Commutative and associative properties of addition (a,c)

59. Additive identity property (e)

61. Finding the difference: $5 - 3 = 2$

63. Computing: $7 + (-2) - 4 = 5 - 4 = 1$

65. Computing: $|-4| - |-3| + |-2| = 4 - 3 + 2 = 3$

67. Finding the differences: $6 - (-3) - 2 - 5 = 6 + 3 - 2 - 5 = 9 - 2 - 5 = 7 - 5 = 2$

69. Finding the differences: $-\frac{1}{12} - \frac{1}{6} - \frac{1}{4} - \frac{1}{3} = -\frac{1}{12} - \frac{2}{12} - \frac{3}{12} - \frac{4}{12} = -\frac{10}{12} = -\frac{5}{6}$

71. Finding the product: $6(-7) = -42$

73. Finding the product: $7(3x) = 21x$

75. Applying the distributive property: $-2(3x - 5) = -2(3x) - (-2)(5) = -6x + 10$

77. Applying the distributive property: $-\frac{1}{2}(2x - 6) = -\frac{1}{2}(2x) - \left(-\frac{1}{2}\right)(6) = -x + 3$

79. Dividing: $-\frac{5}{8} \div \frac{3}{4} = -\frac{5}{8} \cdot \frac{4}{3} = -\frac{5}{6}$

81. Dividing: $\frac{3}{5} \div 6 = \frac{3}{5} \cdot \frac{1}{6} = \frac{1}{10}$

83. Simplifying: $2(-5) - 3 = -10 - 3 = -13$

85. Simplifying: $6 + 3(-2) = 6 + (-6) = 0$

87. Simplifying: $-3(2) - 5(6) = -6 - 30 = -36$

89. Simplifying: $8 - 2(6 - 10) = 8 - 2(-4) = 8 + 8 = 16$

91. Simplifying: $\dfrac{3(-4) - 8}{-5 - 5} = \dfrac{-12 - 8}{-5 - 5} = \dfrac{-20}{-10} = 2$

93. Simplifying: $4 - (-2)\left[\dfrac{6 - 3(-4)}{1 + 5(-2)}\right] = 4 - (-2)\left(\dfrac{6 + 12}{1 - 10}\right) = 4 - (-2)\left(\dfrac{18}{-9}\right) = 4 - (-2)(-2) = 4 - 4 = 0$

95. Simplifying: $4(3x - 1) - 5(6x + 2) = 12x - 4 - 30x - 10 = -18x - 14$

97. The pattern is to add -3, so the next term is $2 + (-3) = -1$. This is an arithmetic sequence.

99. The pattern is to add $-\frac{1}{2}$, so the next term is $-\frac{1}{2} - \frac{1}{2} = -1$. This is an arithmetic sequence.

Chapter 1 Test

1. Written in symbols, the expression is: $2(3x+4y)$

2. Written in symbols, the expression is: $2a-3b < 2a+3b$

3. Using the rule for order of operations: $3 \cdot 2^2 + 5 \cdot 3^2 = 3 \cdot 4 + 5 \cdot 9 = 12 + 45 = 57$

4. Using the rule for order of operations: $6 + 2(4 \cdot 3 - 10) = 6 + 2(12 - 10) = 6 + 2(2) = 6 + 4 = 10$

5. Using the rule for order of operations: $12 - 8 \div 4 + 2 \cdot 3 = 12 - 2 + 6 = 10 + 6 = 16$

6. Using the rule for order of operations:
$$20 - 4\left[3^2 - 2\left(2^3 - 6\right)\right] = 20 - 4\left[9 - 2(8 - 6)\right] = 20 - 4\left[9 - 2(2)\right] = 20 - 4(9 - 4) = 20 - 4(5) = 20 - 20 = 0$$

7. The set is $A \cap B = \{2, 4\}$.

8. The set is $B \cap C = \varnothing$.

9. The integers are: $-5, 0, 1, 4$

10. The rational numbers are: $-5, -4.1, -3.75, -\frac{5}{6}, 0, 1, 1.8, 4$

11. The irrational numbers are: $-\sqrt{2}, \sqrt{3}$

12. Factoring: $585 = 9 \cdot 65 = 3^2 \cdot 5 \cdot 13$

13. The opposite is 3 and the reciprocal is $-\frac{1}{3}$.

14. The opposite is $-\frac{4}{3}$ and the reciprocal is $\frac{3}{4}$.

15. Simplifying: $-(-3) = 3$

16. Simplifying: $-|-2| = -2$

17. Graphing the inequality:

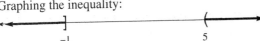

-1 5

18. Graphing the inequality:

-2 4

19. This is the commutative property of addition.

20. This is the multiplicative identity property.

21. These are the commutative and associative properties of multiplication.

22. These are the associative and commutative properties of addition.

23. Simplifying: $5(-4) + 1 = -20 + 1 = -19$

24. Simplifying: $-4(-3) + 2 = 12 + 2 = 14$

25. Simplifying: $3(2-4)^3 - 5(2-7)^2 = 3(-2)^3 - 5(-5)^2 = 3(-8) - 5(25) = -24 - 125 = -149$

26. Simplifying: $-2 + 5\left[7 - 3(-4 - 8)\right] = -2 + 5\left[7 - 3(-12)\right] = -2 + 5(7 + 36) = -2 + 5(43) = -2 + 215 = 213$

27. Simplifying: $\dfrac{-4(-1) - (-10)}{5 - (-2)} = \dfrac{4 + 10}{5 + 2} = \dfrac{14}{7} = 2$

28. Simplifying: $3 - 2\left[\dfrac{8(-1) - 7}{-3(2) - 4}\right] = 3 - 2\left(\dfrac{-8 - 7}{-6 - 4}\right) = 3 - 2\left(\dfrac{-15}{-10}\right) = 3 - 2\left(\dfrac{3}{2}\right) = 3 - 3 = 0$

29. Simplifying: $-\frac{3}{8} + \frac{5}{12} - \left(-\frac{7}{9}\right) = -\frac{3}{8} \cdot \frac{9}{9} + \frac{5}{12} \cdot \frac{6}{6} + \frac{7}{9} \cdot \frac{8}{8} = -\frac{27}{72} + \frac{30}{72} + \frac{56}{72} = \frac{59}{72}$

30. Simplifying: $-\frac{1}{2}(8x) = -4x$

31. Simplifying: $-4(3x + 2) + 7x = -12x - 8 + 7x = -5x - 8$

32. Simplifying: $5(2y - 3) - (6y - 5) = 10y - 15 - 6y + 5 = 4y - 10$

33. Simplifying: $3 + 4(2x - 5) - 5x = 3 + 8x - 20 - 5x = 3x - 17$

34. Simplifying: $2 + 5a + 3(2a - 4) = 2 + 5a + 6a - 12 = 11a - 10$

35. Computing: $-\frac{2}{3} + (-2)\left(\frac{5}{6}\right) = -\frac{2}{3} - \frac{5}{3} = -\frac{7}{3}$

36. Computing: $-4\left(\frac{7}{16}\right) - \frac{3}{4} = -\frac{7}{4} - \frac{3}{4} = -\frac{10}{4} = -\frac{5}{2}$

37. The pattern is to multiply by -4, so the next term is $80(-4) = -320$. This is a geometric sequence.

38. The pattern is to add -1, so the next term is $-1 - 1 = -2$. This is an arithmetic sequence.

39. The pattern is to add 1, then 2, then 3, so the next term is $13 + 4 = 17$.

40. The pattern is to multiply by $\frac{1}{5}$, so the next term is $\frac{1}{25} \cdot \frac{1}{5} = \frac{1}{125}$. This is a geometric sequence.

Chapter 2
Linear Equations and Inequalities in One Variable

2.1 Linear Equations in One Variable

1. Solving the equation:
$$x - 5 = 3$$
$$x = 8$$

3. Solving the equation:
$$2x - 4 = 6$$
$$2x = 10$$
$$x = \frac{10}{2} = 5$$

5. Solving the equation:
$$7 = 4a - 1$$
$$8 = 4a$$
$$a = \frac{8}{4} = 2$$

7. Solving the equation:
$$3 - y = 10$$
$$-y = 7$$
$$y = -7$$

9. Solving the equation:
$$-3 - 4x = 15$$
$$-4x = 18$$
$$x = -\frac{9}{2}$$

11. Solving the equation:
$$-3 = 5 + 2x$$
$$-8 = 2x$$
$$x = -4$$

13. Solving the equation:
$$-300y + 100 = 500$$
$$-300y = 400$$
$$y = -\frac{4}{3}$$

15. Solving the equation:
$$160 = -50x - 40$$
$$200 = -50x$$
$$x = -4$$

17. Solving the equation:
$$-x = 2$$
$$x = -1 \cdot 2 = -2$$

19. Solving the equation:
$$-a = -\frac{3}{4}$$
$$a = -1 \cdot \left(-\frac{3}{4}\right) = \frac{3}{4}$$

21. Solving the equation:
$$\frac{2}{3}x = 8$$
$$x = \frac{3}{2} \cdot 8 = 12$$

23. Solving the equation:
$$-\frac{3}{5}a + 2 = 8$$
$$-\frac{3}{5}a = 6$$
$$a = -\frac{5}{3} \cdot 6 = -10$$

25. Solving the equation:
$$8 = 6 + \tfrac{2}{7}y$$
$$2 = \tfrac{2}{7}y$$
$$y = \tfrac{7}{2} \cdot 2 = 7$$

27. Solving the equation:
$$2x - 5 = 3x + 2$$
$$-x - 5 = 2$$
$$-x = 7$$
$$x = -7$$

29. Solving the equation:
$$-3a + 2 = -2a - 1$$
$$-a + 2 = -1$$
$$-a = -3$$
$$a = 3$$

31. Solving the equation:
$$5 - 2x = 3x + 1$$
$$5 - 5x = 1$$
$$-5x = -4$$
$$x = \tfrac{4}{5}$$

33. Solving the equation:
$$2x - 3 - x = 3x + 5$$
$$x - 3 = 3x + 5$$
$$-2x = 8$$
$$x = -4$$

35. Solving the equation:
$$11x - 5 + 4x - 2 = 8x$$
$$15x - 7 = 8x$$
$$7x = 7$$
$$x = 1$$

37. Solving the equation:
$$5(y + 2) - 4(y + 1) = 3$$
$$5y + 10 - 4y - 4 = 3$$
$$y + 6 = 3$$
$$y = -3$$

39. Solving the equation:
$$6 - 7(m - 3) = -1$$
$$6 - 7m + 21 = -1$$
$$-7m + 27 = -1$$
$$-7m = -28$$
$$m = 4$$

41. Solving the equation:
$$4(a - 3) + 5 = 7(3a - 1)$$
$$4a - 12 + 5 = 21a - 7$$
$$4a - 7 = 21a - 7$$
$$-17a - 7 = -7$$
$$-17a = 0$$
$$a = 0$$

43. Solving the equation:
$$7 + 3(x + 2) = 4(x - 1)$$
$$7 + 3x + 6 = 4x - 4$$
$$3x + 13 = 4x - 4$$
$$-x + 13 = -4$$
$$-x = -17$$
$$x = 17$$

45. Solving the equation:
$$5 = 7 - 2(3x - 1) + 4x$$
$$5 = 7 - 6x + 2 + 4x$$
$$5 = -2x + 9$$
$$-2x = -4$$
$$x = 2$$

47. Solving the equation:
$$\tfrac{1}{2}x + \tfrac{1}{4} = \tfrac{1}{3}x + \tfrac{5}{4}$$
$$12\left(\tfrac{1}{2}x + \tfrac{1}{4}\right) = 12\left(\tfrac{1}{3}x + \tfrac{5}{4}\right)$$
$$6x + 3 = 4x + 15$$
$$2x + 3 = 15$$
$$2x = 12$$
$$x = 6$$

49. Solving the equation:
$$-\tfrac{2}{5}x + \tfrac{2}{15} = \tfrac{2}{3}$$
$$15\left(-\tfrac{2}{5}x + \tfrac{2}{15}\right) = 15\left(\tfrac{2}{3}\right)$$
$$-6x + 2 = 10$$
$$-6x = 8$$
$$x = -\tfrac{4}{3}$$

51. Solving the equation:
$$\tfrac{3}{4}(8x - 4) = \tfrac{2}{3}(6x - 9)$$
$$6x - 3 = 4x - 6$$
$$2x - 3 = -6$$
$$2x = -3$$
$$x = -\tfrac{3}{2}$$

53. Solving the equation:
$$\tfrac{1}{4}(12a + 1) - \tfrac{1}{4} = 5$$
$$3a + \tfrac{1}{4} - \tfrac{1}{4} = 5$$
$$3a = 5$$
$$a = \tfrac{5}{3}$$

55. Solving the equation:
$$\tfrac{1}{2}x + \tfrac{1}{3}x + \tfrac{1}{4}x = 13$$
$$12\left(\tfrac{1}{2}x + \tfrac{1}{3}x + \tfrac{1}{4}x\right) = 12(13)$$
$$6x + 4x + 3x = 156$$
$$13x = 156$$
$$x = 12$$

57. Solving the equation:
$$0.08x + 0.09(9,000 - x) = 750$$
$$0.08x + 810 - 0.09x = 750$$
$$-0.01x + 810 = 750$$
$$-0.01x = -60$$
$$x = 6,000$$

59. Solving the equation:
$$0.12x + 0.10(15,000 - x) = 1,600$$
$$0.12x + 1,500 - 0.10x = 1,600$$
$$0.02x + 1,500 = 1,600$$
$$0.02x = 100$$
$$x = 5,000$$

61. Solving the equation:
$$0.35x - 0.2 = 0.15x + 0.1$$
$$0.2x - 0.2 = 0.1$$
$$0.2x = 0.3$$
$$2x = 3$$
$$x = \tfrac{3}{2}$$

63. Solving the equation:
$$0.42 - 0.18x = 0.48x - 0.24$$
$$0.42 - 0.66x = -0.24$$
$$-0.66x = -0.66$$
$$x = 1$$

65. Solving the equation:
$$3x - 6 = 3(x + 4)$$
$$3x - 6 = 3x + 12$$
$$-6 = 12$$
Since this statement is false, there is no solution.

67. Solving the equation:
$$4y + 2 - 3y + 5 = 3 + y + 4$$
$$y + 7 = y + 7$$
$$7 = 7$$
Since this statement is true, the solution is all real numbers.

69. Solving the equation:
$$2(4t - 1) + 3 = 5t + 4 + 3t$$
$$8t - 2 + 3 = 8t + 4$$
$$8t + 1 = 8t + 4$$
$$1 = 4$$
Since this statement is false, there is no solution.

71. **a.** The equation is $6.60 = 0.4n + 1.80$
  b. Solving the equation:
$$6.60 = 0.4n + 1.80$$
$$4.80 = 0.4n$$
$$n = 12$$
The woman traveled 12 miles.

73. This is the commutative property of multiplication. **75.** This is the associative property of addition.

77. These are the commutative and associative properties of addition.

79. This is the multiplicative identity property. **81.** This is the commutative property of multiplication.

83. This is the additive identity property.

2.2 Formulas

1. Substituting $x = 0$:
$$3(0) - 4y = 12$$
$$-4y = 12$$
$$y = -3$$

3. Substituting $x = 4$:
$$3(4) - 4y = 12$$
$$12 - 4y = 12$$
$$-4y = 0$$
$$y = 0$$

5. Substituting $y = 0$:
$$2x - 3 = 0$$
$$2x = 3$$
$$x = \tfrac{3}{2}$$

7. Substituting $y = 5$:
$$2x - 3 = 5$$
$$2x = 8$$
$$x = 4$$

9. Solving for l:
$$A = lw$$
$$l = \frac{A}{w}$$

11. Solving for t:
$$I = prt$$
$$t = \frac{I}{pr}$$

13. Solving for T:
$$PV = nRT$$
$$T = \frac{PV}{nR}$$

15. Solving for x:
$$y = mx + b$$
$$y - b = mx$$
$$x = \frac{y - b}{m}$$

17. Solving for F:
$$C = \tfrac{5}{9}(F - 32)$$
$$\tfrac{9}{5}C = F - 32$$
$$F = \tfrac{9}{5}C + 32$$

19. Solving for v:
$$h = vt + 16t^2$$
$$h - 16t^2 = vt$$
$$v = \frac{h - 16t^2}{t}$$

21. Solving for d:
$$A = a + (n - 1)d$$
$$A - a = (n - 1)d$$
$$d = \frac{A - a}{n - 1}$$

23. Solving for y:
$$2x + 3y = 6$$
$$3y = -2x + 6$$
$$y = \frac{-2x + 6}{3}$$
$$y = -\tfrac{2}{3}x + 2$$

25. Solving for y:
$$-3x + 5y = 15$$
$$5y = 3x + 15$$
$$y = \frac{3x + 15}{5}$$
$$y = \tfrac{3}{5}x + 3$$

27. Solving for y:
$$2x - 6y + 12 = 0$$
$$-6y = -2x - 12$$
$$y = \frac{-2x - 12}{-6}$$
$$y = \tfrac{1}{3}x + 2$$

29. Solving for x:
$$ax + 4 = bx + 9$$
$$ax - bx + 4 = 9$$
$$ax - bx = 5$$
$$x(a - b) = 5$$
$$x = \frac{5}{a - b}$$

31. Solving for P:
$$A = P + Prt$$
$$A = P(1 + rt)$$
$$P = \frac{A}{1 + rt}$$

33. Solving for y:
$$\frac{x}{8} + \frac{y}{2} = 1$$
$$8\left(\frac{x}{8} + \frac{y}{2}\right) = 8(1)$$
$$x + 4y = 8$$
$$4y = -x + 8$$
$$y = -\tfrac{1}{4}x + 2$$

35. Solving for y:
$$\frac{x}{5} + \frac{y}{-3} = 1$$
$$15\left(\frac{x}{5} + \frac{y}{-3}\right) = 15(1)$$
$$3x - 5y = 15$$
$$-5y = -3x + 15$$
$$y = \tfrac{3}{5}x - 3$$

37. Writing a linear equation and solving:
$$x = 0.54 \cdot 38$$
$$x = 20.52$$

39. Writing a linear equation and solving:
$$x \cdot 36 = 9$$
$$x = \tfrac{1}{4} = 25\%$$

41. Writing a linear equation and solving:
$$0.04 \cdot x = 37$$
$$x = \frac{37}{0.04} = 925$$

43.　The first five terms are: $4, 7, 10, 13, 16$

45.　The first five terms are: $4, 7, 12, 19, 28$

47.　The first five terms are: $\frac{1}{4}, \frac{2}{5}, \frac{1}{2}, \frac{4}{7}, \frac{5}{8}$

49.　The first five terms are: $1, \frac{1}{4}, \frac{1}{9}, \frac{1}{16}, \frac{1}{25}$

51.　The first five terms are: $2, 4, 8, 16, 32$

53.　The first five terms are: $2, \frac{3}{2}, \frac{4}{3}, \frac{5}{4}, \frac{6}{5}$

55.　**a.**　Substituting $A = 30$, $P = 30$, and $N = 4$:　$W = \dfrac{APN}{2000} = \dfrac{(30)(30)4}{2000} = 1.8$ tons

　　b.　Substituting $A = 20$, $P = 30$, and $N = 4$:　$W = \dfrac{APN}{2000} = \dfrac{(20)(30)4}{2000} = 1.2$ tons

57.　Substituting $E = 22071$:
$$22071 = 30366 - 395x$$
$$-8295 = -395x$$
$$x = 21$$
The sulfur dioxide emissions were 22071 thousand tons in the year 1991.

59.　Substituting $P = 500$, $r = 0.05$, $n = 4$, and $t = 6$:　$A = P\left(1 + \dfrac{r}{n}\right)^{nt} = 500\left(1 + \dfrac{0.05}{4}\right)^{4 \cdot 6} \approx \$673.68$

61.　Substituting $P = 5000$, $r = 0.04$, $n = 1$, and $t = 210$:　$A = P\left(1 + \dfrac{r}{n}\right)^{nt} = 5000\left(1 + \dfrac{0.04}{1}\right)^{1 \cdot 210} \approx \$18,878,664$

63.　Substituting $x = 800$:
$$800 = 1,300 - 100p$$
$$-500 = -100p$$
$$p = 5$$
They should charge $5.00 per cartridge.

65.　Substituting $x = 300$:
$$300 = 1,300 - 100p$$
$$-1,000 = -100p$$
$$p = 10$$
They should charge $10.00 per cartridge.

67.　Substituting $V = 308$ and $r = 7$:
$$308 = \tfrac{22}{7}(7)^2 h$$
$$308 = 154h$$
$$h = 2$$
The height is 2 cm.

69.　Substituting $V = 628$ and $r = 10$:
$$628 = 3.14(10)^2 h$$
$$628 = 314h$$
$$h = 2$$
The height is 2 inches.

71.　Substituting $S = 942$ and $r = 10$:
$$942 = 2(3.14)(10)^2 + 2(3.14)(10)h$$
$$942 = 628 + 62.8h$$
$$314 = 62.8h$$
$$h = 5$$
The height is 5 feet.

73.　Translating the expression into symbols: $2(x + 3)$

75.　Translating the expression into symbols: $2(x + 3) = 16$

77.　Translating the expression into symbols: $5(x - 3)$

79.　Translating the expression into symbols: $3x + 2 = x - 4$

81.　Solving the equation for x:
$$\frac{x}{a} + \frac{y}{b} = 1$$
$$ab\left(\frac{x}{a} + \frac{y}{b}\right) = ab(1)$$
$$bx + ay = ab$$
$$bx = -ay + ab$$
$$x = -\frac{a}{b}y + a$$

83.　Solving the equation for a:
$$\frac{1}{a} + \frac{1}{b} = \frac{1}{c}$$
$$abc\left(\frac{1}{a} + \frac{1}{b}\right) = abc\left(\frac{1}{c}\right)$$
$$ac + bc = ab$$
$$bc = ab - ac$$
$$bc = a(b - c)$$
$$a = \frac{bc}{b - c}$$

85.　For Shar, $M = 220 - 46 = 174$ and $R = 60$:　$T = R + 0.6(M - R) = 60 + 0.6(174 - 60) = 128.4$ beats per minute

　　For Sara, $M = 220 - 26 = 194$ and $R = 60$:　$T = R + 0.6(M - R) = 60 + 0.6(194 - 60) = 140.4$ beats per minute

2.3 Applications

1. Let w represent the width and $2w$ represent the length. Using the perimeter formula:
$$2w + 2(2w) = 60$$
$$2w + 4w = 60$$
$$6w = 60$$
$$w = 10$$
The dimensions are 10 feet by 20 feet.

3. Let s represent the side of the square. Using the perimeter formula:
$$4s = 28$$
$$s = 7$$
The length of each side is 7 feet.

5. Let x represent the shortest side, $x + 3$ represent the medium side, and $2x$ represent the longest side. Using the perimeter formula:
$$x + x + 3 + 2x = 23$$
$$4x + 3 = 23$$
$$4x = 20$$
$$x = 5$$
The shortest side is 5 inches.

7. Let w represent the width and $2w - 3$ represent the length. Using the perimeter formula:
$$2w + 2(2w - 3) = 18$$
$$2w + 4w - 6 = 18$$
$$6w - 6 = 18$$
$$6w = 24$$
$$w = 4$$
The width is 4 meters.

9. Let w represent the width and $2w$ represent the length. Using the perimeter formula:
$$2w + 2(2w) = 48$$
$$2w + 4w = 48$$
$$6w = 48$$
$$w = 8$$
The width is 8 feet and the length is 16 feet. Finding the cost: $C = 1.75(32) + 2.25(16) = 56 + 36 = 92$
The cost to build the pen is $92.00.

11. Let b represent the amount of money Shane had at the beginning of the trip. Using the percent increase:
$$b + 0.50b = 300$$
$$1.5b = 300$$
$$b = 200$$
Shane had $200.00 at the beginning of the trip.

13. Let w represent the number of employees before the layoffs. The equation is:
$$0.02n = 150$$
$$n = 7500$$
Amazon.com had 7,500 employees before the layoffs.

15. Let I represent the accountant's monthly income before the raise. The equation is:
$$I + 0.055I = 3440$$
$$1.055I = 3440$$
$$I \approx 3260.66$$
The accountant's monthly income before the raise was approximately $3,260.66.

17. The light milk contains 20% less calories, so its calories represent 80% of that in whole milk. The equation is:
$$0.80x = 120$$
$$x = 150$$
One serving of whole milk contains 150 calories.

19. Let x represent one angle and $8x$ represent the other angle. Since the angles are supplementary:

$$x + 8x = 180$$
$$9x = 180$$
$$x = 20$$

The two angles are 20° and 160°.

21. **a.** Let x represent one angle and $4x - 12$ represent the other angle. Since the angles are complementary:

$$x + 4x - 12 = 90$$
$$5x - 12 = 90$$
$$5x = 102$$
$$x = 20.4$$
$$4x - 12 = 4(20.4) - 12 = 69.6$$

The two angles are 20.4° and 69.6°.

b. Let x represent one angle and $4x - 12$ represent the other angle. Since the angles are supplementary:

$$x + 4x - 12 = 180$$
$$5x - 12 = 180$$
$$5x = 192$$
$$x = 38.4$$
$$4x - 12 = 4(38.4) - 12 = 141.6$$

The two angles are 38.4° and 141.6°.

23. Let x represent the smallest angle, $3x$ represent the largest angle, and $3x - 9$ represent the third angle. The equation is:

$$x + 3x + 3x - 9 = 180$$
$$7x - 9 = 180$$
$$7x = 189$$
$$x = 27$$

The three angles are 27°, 72° and 81°.

25. Let x represent the largest angle, $\frac{1}{3}x$ represent the smallest angle, and $\frac{1}{3}x + 10$ represent the third angle.

The equation is:

$$\frac{1}{3}x + x + \frac{1}{3}x + 10 = 180$$
$$\frac{5}{3}x + 10 = 180$$
$$\frac{5}{3}x = 170$$
$$x = 102$$

The three angles are 34°, 44° and 102°.

27. Let x represent the measure of the two base angles, and $2x + 8$ represent the third angle. The equation is:

$$x + x + 2x + 8 = 180$$
$$4x + 8 = 180$$
$$4x = 172$$
$$x = 43$$

The three angles are 43°, 43° and 94°.

29. Let x represent the amount invested at 8% and $9000 - x$ represent the amount invested at 9%. The equation is:

$$0.08x + 0.09(9000 - x) = 750$$
$$0.08x + 810 - 0.09x = 750$$
$$-0.01x + 810 = 750$$
$$-0.01x = -60$$
$$x = 6000$$

She invested $6,000 at 8% and $3,000 at 9%.

31. Let x represent the amount invested at 12% and $15000 - x$ represent the amount invested at 10%. The equation is:
$$0.12x + 0.10(15000 - x) = 1600$$
$$0.12x + 1500 - 0.10x = 1600$$
$$0.02x + 1500 = 1600$$
$$0.02x = 100$$
$$x = 5000$$
The investment was $5,000 at 12% and $10,000 at 10%.

33. Let x represent the amount invested at 8% and $6000 - x$ represent the amount invested at 9%. The equation is:
$$0.08x + 0.09(6000 - x) = 500$$
$$0.08x + 540 - 0.09x = 500$$
$$-0.01x + 540 = 500$$
$$-0.01x = -40$$
$$x = 4000$$
Stacey invested $4,000 at 8% and $2,000 at 9%.

35. Let x represent the amount of her Shell purchases and $2850 - x$ represent the amount of her other purchases. The equation is:
$$0.03x + 0.02(2850 - x) = 64$$
$$0.03x + 57 - 0.02x = 64$$
$$0.01x + 57 = 64$$
$$0.01x = 7$$
$$x = 700$$
Her Shell purchases were $700 and her other purchases were $2,150.

37. Let x represent the number of father tickets and $75 - x$ represent the number of son tickets. The equation is:
$$2x + 1.5(75 - x) = 127.5$$
$$2x + 112.5 - 1.5x = 127.5$$
$$0.5x + 112.5 = 127.5$$
$$0.5x = 15$$
$$x = 30$$
So 30 fathers and 45 sons attended the breakfast.

39. The total money collected is: $1204 - $250 = 954
Let x represent the amount of her sales (not including tax). Since this amount includes the tax collected, the equation is:
$$x + 0.06x = 954$$
$$1.06x = 954$$
$$x = 900$$
Her sales were $900, so the sales tax is: $0.06(900) = 54

41. Let x represent the length of Patrick's call. He talks 1 minute at 40 cents and $x - 1$ minutes at 30 cents, so the equation is:
$$40(1) + 30(x - 1) + 50 = 1380$$
$$40 + 30x - 30 + 50 = 1380$$
$$30x + 60 = 1380$$
$$30x = 1320$$
$$x = 44$$
Patrick talked for 44 minutes.

43. Completing the table:

Age (years)	Maximum Heart Rate (beats per minute)
18	202
19	201
20	200
21	199
22	198
23	197

45. Completing the table:

Resting Heart Rate (beats per minute)	Training Heart Rate (beats per minute)
60	144
62	145
64	146
68	147
70	148
72	149

47. Completing the table:

w	l	A
2	22	44
4	20	80
6	18	108
8	16	128
10	14	140
12	12	144

49. The largest area produced is 9 square inches. Completing the table:

Width	Length	Area
0	6	0
0.5	5.5	2.75
1	5	5
1.5	4.5	6.75
2	4	8
2.5	3.5	8.75
3	3	9
3.5	2.5	8.75
4	2	8
4.5	1.5	6.75
5	1	5
5.5	0.5	2.75
6	0	0

51. Graphing the inequality:

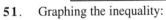

53. Graphing the inequality:

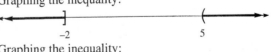

55. Graphing the inequality:

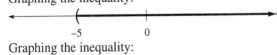

57. Graphing the inequality:

2.4 Linear Inequalities in One Variable

1. Solving the inequality:

$$2x \le 3$$

$$x \le \frac{3}{2}$$

Graphing the solution set:

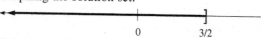

3. Solving the inequality:

$$\frac{1}{2}x > 2$$

$$x > 4$$

Graphing the solution set:

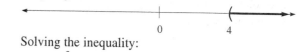

5. Solving the inequality:

$$-5x \le 25$$

$$x \ge -5$$

Graphing the solution set:

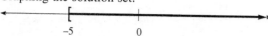

7. Solving the inequality:

$$-\frac{3}{2}x > -6$$

$$-3x > -12$$

$$x < 4$$

Graphing the solution set:

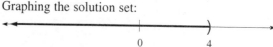

9. Solving the inequality:

$$-12 \le 2x$$

$$x \ge -6$$

Graphing the solution set:

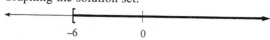

11. Solving the inequality:

$$-1 \ge -\frac{1}{4}x$$

$$x \ge 4$$

Graphing the solution set:

13. Solving the inequality:

$$-3x + 1 > 10$$

$$-3x > 9$$

$$x < -3$$

Graphing the solution set:

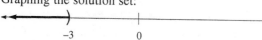

15. Solving the inequality:

$$\frac{1}{2} - \frac{m}{12} \le \frac{7}{12}$$

$$12\left(\frac{1}{2} - \frac{m}{12}\right) \le 12\left(\frac{7}{12}\right)$$

$$6 - m \le 7$$

$$-m \le 1$$

$$m \ge -1$$

Graphing the solution set:

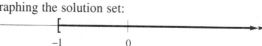

17. Solving the inequality:

$$\frac{1}{2} \ge -\frac{1}{6} - \frac{2}{9}x$$

$$18\left(\frac{1}{2}\right) \ge 18\left(-\frac{1}{6} - \frac{2}{9}x\right)$$

$$9 \ge -3 - 4x$$

$$12 \ge -4x$$

$$x \ge -3$$

Graphing the solution set:

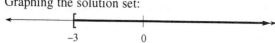

19. Solving the inequality:

$$-40 \le 30 - 20y$$

$$-70 \le -20y$$

$$y \le \frac{7}{2}$$

Graphing the solution set:

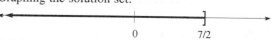

21. Solving the inequality:

$$\frac{2}{3}x - 3 < 1$$

$$\frac{2}{3}x < 4$$

$$2x < 12$$

$$x < 6$$

Graphing the solution set:

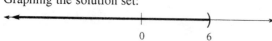

23. Solving the inequality:

$$10 - \frac{1}{2}y \le 36$$

$$-\frac{1}{2}y \le 26$$

$$y \ge -52$$

Graphing the solution set:

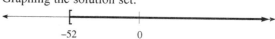

25. Solving the inequality:
$$2(3y+1) \le -10$$
$$6y+2 \le -10$$
$$6y \le -12$$
$$y \le -2$$
The solution set is $(-\infty, -2]$.

29. Solving the inequality:
$$\tfrac{1}{3}t - \tfrac{1}{2}(5-t) < 0$$
$$6\left(\tfrac{1}{3}t - \tfrac{1}{2}(5-t)\right) < 6(0)$$
$$2t - 3(5-t) < 0$$
$$2t - 15 + 3t < 0$$
$$5t - 15 < 0$$
$$5t < 15$$
$$t < 3$$
The solution set is $(-\infty, 3)$.

33. Solving the inequality:
$$-\tfrac{1}{3}(x+5) \le -\tfrac{2}{9}(x-1)$$
$$9\left[-\tfrac{1}{3}(x+5)\right] \le 9\left[-\tfrac{2}{9}(x-1)\right]$$
$$-3(x+5) \le -2(x-1)$$
$$-3x - 15 \le -2x + 2$$
$$-x - 15 \le 2$$
$$-x \le 17$$
$$x \ge -17$$
The solution set is $[-17, \infty)$.

37. Solving the inequality:
$$-2 \le m - 5 \le 2$$
$$3 \le m \le 7$$
The solution set is $[3,7]$. Graphing the solution set:

41. Solving the inequality:
$$0.5 \le 0.3a - 0.7 \le 1.1$$
$$1.2 \le 0.3a \le 1.8$$
$$4 \le a \le 6$$
The solution set is $[4,6]$. Graphing the solution set:

45. Solving the inequality:
$$4 < 6 + \tfrac{2}{3}x < 8$$
$$-2 < \tfrac{2}{3}x < 2$$
$$-6 < 2x < 6$$
$$-3 < x < 3$$
The solution set is $(-3,3)$. Graphing the solution set:

27. Solving the inequality:
$$-(a+1) - 4a \le 2a - 8$$
$$-a - 1 - 4a \le 2a - 8$$
$$-5a - 1 \le 2a - 8$$
$$-7a \le -7$$
$$a \ge 1$$
The solution set is $[1, \infty)$.

31. Solving the inequality:
$$-2 \le 5 - 7(2a+3)$$
$$-2 \le 5 - 14a - 21$$
$$-2 \le -16 - 14a$$
$$14 \le -14a$$
$$a \le -1$$
The solution set is $(-\infty, -1]$.

35. Solving the inequality:
$$5(y+3) + 4 < 6y - 1 - 5y$$
$$5y + 15 + 4 < y - 1$$
$$5y + 19 < y - 1$$
$$4y + 19 < -1$$
$$4y < -20$$
$$y < -5$$
The solution set is $(-\infty, -5)$.

39. Solving the inequality:
$$-60 < 20a + 20 < 60$$
$$-80 < 20a < 40$$
$$-4 < a < 2$$
The solution set is $(-4, 2)$. Graphing the solution set:

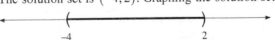

43. Solving the inequality:
$$3 < \tfrac{1}{2}x + 5 < 6$$
$$-2 < \tfrac{1}{2}x < 1$$
$$-4 < x < 2$$
The solution set is $(-4, 2)$. Graphing the solution set:

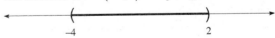

47.   Solving the inequality:
$$x+5 \le -2 \qquad \text{or} \qquad x+5 \ge 2$$
$$x \le -7 \qquad \text{or} \qquad x \ge -3$$
The solution set is $(-\infty, -7] \cup [-3, \infty)$. Graphing the solution set:

$$-7 \qquad\qquad -3$$

49.   Solving the inequality:
$$5y+1 \le -4 \qquad \text{or} \qquad 5y+1 \ge 4$$
$$5y \le -5 \qquad \text{or} \qquad 5y \ge 3$$
$$y \le -1 \qquad \text{or} \qquad y \ge \tfrac{3}{5}$$
The solution set is $(-\infty, -1] \cup \left[\tfrac{3}{5}, \infty\right)$. Graphing the solution set:

$$-1 \qquad\qquad 3/5$$

51.   Solving the inequality:
$$2x+5 < 3x-1 \qquad \text{or} \qquad x-4 > 2x+6$$
$$-x+5 < -1 \qquad \text{or} \qquad -x-4 > 6$$
$$-x < -6 \qquad \text{or} \qquad -x > 10$$
$$x > 6 \qquad \text{or} \qquad x < -10$$
The solution set is $(-\infty, -10) \cup (6, \infty)$. Graphing the solution set:

$$-10 \qquad\qquad 6$$

53.   Solving the inequality:
$$900 - 300p \ge 300$$
$$-300p \ge -600$$
$$p \le 2$$
They should charge $2.00 per pad or less.

55.   Solving the inequality:
$$900 - 300p < 525$$
$$-300p < -375$$
$$p > 1.25$$
They should charge more than $1.25 per pad.

57.   Solving the inequality:
$$22149 - 399x > 20500$$
$$-399x > -1649$$
$$x < 4.13$$
In the years 1990 through 1994 Amtrak had more than 20,500 million passengers.

59.   Let x represent the number of minutes you talk. The inequality is:
$$20 + 0.40(x - 30) < 30$$
$$20 + 0.4x - 12 < 30$$
$$0.4x + 8 < 30$$
$$0.4x < 22$$
$$x < 55$$
You may use less than 55 minutes to keep your bill under $30.

61.   Solving the inequality:
$$95 \le \tfrac{9}{5}C + 32 \le 113$$
$$63 \le \tfrac{9}{5}C \le 81$$
$$315 \le 9C \le 405$$
$$35° \le C \le 45°$$

63.   Solving the inequality:
$$-13 \le \tfrac{9}{5}C + 32 \le 14$$
$$-45 \le \tfrac{9}{5}C \le -18$$
$$-225 \le 9C \le -90$$
$$-25° \le C \le -10°$$

65.   Simplifying: $|-3| = 3$

67.   Simplifying: $-|-3| = -3$

69. $|x|$ represents the distance from x to 0 on the number line.

71.   Solving the inequality:
$$ax + b < c$$
$$ax < c - b$$
$$x < \frac{c-b}{a}$$

73.   Solving the inequality:
$$-c < ax + b < c$$
$$-c - b < ax < c - b$$
$$\frac{-c-b}{a} < x < \frac{c-b}{a}$$

2.5   Equations with Absolute Value

1. Solving the equation:
$$|x| = 4$$
$$x = -4, 4$$

3. Solving the equation:
$$2 = |a|$$
$$a = -2, 2$$

5. The equation $|x| = -3$ has no solution, or $\varnothing$.

7. Solving the equation:
$$|a| + 2 = 3$$
$$|a| = 1$$
$$a = -1, 1$$

9. Solving the equation:
$$|y| + 4 = 3$$
$$|y| = -1$$
The equation $|y| = -1$ has no solution, or $\varnothing$.

11. Solving the equation:
$$4 = |x| - 2$$
$$|x| = 6$$
$$x = -6, 6$$

13. Solving the equation:
$$|x - 2| = 5$$
$$x - 2 = -5, 5$$
$$x = -3, 7$$

15. Solving the equation:
$$|a - 4| = \tfrac{5}{3}$$
$$a - 4 = -\tfrac{5}{3}, \tfrac{5}{3}$$
$$a = \tfrac{7}{3}, \tfrac{17}{3}$$

17. Solving the equation:
$$1 = |3 - x|$$
$$3 - x = -1, 1$$
$$-x = -4, -2$$
$$x = 2, 4$$

19. Solving the equation:
$$\left|\tfrac{3}{5} a + \tfrac{1}{2}\right| = 1$$
$$\tfrac{3}{5} a + \tfrac{1}{2} = -1, 1$$
$$\tfrac{3}{5} a = -\tfrac{3}{2}, \tfrac{1}{2}$$
$$a = -\tfrac{5}{2}, \tfrac{5}{6}$$

21. Solving the equation:
$$60 = |20x - 40|$$
$$20x - 40 = -60, 60$$
$$20x = -20, 100$$
$$x = -1, 5$$

23. Since $|2x + 1| = -3$ is impossible, there is no solution, or $\varnothing$.

25. Solving the equation:
$$\left|\tfrac{3}{4} x - 6\right| = 9$$
$$\tfrac{3}{4} x - 6 = -9, 9$$
$$\tfrac{3}{4} x = -3, 15$$
$$3x = -12, 60$$
$$x = -4, 20$$

27. Solving the equation:
$$\left|1 - \tfrac{1}{2} a\right| = 3$$
$$1 - \tfrac{1}{2} a = -3, 3$$
$$-\tfrac{1}{2} a = -4, 2$$
$$a = -4, 8$$

29. Solving the equation:
$$|3x + 4| + 1 = 7$$
$$|3x + 4| = 6$$
$$3x + 4 = -6, 6$$
$$3x = -10, 2$$
$$x = -\tfrac{10}{3}, \tfrac{2}{3}$$

31. Solving the equation:
$$|3 - 2y| + 4 = 3$$
$$|3 - 2y| = -1$$
Since this equation is impossible, there is no solution, or $\varnothing$.

33.   Solving the equation:

$$3 + |4t - 1| = 8$$
$$|4t - 1| = 5$$
$$4t - 1 = -5, 5$$
$$4t = -4, 6$$
$$t = -1, \tfrac{3}{2}$$

35.   Solving the equation:

$$\left|9 - \tfrac{3}{5}x\right| + 6 = 12$$
$$\left|9 - \tfrac{3}{5}x\right| = 6$$
$$9 - \tfrac{3}{5}x = -6, 6$$
$$-\tfrac{3}{5}x = -15, -3$$
$$-3x = -75, -15$$
$$x = 5, 25$$

37.   Solving the equation:

$$5 = \left|\tfrac{2x}{7} + \tfrac{4}{7}\right| - 3$$
$$\left|\tfrac{2x}{7} + \tfrac{4}{7}\right| = 8$$
$$\tfrac{2x}{7} + \tfrac{4}{7} = -8, 8$$
$$2x + 4 = -56, 56$$
$$2x = -60, 52$$
$$x = -30, 26$$

39.   Solving the equation:

$$2 = -8 + \left|4 - \tfrac{1}{2}y\right|$$
$$\left|4 - \tfrac{1}{2}y\right| = 10$$
$$4 - \tfrac{1}{2}y = -10, 10$$
$$-\tfrac{1}{2}y = -14, 6$$
$$y = -12, 28$$

41.   Solving the equation:

$$|3a + 1| = |2a - 4|$$

$3a + 1 = 2a - 4$	or	$3a + 1 = -2a + 4$
$a + 1 = -4$		$5a = 3$
$a = -5$		$a = \tfrac{3}{5}$

43.   Solving the equation:

$$\left|x - \tfrac{1}{3}\right| = \left|\tfrac{1}{2}x + \tfrac{1}{6}\right|$$

$x - \tfrac{1}{3} = \tfrac{1}{2}x + \tfrac{1}{6}$	or	$x - \tfrac{1}{3} = -\tfrac{1}{2}x - \tfrac{1}{6}$
$6x - 2 = 3x + 1$		$6x - 2 = -3x - 1$
$3x - 2 = 1$		$9x - 2 = -1$
$3x = 3$		$9x = 1$
$x = 1$		$x = \tfrac{1}{9}$

45.   Solving the equation:

$$|y - 2| = |y + 3|$$

$y - 2 = y + 3$	or	$y - 2 = -y - 3$
$-2 = -3$		$2y = -1$
$y = $ impossible		$y = -\tfrac{1}{2}$

47.   Solving the equation:

$$|3x - 1| = |3x + 1|$$

$3x - 1 = 3x + 1$	or	$3x - 1 = -3x - 1$
$-1 = 1$		$6x = 0$
$x = $ impossible		$x = 0$

49.   Solving the equation:

$$|3 - m| = |m + 4|$$

$3 - m = m + 4$	or	$3 - m = -m - 4$
$-2m = 1$		$3 = -4$
$m = -\tfrac{1}{2}$		$m = $ impossible

51. Solving the equation:
$$|0.03 - 0.01x| = |0.04 + 0.05x|$$
$$0.03 - 0.01x = 0.04 + 0.05x \qquad \text{or} \qquad 0.03 - 0.01x = -0.04 - 0.05x$$
$$-0.06x = 0.01 \qquad\qquad\qquad\qquad 0.04x = -0.07$$
$$x = -\frac{1}{6} \qquad\qquad\qquad\qquad x = -\frac{7}{4}$$

53. Since $|x - 2| = |2 - x|$ is always true, the solution set is all real numbers.

55. Since $\left|\dfrac{x}{5} - 1\right| = \left|1 - \dfrac{x}{5}\right|$ is always true, the solution set is all real numbers.

57. Setting $R = 722$:
$$-60|x - 11| + 962 = 722$$
$$-60|x - 11| = -240$$
$$|x - 11| = 4$$
$$x - 11 = -4, 4$$
$$x = 7, 15$$
The revenue was 722 million dollars in the years 1987 and 1995.

59. Graphing the inequality:

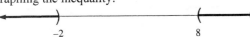

$$-2 \qquad\qquad\qquad 8$$

61. Graphing the inequality:

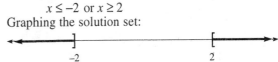

$$-2 \qquad\qquad\qquad 1$$

63. Solving the inequality:
$$4t - 3 \le -9$$
$$4t \le -6$$
$$t \le -\frac{3}{2}$$

65. Solving the inequality:
$$-3x > 15$$
$$x < -5$$

2.6 Inequalities Involving Absolute Value

1. Solving the inequality:
$$|x| < 3$$
$$-3 < x < 3$$
Graphing the solution set:

$$-3 \qquad\qquad\qquad 3$$

3. Solving the inequality:
$$|x| \ge 2$$
$$x \le -2 \text{ or } x \ge 2$$
Graphing the solution set:

$$-2 \qquad\qquad\qquad 2$$

5. Solving the inequality:
$$|x| + 2 < 5$$
$$|x| < 3$$
$$-3 < x < 3$$
Graphing the solution set:

$$-3 \qquad\qquad\qquad 3$$

7. Solving the inequality:
$$|t| - 3 > 4$$
$$|t| > 7$$
$$t < -7 \text{ or } t > 7$$
Graphing the solution set:

$$-7 \qquad\qquad\qquad 7$$

9. Since the inequality $|y| < -5$ is never true, there is no solution, or $\varnothing$. Graphing the solution set:

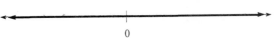

$$0$$

11. Since the inequality $|x| \ge -2$ is always true, the solution set is all real numbers.
Graphing the solution set:

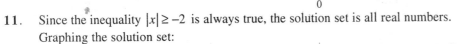

$$0$$

13. Solving the inequality:
$$|x-3|<7$$
$$-7<x-3<7$$
$$-4<x<10$$
Graphing the solution set:

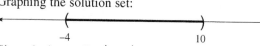

15. Solving the inequality:
$$|a+5|\geq 4$$
$$a+5\leq -4 \text{ or } a+5\geq 4$$
$$a\leq -9 \text{ or } \quad a\geq -1$$
Graphing the solution set:

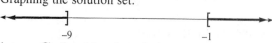

17. Since the inequality $|a-1|<-3$ is never true, there is no solution, or $\varnothing$. Graphing the solution set:

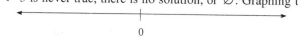

19. Solving the inequality:
$$|2x-4|<6$$
$$-6<2x-4<6$$
$$-2<2x<10$$
$$-1<x<5$$
Graphing the solution set:

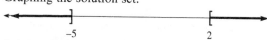

21. Solving the inequality:
$$|3y+9|\geq 6$$
$$3y+9\leq -6 \qquad \text{or} \qquad 3y+9\geq 6$$
$$3y\leq -15 \qquad\qquad 3y\geq -3$$
$$y\leq -5 \qquad\qquad y\geq -1$$
Graphing the solution set:

23. Solving the inequality:
$$|2k+3|\geq 7$$
$$2k+3\leq -7 \qquad \text{or} \qquad 2k+3\geq 7$$
$$2k\leq -10 \qquad\qquad 2k\geq 4$$
$$k\leq -5 \qquad\qquad k\geq 2$$
Graphing the solution set:

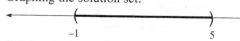

25. Solving the inequality:
$$|x-3|+2<6$$
$$|x-3|<4$$
$$-4<x-3<4$$
$$-1<x<7$$
Graphing the solution set:

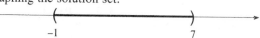

27. Solving the inequality:
$$|2a+1|+4\geq 7$$
$$|2a+1|\geq 3$$
$$2a+1\leq -3 \qquad \text{or} \qquad 2a+1\geq 3$$
$$2a\leq -4 \qquad\qquad 2a\geq 2$$
$$a\leq -2 \qquad\qquad a\geq 1$$
Graphing the solution set:

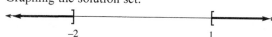

29. Solving the inequality:
$$|3x+5|-8<5$$
$$|3x+5|<13$$
$$-13<3x+5<13$$
$$-18<3x<8$$
$$-6<x<\frac{8}{3}$$
Graphing the solution set:

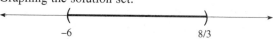

31. Solving the inequality:
$$|5-x|>3$$
$$5-x<-3 \qquad \text{or} \qquad 5-x>3$$
$$-x<-8 \qquad\qquad -x>-2$$
$$x>8 \qquad\qquad x<2$$
Graphing the solution set:

33. Solving the inequality:
$$\left|3-\tfrac{2}{3}x\right|\geq 5$$
$$3-\tfrac{2}{3}x\leq -5 \qquad \text{or} \qquad 3-\tfrac{2}{3}x\geq 5$$
$$-\tfrac{2}{3}x\leq -8 \qquad\qquad -\tfrac{2}{3}x\geq 2$$
$$-2x\leq -24 \qquad\qquad -2x\geq 6$$
$$x\geq 12 \qquad\qquad x\leq -3$$
Graphing the solution set:

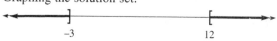

35. Solving the inequality:
$$\left|2 - \tfrac{1}{2}x\right| > 1$$

$$2 - \tfrac{1}{2}x < -1 \qquad \text{or} \qquad 2 - \tfrac{1}{2}x > 1$$
$$-\tfrac{1}{2}x < -3 \qquad\qquad\qquad -\tfrac{1}{2}x > -1$$
$$x > 6 \qquad\qquad\qquad\qquad x < 2$$

Graphing the solution set:

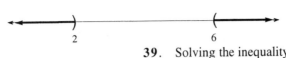

37. Solving the inequality:

$$|x - 1| < 0.01$$
$$-0.01 < x - 1 < 0.01$$
$$0.99 < x < 1.01$$

39. Solving the inequality:
$$|2x + 1| \geq \tfrac{1}{5}$$

$$2x + 1 \leq -\tfrac{1}{5} \qquad \text{or} \qquad 2x + 1 \geq \tfrac{1}{5}$$
$$2x \leq -\tfrac{6}{5} \qquad\qquad\qquad 2x \geq -\tfrac{4}{5}$$
$$x \leq -\tfrac{3}{5} \qquad\qquad\qquad x \geq -\tfrac{2}{5}$$

41. Solving the inequality:
$$\left|\frac{3x - 2}{5}\right| \leq \tfrac{1}{2}$$

$$-\tfrac{1}{2} \leq \frac{3x - 2}{5} \leq \tfrac{1}{2}$$
$$-\tfrac{5}{2} \leq 3x - 2 \leq \tfrac{5}{2}$$
$$-\tfrac{1}{2} \leq 3x \leq \tfrac{9}{2}$$
$$-\tfrac{1}{6} \leq x \leq \tfrac{3}{2}$$

43. Solving the inequality:

$$\left|2x - \tfrac{1}{5}\right| < 0.3$$
$$-0.3 < 2x - 0.2 < 0.3$$
$$-0.1 < 2x < 0.5$$
$$-0.05 < x < 0.25$$

45. Writing as an absolute value inequality: $|x| \leq 4$

47. Writing as an absolute value inequality: $|x - 5| \leq 1$

49. The absolute value inequality is: $|x - 65| \leq 10$

51. Solving the inequality:
$$|v - 455| < 23$$
$$-23 < v - 455 < 23$$
$$432 < v < 478$$

The color of the fireworks display is blue.

53. Simplifying: $-9 \div \tfrac{3}{2} = -9 \cdot \tfrac{2}{3} = -6$

55. Simplifying: $3 - 7(-6 - 3) = 3 - 7(-9) = 3 + 63 = 66$

57. Simplifying: $-4(-2)^3 - 5(-3)^2 = -4(-8) - 5(9) = 32 - 45 = -13$

59. Simplifying: $-2(-3 + 8) - 7(-9 + 6) = -2(5) - 7(-3) = -10 + 21 = 11$

61. Simplifying: $\dfrac{2(-3) - 5(-6)}{-1 - 2 - 3} = \dfrac{-6 + 30}{-1 - 2 - 3} = \dfrac{24}{-6} = -4$

Chapter 2 Review

1. Solving the equation:
$$x - 3 = 7$$
$$x = 10$$

3. Solving the equation:
$$400 - 100a = 200$$
$$-100a = -200$$
$$a = 2$$

5. Solving the equation:
$$4x - 2 = 7x + 7$$
$$-3x - 2 = 7$$
$$-3x = 9$$
$$x = -3$$

7. Solving the equation:
$$7y - 5 - 2y = 2y - 3$$
$$5y - 5 = 2y - 3$$
$$3y - 5 = -3$$
$$3y = 2$$
$$y = \tfrac{2}{3}$$

9.   Solving the equation:

$$3(2x+1) = 18$$
$$6x + 3 = 18$$
$$6x = 15$$
$$x = \frac{5}{2}$$

11.   Solving the equation:
$$8 - 3(2t+1) = 5(t+2)$$
$$8 - 6t - 3 = 5t + 10$$
$$-6t + 5 = 5t + 10$$
$$-11t + 5 = 10$$
$$-11t = 5$$
$$t = -\frac{5}{11}$$

13.   Solving for h:
$$40 = 2(3) + 2h$$
$$40 = 6 + 2h$$
$$2h = 34$$
$$h = 17$$

15.   Solving for t:

$$I = prt$$
$$p = \frac{I}{rt}$$

17.   Solving for y:
$$4x - 3y = 12$$
$$-3y = -4x + 12$$
$$y = \frac{4}{3}x - 4$$

19.   The first four terms are: 3,6,9,12. This sequence is increasing.

21.   The first four terms are: –2,4,–8,16. This sequence is alternating.

23.   Let w represent the width and $3w$ represent the length. Using the perimeter formula:
$$2w + 2(3w) = 32$$
$$2w + 6w = 32$$
$$8w = 32$$
$$w = 4$$
The dimensions are 4 feet by 12 feet.

25.   **a.**   Substituting $L = 45$ and $H = 12$: $N = 7(45)(12) = 3,780$ bricks

   b.   Substituting $N = 35,000$ and $H = 8$:
$$35000 = 7L(8)$$
$$35000 = 56L$$
$$L = 625$$
The wall could be 625 feet long.

27.   Solving the inequality:

$$-8a > -4$$
$$a < \frac{1}{2}$$

The solution set is $\left(-\infty, \frac{1}{2}\right)$.

29.   Solving the inequality:
$$\frac{3}{4}x + 1 \le 10$$
$$\frac{3}{4}x \le 9$$
$$3x \le 36$$
$$x \le 12$$
The solution set is $(-\infty, 12]$.

31.   Solving the inequality:

$$\frac{1}{3} \le \frac{1}{6}x \le 1$$
$$2 \le x \le 6$$

The solution set is $[2, 6]$.

33.   Solving the inequality:
$$5t + 1 \le 3t - 2 \quad \text{or} \quad -7t \le -21$$
$$2t \le -3 \qquad\qquad t \ge 3$$
$$t \le -\frac{3}{2} \qquad\qquad t \ge 3$$
The solution set is $\left(-\infty, -\frac{3}{2}\right] \cup [3, \infty)$.

35.   Solving the equation:

$$|x| = 2$$
$$x = -2, 2$$

37.   Solving the equation:
$$|x - 3| = 1$$
$$x - 3 = -1, 1$$
$$x = 2, 4$$

39. Solving the equation:
$$|4x - 3| + 2 = 11$$
$$|4x - 3| = 9$$
$$4x - 3 = -9, 9$$
$$4x = -6, 12$$
$$x = -\tfrac{3}{2}, 3$$

41. Solving the equation:
$$|5t - 3| = |3t - 5|$$

$5t - 3 = 3t - 5$ or $5t - 3 = -3t + 5$

$\quad\quad 2t = -2$ $8t = 8$

$\quad\quad t = -1$ $t = 1$

43. Solving the inequality:
$$|x| < 5$$
$$-5 < x < 5$$
Graphing the solution set:

$\quad\quad\quad\quad -5 \quad\quad\quad\quad\quad 5$

45. Since the inequality $|x| < 0$ is never true, there is no solution, or $\varnothing$. Graphing the solution set:

$\quad\quad\quad\quad\quad\quad 0$

47. Solving the equation:
$$2x - 3 = 2(x - 3)$$
$$2x - 3 = 2x - 6$$
$$-3 = -6$$
Since this statement is false, there is no solution, or $\varnothing$.

49. Since $|4y + 8| = -1$ is never true, there is no solution, or $\varnothing$.

51. Solving the inequality:
$$|5 - 8t| + 4 \le 1$$
$$|5 - 8t| \le -3$$
Since this statement is never true, there is no solution, or $\varnothing$.

Chapters 1-2 Cumulative Review

1. Simplifying: $7^2 = 7 \cdot 7 = 49$

3. Simplifying: $-5 + 3 = -2$

5. Simplifying: $-5 - 6 = -11$

7. Simplifying: $3 + 2(7 + 4) = 3 + 2(11) = 3 + 22 = 25$

9. Simplifying: $-(6 - 4) - (3 - 8) = -2 - (-5) = -2 + 5 = 3$

11. Simplifying: $\dfrac{5(-3) + 6}{8(-3) + 7(3)} = \dfrac{-15 + 6}{-24 + 21} = \dfrac{-9}{-3} = 3$

13. Simplifying: $5(2x + 4) + 6 = 10x + 20 + 6 = 10x + 26$

15. Simplifying: $2x + 6 + 3x + 5 = (2x + 3x) + (6 + 5) = 5x + 11$

17. Computing the value: $-4 + 5(-6) = -4 - 30 = -34$

19. The set is $A \cup C = \{2, 3, 4, 6, 8, 9\}$.

21. The set is $\{4, 5\}$.

23. The rational numbers are: $-13, -6.7, 0, \tfrac{1}{2}, 2, \tfrac{5}{2}$

25. The irrational numbers are: $-\sqrt{5}, \pi, \sqrt{13}$

27. The pattern is to multiply by -3, so the next number is $18(-3) = -54$.

29. Solving the equation:
$$3 - \tfrac{4}{5}a = -5$$
$$5\left(3 - \tfrac{4}{5}a\right) = 5(-5)$$
$$15 - 4a = -25$$
$$-4a = -40$$
$$a = 10$$

31. Solving the equation:
$$\tfrac{y}{4} - 1 + \tfrac{3y}{8} = \tfrac{3}{4} - y$$
$$8\left(\tfrac{y}{4} - 1 + \tfrac{3y}{8}\right) = 8\left(\tfrac{3}{4} - y\right)$$
$$2y - 8 + 3y = 6 - 8y$$
$$5y - 8 = -8y + 6$$
$$13y = 14$$
$$y = \tfrac{14}{13}$$

33. Solving the equation:
$$7 - 2(8t - 3) = 4(t - 2)$$
$$7 - 16t + 6 = 4t - 8$$
$$-16t + 13 = 4t - 8$$
$$-20t = -21$$
$$t = \frac{21}{20}$$

35. Solving for t:
$$40 = 4 + (n - 1)(9)$$
$$40 = 4 + 9n - 9$$
$$40 = 9n - 5$$
$$9n = 45$$
$$n = 5$$

37. Solving for C:
$$F = \frac{9}{5}C + 32$$
$$F - 32 = \frac{9}{5}C$$
$$C = \frac{5}{9}(F - 32)$$

39. The first four terms are: $1, \frac{1}{4}, \frac{1}{9}, \frac{1}{16}$. The sequence is decreasing.

41. Let x and $2x + 15$ represent the two angles. Since they are complementary, the equation is:
$$x + 2x + 15 = 90$$
$$3x + 15 = 90$$
$$3x = 75$$
$$x = 25$$
The angles are $25°$ and $65°$.

43. Solving the inequality:
$$-\frac{1}{2} \le \frac{1}{6}x \le \frac{1}{3}$$
$$-3 \le x \le 2$$

The solution set is $[-3, 2]$.

45. Solving the inequality:
$$|a| - 2 = 3$$
$$|a| = 5$$
$$a = -5, 5$$

47. Solving the equation:
$$|6x - 2| + 4 = 16$$
$$|6x - 2| = 12$$
$$6x - 2 = -12, 12$$
$$6x = -10, 14$$
$$x = -\frac{5}{3}, \frac{7}{3}$$

49. Solving the inequality:
$$|5x - 1| > 3$$

$$5x - 1 < -3 \qquad \text{or} \qquad 5x - 1 > 3$$
$$5x < -2 \qquad\qquad\qquad 5x > 4$$
$$x < -\frac{2}{5} \qquad\qquad\qquad x > \frac{4}{5}$$

Graphing the solution set:

$$-2/5 \qquad\qquad\qquad 4/5$$

Chapter 2 Test

1. Solving the equation:
$$x - 5 = 7$$
$$x = 12$$

2. Solving the equation:
$$3y = -4$$
$$y = -\frac{4}{3}$$

3. Solving the equation:
$$5 - \frac{4}{7}a = -11$$
$$7\left(5 - \frac{4}{7}a\right) = 7(-11)$$
$$35 - 4a = -77$$
$$-4a = -112$$
$$a = 28$$

4. Solving the equation:
$$\frac{1}{5}x - \frac{1}{2} - \frac{1}{10}x + \frac{2}{5} = \frac{3}{10}x + \frac{1}{2}$$
$$10\left(\frac{1}{5}x - \frac{1}{2} - \frac{1}{10}x + \frac{2}{5}\right) = 10\left(\frac{3}{10}x + \frac{1}{2}\right)$$
$$2x - 5 - x + 4 = 3x + 5$$
$$x - 1 = 3x + 5$$
$$-2x = 6$$
$$x = -3$$

5. Solving the equation:
$$5(x-1) - 2(2x+3) = 5x - 4$$
$$5x - 5 - 4x - 6 = 5x - 4$$
$$x - 11 = 5x - 4$$
$$-4x = 7$$
$$x = -\frac{7}{4}$$

6. Solving the equation:
$$0.07 - 0.02(3x+1) = -0.04x + 0.01$$
$$0.07 - 0.06x - 0.02 = -0.04x + 0.01$$
$$-0.06x + 0.05 = -0.04x + 0.01$$
$$-0.02x = -0.04$$
$$x = 2$$

7. Solving for w:
$$P = 2l + 2w$$
$$P - 2l = 2w$$
$$w = \frac{P - 2l}{2}$$

8. Solving for B:
$$A = \frac{1}{2}h(b + B)$$
$$2A = h(b + B)$$
$$2A = hb + hB$$
$$2A - hb = hB$$
$$B = \frac{2A - hb}{h}$$

9. The first four terms are: $-1, 1, 3, 5$. The sequence is increasing.

10. The first four terms are: $-3, 9, -27, 81$. The sequence is alternating.

11. Substituting $R = 4.5$ and $r = 4.45$ into the volume formula: $V = \frac{4}{3}\pi\left(4.5^3 - 4.45^3\right) \approx 12.6$ cubic inches

12. Substituting $W = 11$ and $L = 6$: $P = \frac{11}{11 + 6} = \frac{11}{17} \approx 64.7\%$

13. The amount collected for the day is $881.25 - \$75 = \$806.25$. Since this amount includes the sales s plus the tax, the equation is:
$$s + 0.075s = 806.25$$
$$1.075s = 806.25$$
$$s = 750$$
The portion that is sales tax is: $0.075(750) = 56.25$. The sales tax collected is $56.25.

14. Let x and $2x + 15$ represent the two angles. Since they are supplementary:
$$x + 2x + 15 = 180$$
$$3x + 15 = 180$$
$$3x = 165$$
$$x = 55$$
The angles are $55°$ and $125°$.

15.   Solving the inequality:

$$-5t \leq 30$$
$$t \geq -6$$

The solution set is $[-6, \infty)$. Graphing the solution set:

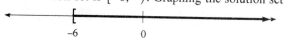

16.   Solving the inequality:

$$5 - \frac{3}{2}x > -1$$
$$-\frac{3}{2}x > -6$$
$$-3x > -12$$
$$x < 4$$

The solution set is $(-\infty, 4)$. Graphing the solution set:

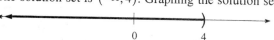

17.   Solving the inequality:

$$1.6x - 2 < 0.8x + 2.8$$
$$0.8x - 2 < 2.8$$
$$0.8x < 4.8$$
$$x < 6$$

The solution set is $(-\infty, 6)$. Graphing the solution set:

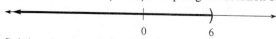

18.   Solving the inequality:

$$3(2y + 4) \geq 5(y - 8)$$
$$6y + 12 \geq 5y - 40$$
$$y + 12 \geq -40$$
$$y \geq -52$$

The solution set is $[-52, \infty)$. Graphing the solution set:

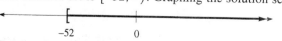

19.   Solving the equation:

$$\left|\frac{1}{4}x - 1\right| = \frac{1}{2}$$
$$\frac{1}{4}x - 1 = -\frac{1}{2}, \frac{1}{2}$$
$$\frac{1}{4}x = \frac{1}{2}, \frac{3}{2}$$
$$x = 2, 6$$

20.   Solving the equation:

$$\left|\frac{2}{3}a + 4\right| = 6$$
$$\frac{2}{3}a + 4 = -6, 6$$
$$\frac{2}{3}a = -10, 2$$
$$2a = -30, 6$$
$$a = -15, 3$$

21.   Solving the equation:

$$|3 - 2x| + 5 = 2$$
$$|3 - 2x| = -3$$

Since this statement is false, there is no solution, or $\varnothing$.

22.   Solving the equation:

$$5 = |3y + 6| - 4$$
$$9 = |3y + 6|$$
$$3y + 6 = -9, 9$$
$$3y = -15, 3$$
$$y = -5, 1$$

23.   Solving the inequality:

$$|6x - 1| > 7$$
$$6x - 1 < -7 \quad \text{or} \quad 6x - 1 > 7$$
$$6x < -6 \qquad\qquad 6x > 8$$
$$x < -1 \qquad\qquad x > \frac{4}{3}$$

Graphing the solution set:

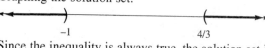

24.   Solving the inequality:

$$|3x - 5| - 4 \leq 3$$
$$|3x - 5| \leq 7$$
$$-7 \leq 3x - 5 \leq 7$$
$$-2 \leq 3x \leq 12$$
$$-\frac{2}{3} \leq x \leq 4$$

Graphing the solution set:

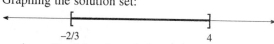

25.   Since the inequality is always true, the solution set is all real numbers. Graphing the solution set:

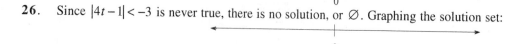

26.   Since $|4t - 1| < -3$ is never true, there is no solution, or $\varnothing$. Graphing the solution set:

―――――――――――――――――
0

Chapter 3
Equations and Inequalities in Two Variables

3.1 Paired Data and the Rectangular Coordinate System

1. Plotting the points:

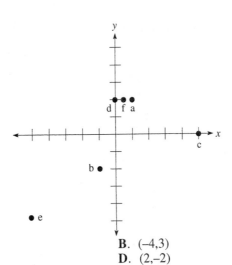

3.
A. (4,1)
C. (−2,−5)
E. (0,5)
G. (1,0)

B. (−4,3)
D. (2,−2)
F. (−4,0)

5. The *x*-intercept is 3 and the *y*-intercept is −2.

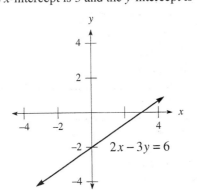

7. The *x*-intercept is 5 and the *y*-intercept is −4.

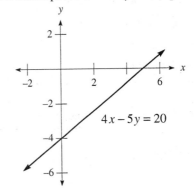

9.   The x-intercept is $-\frac{3}{2}$ and the y-intercept is 3.

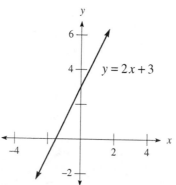

$y = 2x + 3$

11.   Table b, since its values match the equation.

13.   Graphing the line:

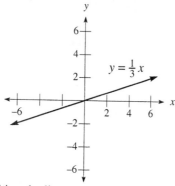

$y = \frac{1}{3}x$

15.   Graphing the line:

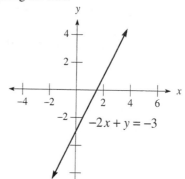

$-2x + y = -3$

17.   Graphing the line:

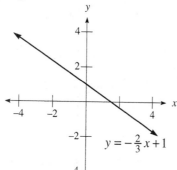

$y = -\frac{2}{3}x + 1$

19.   Graphing the line:

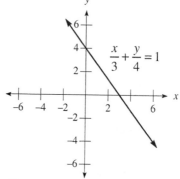

$\frac{x}{3} + \frac{y}{4} = 1$

21.   Since the x-intercept is 3 and the y-intercept is -2, this is the graph of b.

23. **a.** Graphing the line:

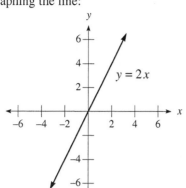

b. Graphing the line:

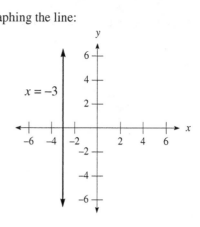

c. Graphing the line:

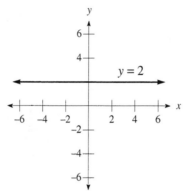

25. **a.** Graphing the line:

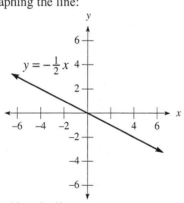

b. Graphing the line:

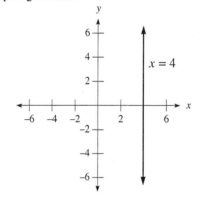

c. Graphing the line:

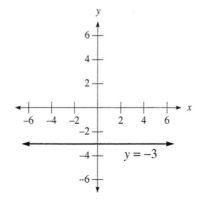

27. Graphing the line:

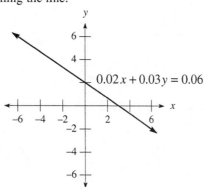

$0.02x + 0.03y = 0.06$

29. Graphing the line:

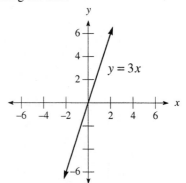

$y = 3x$

31. Graphing the distance:

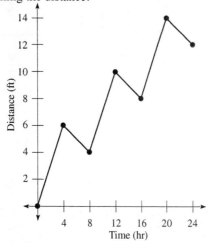

33. Sketching the line graph:

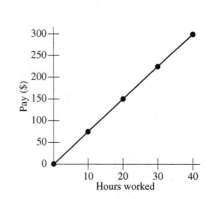

35. Sketching the bar chart:

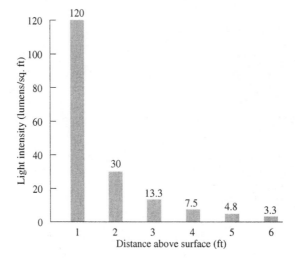

37. **a.**   Graphing the line:

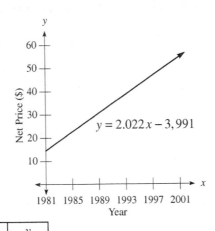

b.   Completing the table:

x	y
1982	16.60
1985	22.67
1989	30.76
1993	38.85
1997	46.93
2001	55.02

39. **a.**   There were 159 fewer deaths each successive year.
   b.   Substituting $y = 1930$: $D = 325{,}870 - 159(1930) = 19{,}000$ deaths
   c.   Substituting $y = 1975$: $D = 325{,}870 - 159(1975) = 11{,}845$ deaths
   d.   No, because for the year 2050, $D = -80$, and there cannot be a negative number of accident-related deaths.

41. Solving the equation:

$$5x - 4 = -3x + 12$$
$$8x - 4 = 12$$
$$8x = 16$$
$$x = 2$$

43. Solving the equation:

$$\tfrac{1}{2} - \tfrac{1}{8}(3t - 4) = -\tfrac{7}{8}t$$
$$8\left(\tfrac{1}{2} - \tfrac{1}{8}(3t - 4)\right) = 8\left(-\tfrac{7}{8}t\right)$$
$$4 - (3t - 4) = -7t$$
$$4 - 3t + 4 = -7t$$
$$4t + 8 = 0$$
$$4t = -8$$
$$t = -2$$

45. **a.**   Constructing a line graph:

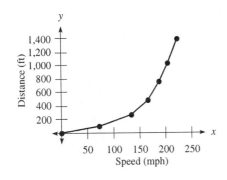

b.   Constructing a line graph:

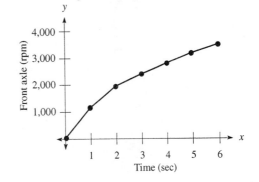

3.2   The Slope of a Line

1.   The slope is $\frac{3}{2}$.

3.   There is no slope (undefined).

5.   The slope is $\frac{2}{3}$.

7.   Finding the slope: $m = \dfrac{4-1}{4-2} = \dfrac{3}{2}$

Sketching the graph:

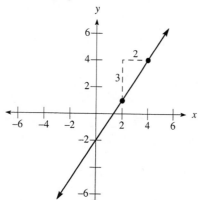

9.   Finding the slope: $m = \dfrac{2-4}{5-1} = \dfrac{-2}{4} = -\dfrac{1}{2}$

Sketching the graph:

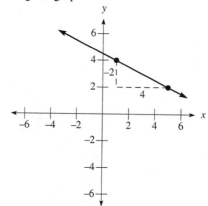

11.   Finding the slope: $m = \dfrac{2-(-3)}{4-1} = \dfrac{2+3}{3} = \dfrac{5}{3}$

Sketching the graph:

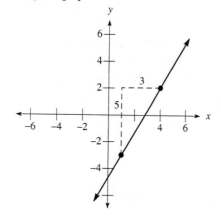

13.   Finding the slope: $m = \dfrac{3-(-2)}{1-(-3)} = \dfrac{3+2}{1+3} = \dfrac{5}{4}$

Sketching the graph:

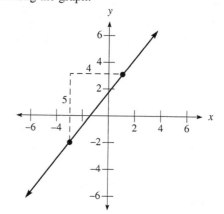

15. Finding the slope: $m = \dfrac{-2-2}{3-(-3)} = \dfrac{-4}{3+3} = -\dfrac{2}{3}$

Sketching the graph:

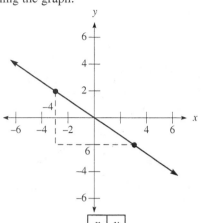

17. Finding the slope: $m = \dfrac{-2-(-5)}{3-2} = \dfrac{-2+5}{1} = 3$

Sketching the graph:

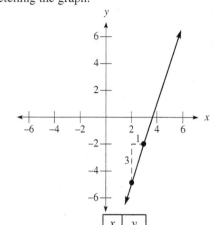

19. Completing the table:

x	y
0	2
3	0

Finding the slope: $m = \dfrac{2-0}{0-3} = -\dfrac{2}{3}$

21. Completing the table:

x	y
0	-5
3	-3

Finding the slope: $m = \dfrac{-5-(-3)}{0-3} = \dfrac{-5+3}{-3} = \dfrac{2}{3}$

23. Graphing the line:

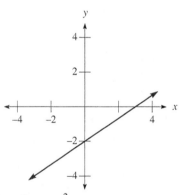

The slope is $m = \dfrac{2}{3}$.

25. Graphing the line:

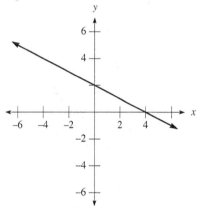

The slope is $m = -\dfrac{1}{2}$.

27. **a.** Since the slopes between each successive pairs of points is 2, this could represent ordered pairs from a line.
  b. Since the slopes between each successive pairs of points is not the same, this could not represent ordered pairs from a line.

29. Finding the slope of this line: $m = \dfrac{1-3}{-8-2} = \dfrac{-2}{-10} = \dfrac{1}{5}$. Since the parallel slope is the same, its slope is $\dfrac{1}{5}$.

31. Finding the slope of this line: $m = \dfrac{2-(-6)}{5-5} = \dfrac{8}{0} =$ undefined

Since the perpendicular slope is a horizontal line, its slope is 0.

33. Using the slope formula:
$$\dfrac{\Delta y}{12} = \dfrac{2}{3}$$
$$3\Delta y = 24$$
$$\Delta y = 8$$

35. **a.** Let d represent the distance. Using the slope formula:
$$\frac{0-1106}{d-0}=-\frac{7}{100}$$
$$-7d=-110600$$
$$d=15800$$
The distance to point A is 15,800 feet.

b. The slope is $-\frac{7}{100}=-0.07$.

37. **a.** It takes 10 minutes for all the ice to melt.   **b.** It takes 20 minutes before the water boils.
c. The slope of A is 20°C per minute.   **d.** The slope of C is 10°C per minute.
e. It is changing faster during the first minute, since its slope is greater.

39. Substituting $x=4$:
$$3(4)+2y=12$$
$$12+2y=12$$
$$2y=0$$
$$y=0$$

41. Solving for y:
$$3x+2y=12$$
$$2y=-3x+12$$
$$y=-\frac{3}{2}x+6$$

43. Solving for t:
$$A=P+Prt$$
$$A-P=Prt$$
$$t=\frac{A-P}{Pr}$$

45. Graphing the curves:

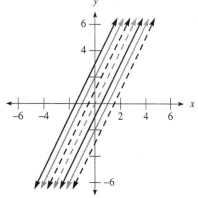

47. Graphing the curves:

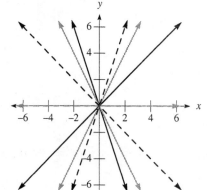

49. Graphing the curves:

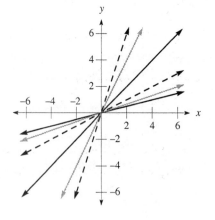

3.3   The Equation of a Line

1. Using the slope-intercept formula: $y = 2x + 3$

3. Using the slope-intercept formula: $y = x - 5$

5. Using the slope-intercept formula: $y = \frac{1}{2}x + \frac{3}{2}$

7. Using the slope-intercept formula: $y = 4$

9. The slope is 3, the y-intercept is -2, and the perpendicular slope is $-\frac{1}{3}$.

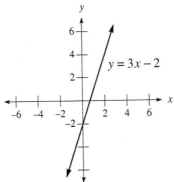

11. The slope is $\frac{2}{3}$, the y-intercept is -4, and the perpendicular slope is $-\frac{3}{2}$.

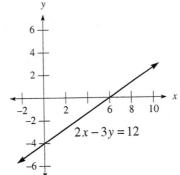

13. The slope is $-\frac{4}{5}$, the y-intercept is 4, and the perpendicular slope is $\frac{5}{4}$.

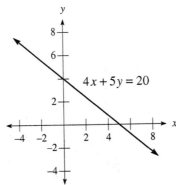

15. The slope is $\frac{1}{2}$ and the y-intercept is -4. Using the slope-intercept form, the equation is $y = \frac{1}{2}x - 4$.

17. The slope is $-\frac{2}{3}$ and the y-intercept $= 3$. Using the slope-intercept form, the equation is $y = -\frac{2}{3}x + 3$.

19.   Using the point-slope formula:
$$y - (-5) = 2(x - (-2))$$
$$y + 5 = 2(x + 2)$$
$$y + 5 = 2x + 4$$
$$y = 2x - 1$$

21.   Using the point-slope formula:
$$y - 1 = -\tfrac{1}{2}(x - (-4))$$
$$y - 1 = -\tfrac{1}{2}(x + 4)$$
$$y - 1 = -\tfrac{1}{2}x - 2$$
$$y = -\tfrac{1}{2}x - 1$$

23.   Using the point-slope formula:
$$y - 2 = -3\left(x - \left(-\tfrac{1}{3}\right)\right)$$
$$y - 2 = -3\left(x + \tfrac{1}{3}\right)$$
$$y - 2 = -3x - 1$$
$$y = -3x + 1$$

25.   First find the slope: $m = \dfrac{-1 - (-4)}{1 - (-2)} = \dfrac{-1 + 4}{1 + 2} = \dfrac{3}{3} = 1$. Using the point-slope formula:
$$y - (-1) = 1(x - 1)$$
$$y + 1 = x - 1$$
$$-x + y = -2$$
$$x - y = 2$$

27.   First find the slope: $m = \dfrac{1 - (-5)}{2 - (-1)} = \dfrac{1 + 5}{2 + 1} = \dfrac{6}{3} = 2$. Using the point-slope formula:
$$y - 1 = 2(x - 2)$$
$$y - 1 = 2x - 4$$
$$-2x + y = -3$$
$$2x - y = 3$$

29.   First find the slope: $m = \dfrac{-1 - \left(-\tfrac{1}{5}\right)}{-\tfrac{1}{3} - \tfrac{1}{3}} = \dfrac{-1 + \tfrac{1}{5}}{-\tfrac{2}{3}} = \dfrac{-\tfrac{4}{5}}{-\tfrac{2}{3}} = \tfrac{4}{5} \cdot \tfrac{3}{2} = \tfrac{6}{5}$. Using the point-slope formula:
$$y - (-1) = \tfrac{6}{5}\left(x - \left(-\tfrac{1}{3}\right)\right)$$
$$y + 1 = \tfrac{6}{5}\left(x + \tfrac{1}{3}\right)$$
$$5(y + 1) = 6\left(x + \tfrac{1}{3}\right)$$
$$5y + 5 = 6x + 2$$
$$6x - 5y = 3$$

31.   Two points on the line are (0,–4) and (2,0). Finding the slope: $m = \dfrac{0 - (-4)}{2 - 0} = \dfrac{4}{2} = 2$

Using the slope-intercept form, the equation is $y = 2x - 4$.

33.   Two points on the line are (0,4) and (–2,0). Finding the slope: $m = \dfrac{0 - 4}{-2 - 0} = \dfrac{-4}{-2} = 2$

Using the slope-intercept form, the equation is $y = 2x + 4$.

35. The slope is 0 and the y-intercept is –2.

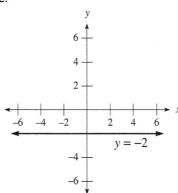

$$y = -2$$

37. First find the slope:
$$3x - y = 5$$
$$-y = -3x + 5$$
$$y = 3x - 5$$
So the slope is 3. Using $(-1,4)$ in the point-slope formula:
$$y - 4 = 3(x - (-1))$$
$$y - 4 = 3(x + 1)$$
$$y - 4 = 3x + 3$$
$$y = 3x + 7$$

39. First find the slope:
$$2x - 5y = 10$$
$$-5y = -2x + 10$$
$$y = \tfrac{2}{5}x - 2$$
So the perpendicular slope is $-\tfrac{5}{2}$. Using $(-4,-3)$ in the point-slope formula:
$$y - (-3) = -\tfrac{5}{2}(x - (-4))$$
$$y + 3 = -\tfrac{5}{2}(x + 4)$$
$$y + 3 = -\tfrac{5}{2}x - 10$$
$$y = -\tfrac{5}{2}x - 13$$

41. The perpendicular slope is $\tfrac{1}{4}$. Using $(-1,0)$ in the point-slope formula:
$$y - 0 = \tfrac{1}{4}(x - (-1))$$
$$y = \tfrac{1}{4}(x + 1)$$
$$y = \tfrac{1}{4}x + \tfrac{1}{4}$$

43. Using the points $(3,0)$ and $(0,2)$, first find the slope: $m = \dfrac{2 - 0}{0 - 3} = -\tfrac{2}{3}$

Using the slope-intercept formula, the equation is: $y = -\tfrac{2}{3}x + 2$

45. **a.** Using the points $(0,32)$ and $(25,77)$, first find the slope: $m = \dfrac{77 - 32}{25 - 0} = \dfrac{45}{25} = \dfrac{9}{5}$

Using the slope-intercept formula, the equation is: $F = \tfrac{9}{5}C + 32$

b. Substituting $C = 30$: $F = \tfrac{9}{5}(30) + 32 = 54 + 32 = 86°$

47.   First find the slope: $m = \dfrac{1.5-1}{155-98} = \dfrac{\frac{1}{2}}{57} = \dfrac{1}{114}$. Using $(98,1)$ in the point-slope formula:

$$y - 1 = \tfrac{1}{114}(x - 98)$$
$$y - 1 = \tfrac{1}{114}x - \tfrac{49}{57}$$
$$y = \tfrac{1}{114}x + \tfrac{8}{57}$$

49.   **a.**   First finding the slope: $m = \dfrac{10-9.3}{5-3} = \dfrac{0.7}{2} = 0.35$. Using $(5,10)$ in the point-slope formula:

$$C - 10 = 0.35(x - 5)$$
$$C - 10 = 0.35x - 1.75$$
$$C = 0.35x + 8.25$$

b.   The slope is 0.35; this means the variable cost increases by 0.35 cents/mile each year.

51.   Let w represent the width and $4w + 3$ represent the length. Using the perimeter formula:

$$2w + 2(4w + 3) = 56$$
$$2w + 8w + 6 = 56$$
$$10w = 50$$
$$w = 5$$

The width is 5 inches and the length is 23 inches.

53.   The total amount collected is: \$732.50 – \$66 = \$666.50

Let x represent the sales. Since this amount includes the sales tax:

$$x + 0.075x = 666.50$$
$$1.075x = 666.50$$
$$x = 620$$

The amount which is sales tax is therefore: \$666.50 – \$620 = \$46.50

55.   Writing the equation in slope-intercept form:

$$\frac{x}{2} + \frac{y}{3} = 1$$
$$\frac{y}{3} = -\frac{x}{2} + 1$$
$$y = -\tfrac{3}{2}x + 3$$

The slope is $-\tfrac{3}{2}$, the x-intercept is 2, and the y-intercept is 3.

57.   Writing the equation in slope-intercept form:

$$\frac{x}{-2} + \frac{y}{3} = 1$$
$$\frac{y}{3} = \frac{x}{2} + 1$$
$$y = \tfrac{3}{2}x + 3$$

The slope is $\tfrac{3}{2}$, the x-intercept is –2, and the y-intercept is 3.

59.   The x-intercept is a, the y-intercept is b, and the slope is $-\dfrac{b}{a}$.

3.4   Linear Inequalities in Two Variables

1. Graphing the solution set:

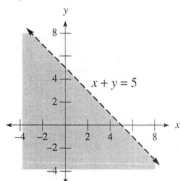

3. Graphing the solution set:

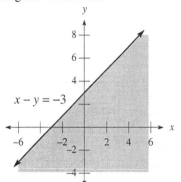

5. Graphing the solution set:

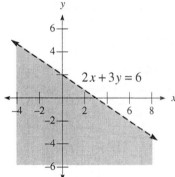

7. Graphing the solution set:

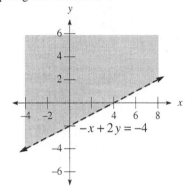

9. Graphing the solution set:

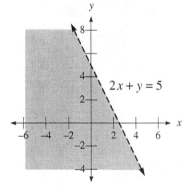

11. Graphing the solution set:

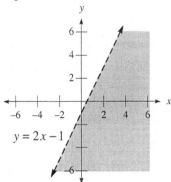

13. The inequality is $x + y > 4$.

15. The inequality is $-x + 2y \leq 4$ or $y \leq \frac{1}{2}x + 2$.

17.   Graphing the inequality:

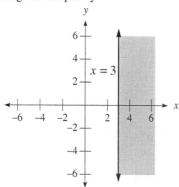

19.   Graphing the inequality:

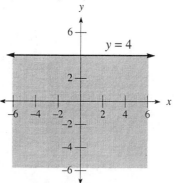

21.   Graphing the inequality:

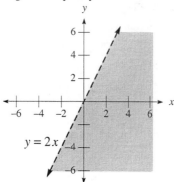

23.   Graphing the inequality:

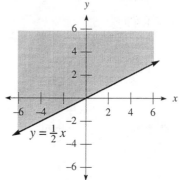

25.   Graphing the inequality:

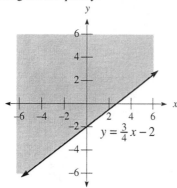

27.   Graphing the inequality:

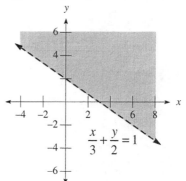

29.   Graphing the region:

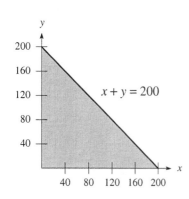

31.   Graphing the region:

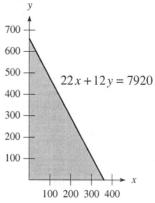

33. Graphing the region:

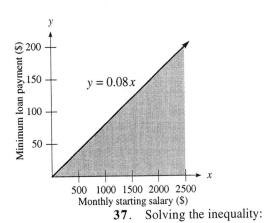

$y = 0.08x$

Minimum loan payment ($)

Monthly starting salary ($)

35. Solving the inequality:

$$\frac{1}{3} + \frac{y}{5} \le \frac{26}{15}$$

$$15\left(\frac{1}{3} + \frac{y}{5}\right) \le 15\left(\frac{26}{15}\right)$$

$$5 + 3y \le 26$$

$$3y \le 21$$

$$y \le 7$$

37. Solving the inequality:

$$5t - 4 > 3t - 8$$

$$2t - 4 > -8$$

$$2t > -4$$

$$t > -2$$

39. Solving the inequality:

$$-9 < -4 + 5t < 6$$

$$-5 < 5t < 10$$

$$-1 < t < 2$$

41. Graphing the inequality:

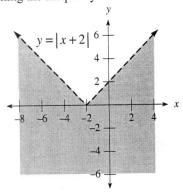

$y = |x + 2|$

43. Graphing the inequality:

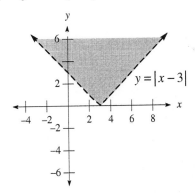

$y = |x - 3|$

3.5   Introduction to Functions

1. The domain is {1,2,4} and the range is {1,3,5}. This is a function.
3. The domain is {−1,1,2} and the range is {−5,3}. This is a function.
5. The domain is {3,7} and the range is {−1,4}. This is not a function.
7. Yes, since it passes the vertical line test.
9. No, since it fails the vertical line test.
11. No, since it fails the vertical line test.
13. Yes, since it passes the vertical line test.
15. Yes, since it passes the vertical line test.
17. The domain is all real numbers and the range is $\{y \mid y \geq -1\}$. This is a function.

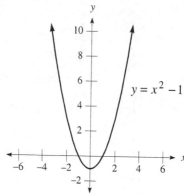

19. The domain is all real numbers and the range is $\{y \mid y \geq 4\}$. This is a function.

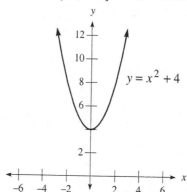

21. The domain is $\{x \mid x \geq -1\}$ and the range is all real numbers. This is not a function.

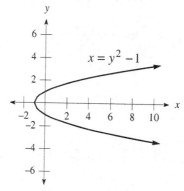

23.   The domain is $\{x \mid x \geq 4\}$ and the range is all real numbers. This is not a function.

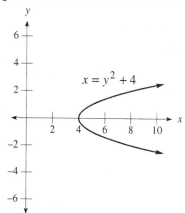

$$x = y^2 + 4$$

25.   The domain is all real numbers and the range is $\{y \mid y \geq 0\}$. This is a function.

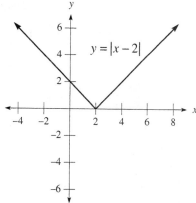

$$y = |x - 2|$$

27.   The domain is all real numbers and the range is $\{y \mid y \geq -2\}$. This is a function.

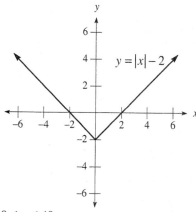

$$y = |x| - 2$$

29. a.   The equation is $y = 8.5x$ for $10 \leq x \leq 40$.

b.   Completing the table:

Hours Worked	Function Rule	Gross Pay (\$)
x	$y = 8.5x$	y
10	$y = 8.5(10) = 85$	85
20	$y = 8.5(20) = 170$	170
30	$y = 8.5(30) = 255$	255
40	$y = 8.5(40) = 340$	340

c.   Constructing a line graph:

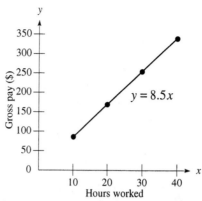

d.   The domain is $\{x \mid 10 \le x \le 40\}$ and the range is $\{y \mid 85 \le y \le 340\}$.

e.   The minimum is \$85 and the maximum is \$340.

Time (sec)	Function Rule	Distance (ft)
t	$h = 16t - 16t^2$	h
0	$h = 16(0) - 16(0)^2$	0
0.1	$h = 16(0.1) - 16(0.1)^2$	1.44
0.2	$h = 16(0.2) - 16(0.2)^2$	2.56
0.3	$h = 16(0.3) - 16(0.3)^2$	3.36
0.4	$h = 16(0.4) - 16(0.4)^2$	3.84
0.5	$h = 16(0.5) - 16(0.5)^2$	4
0.6	$h = 16(0.6) - 16(0.6)^2$	3.84
0.7	$h = 16(0.7) - 16(0.7)^2$	3.36
0.8	$h = 16(0.8) - 16(0.8)^2$	2.56
0.9	$h = 16(0.9) - 16(0.9)^2$	1.44
1	$h = 16(1) - 16(1)^2$	0

31. **a.**   Completing the table:

b.   The domain is $\{t \mid 0 \le t \le 1\}$ and the range is $\{h \mid 0 \le h \le 4\}$.

c.   Graphing the function:

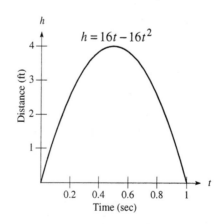

33. **a.** Graphing the function:

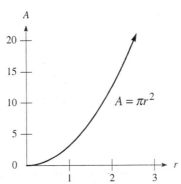

b. The domain is $\{r \mid 0 \le r \le 3\}$ and the range is $\{A \mid 0 \le A \le 9\pi\}$.

35. **a.** Yes, since it passes the vertical line test.
b. The domain is $\{t \mid 0 \le t \le 6\}$ and the range is $\{h \mid 0 \le h \le 60\}$.
c. At time $t = 3$ the ball reaches its maximum height.
d. The maximum height is $h = 60$.   **e.** At time $t = 6$ the ball hits the ground.

37. Substituting $x = 4$: $y = 3(4) - 2 = 12 - 2 = 10$ **39.** Substituting $x = -4$: $y = 3(-4) - 2 = -12 - 2 = -14$

41. Substituting $x = 2$: $y = (2)^2 - 3 = 4 - 3 = 1$ **43.** Substituting $x = 0$: $y = (0)^2 - 3 = 0 - 3 = -3$

45. **a.** Figure 3   **b.** Figure 4
    c. Figure 2   **d.** Figure 1

3.6 Function Notation

1. Evaluating the function: $f(2) = 2(2) - 5 = 4 - 5 = -1$
3. Evaluating the function: $f(-3) = 2(-3) - 5 = -6 - 5 = -11$
5. Evaluating the function: $g(-1) = (-1)^2 + 3(-1) + 4 = 1 - 3 + 4 = 2$
7. Evaluating the function: $g(-3) = (-3)^2 + 3(-3) + 4 = 9 - 9 + 4 = 4$
9. First evaluate each function:
    $g(4) = (4)^2 + 3(4) + 4 = 16 + 12 + 4 = 32$ $f(4) = 2(4) - 5 = 8 - 5 = 3$
    Now evaluating: $g(4) + f(4) = 32 + 3 = 35$
11. First evaluate each function:
    $f(3) = 2(3) - 5 = 6 - 5 = 1$ $g(2) = (2)^2 + 3(2) + 4 = 4 + 6 + 4 = 14$
    Now evaluating: $f(3) - g(2) = 1 - 14 = -13$
13. Evaluating the function: $f(0) = 3(0)^2 - 4(0) + 1 = 0 - 0 + 1 = 1$
15. Evaluating the function: $g(-4) = 2(-4) - 1 = -8 - 1 = -9$
17. Evaluating the function: $f(-1) = 3(-1)^2 - 4(-1) + 1 = 3 + 4 + 1 = 8$
19. Evaluating the function: $g(10) = 2(10) - 1 = 20 - 1 = 19$
21. Evaluating the function: $f(3) = 3(3)^2 - 4(3) + 1 = 27 - 12 + 1 = 16$
23. Evaluating the function: $g\left(\tfrac{1}{2}\right) = 2\left(\tfrac{1}{2}\right) - 1 = 1 - 1 = 0$ **25.** Evaluating the function: $f(a) = 3a^2 - 4a + 1$
27. $f(1) = 4$ **29.** $g\left(\tfrac{1}{2}\right) = 0$
31. $g(-2) = 2$ **33.** Evaluating the function: $f(0) = 2(0)^2 - 8 = 0 - 8 = -8$
35. Evaluating the function: $g(-4) = \tfrac{1}{2}(-4) + 1 = -2 + 1 = -1$
37. Evaluating the function: $f(a) = 2a^2 - 8$ **39.** Evaluating the function: $f(b) = 2b^2 - 8$
41. Evaluating the function: $f[g(2)] = f\left[\tfrac{1}{2}(2) + 1\right] = f(2) = 2(2)^2 - 8 = 8 - 8 = 0$

43. Evaluating the function: $g\left[f(-1)\right]=g\left[2(-1)^2-8\right]=g(-6)=\frac{1}{2}(-6)+1=-3+1=-2$

45. Evaluating the function: $g\left[f(0)\right]=g\left[2(0)^2-8\right]=g(-8)=\frac{1}{2}(-8)+1=-4+1=-3$

47. Graphing the function:

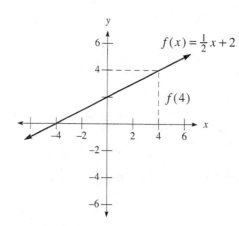

49. Finding where $f(x)=x$:
$$\frac{1}{2}x+2=x$$
$$2=\frac{1}{2}x$$
$$x=4$$

51. Graphing the function:

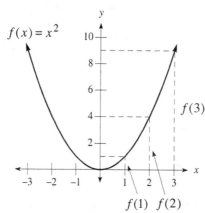

53. Evaluating: $V(3)=150\cdot2^{3/3}=150\cdot2=300$; The painting is worth \$300 in 3 years.

Evaluating: $V(6)=150\cdot2^{6/3}=150\cdot4=600$; The painting is worth \$600 in 6 years.

55. Let x represent the width and $2x+3$ represent the length. Then the perimeter is given by:
$$P(x)=2(x)+2(2x+3)=2x+4x+6=6x+6,\text{ where }x>0$$

57. Finding the values:
$$A(2)=3.14(2)^2=3.14(4)=12.56$$
$$A(5)=3.14(5)^2=3.14(25)=78.5$$
$$A(10)=3.14(10)^2=3.14(100)=314$$

59. **a.** Evaluating: $V(3.75) = -3300(3.75) + 18000 = \$5,625$

b. Evaluating: $V(5) = -3300(5) + 18000 = \$1,500$

c. The domain of this function is $\{t \mid 0 \le t \le 5\}$.

d. Sketching the graph:

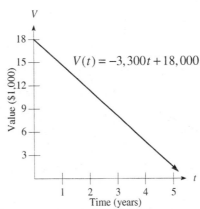

$V(t) = -3,300t + 18,000$

Value (\$1,000)

Time (years)

e. The range of this function is $\{V(t) \mid 1,500 \le V(t) \le 18,000\}$.

f. Solving $V(t) = 10000$:
$$-3300t + 18000 = 10000$$
$$-3300t = -8000$$
$$t \approx 2.42$$
The copier will be worth \$10,000 after approximately 2.42 years.

61. Solving the equation:
$$|3x - 5| = 7$$
$$3x - 5 = -7, 7$$
$$3x = -2, 12$$
$$x = -\tfrac{2}{3}, 4$$

63. Solving the equation:
$$|4y + 2| - 8 = -2$$
$$|4y + 2| = 6$$
$$4y + 2 = -6, 6$$
$$4y = -8, 4$$
$$y = -2, 1$$

65. Solving the equation:
$$5 + |6t + 2| = 3$$
$$|6t + 2| = -2$$
Since this last equation is impossible, there is no solution, or $\varnothing$.

67. **a.** $f(2) = 2$ **b.** $f(-4) = 0$

c. $g(0) = 1$ **d.** $g(3) = 4$

69. **a.** Completing the table:

Weight (ounces)	0.6	1.0	1.1	2.5	3.0	4.8	5.0	5.3
Cost (cents)	32	32	55	78	78	124	124	147

b. The letter weighs over 2 ounces, but not over 3 ounces. As an inequality, this can be written as $2 < x \le 3$.

c. The domain is $\{x \mid 0 < x \le 6\}$.   **d.** The range is $\{32, 55, 78, 101, 124, 147\}$.

3.7   Algebra and Composition with Functions

1. Writing the formula: $f + g = f(x) + g(x) = (4x - 3) + (2x + 5) = 6x + 2$

3. Writing the formula: $g - f = g(x) - f(x) = (2x + 5) - (4x - 3) = -2x + 8$

5. Writing the formula: $fg = f(x) \cdot g(x) = (4x - 3)(2x + 5) = 8x^2 + 14x - 15$

7. Writing the formula: $g / f = \dfrac{g(x)}{f(x)} = \dfrac{2x + 5}{4x - 3}$

9. Writing the formula: $f + g = f(x) + g(x) = (3x - 5) + (x - 2) = 4x - 7$

11. Writing the formula: $g + h = g(x) + h(x) = (x - 2) + \left(3x^2 - 11x + 10\right) = 3x^2 - 10x + 8$

13. Writing the formula: $g - f = g(x) - f(x) = (x - 2) - (3x - 5) = -2x + 3$

15. Writing the formula: $fg = f(x) \cdot g(x) = (3x - 5)(x - 2) = 3x^2 - 11x + 10$

17. Writing the formula:

$$fh = f(x) \cdot h(x) = (3x - 5)\left(3x^2 - 11x + 10\right) = 9x^3 - 33x^2 + 30x - 15x^2 + 55x - 50 = 9x^3 - 48x^2 + 85x - 50$$

19. Writing the formula: $h / f = \dfrac{h(x)}{f(x)} = \dfrac{3x^2 - 11x + 10}{3x - 5} = \dfrac{(3x - 5)(x - 2)}{3x - 5} = x - 2$

21. Writing the formula: $f / h = \dfrac{f(x)}{h(x)} = \dfrac{3x - 5}{3x^2 - 11x + 10} = \dfrac{3x - 5}{(3x - 5)(x - 2)} = \dfrac{1}{x - 2}$

23. Writing the formula: $f + g + h = f(x) + g(x) + h(x) = (3x - 5) + (x - 2) + \left(3x^2 - 11x + 10\right) = 3x^2 - 7x + 3$

25. Writing the formula:

$$\begin{aligned} h + fg &= h(x) + f(x)g(x) \\ &= \left(3x^2 - 11x + 10\right) + (3x - 5)(x - 2) \\ &= 3x^2 - 11x + 10 + 3x^2 - 11x + 10 \\ &= 6x^2 - 22x + 20 \end{aligned}$$

27. Evaluating: $(f + g)(2) = f(2) + g(2) = (2 \cdot 2 + 1) + (4 \cdot 2 + 2) = 5 + 10 = 15$

29. Evaluating: $(fg)(3) = f(3) \cdot g(3) = (2 \cdot 3 + 1)(4 \cdot 3 + 2) = 7 \cdot 14 = 98$

31. Evaluating: $(h / g)(1) = \dfrac{h(1)}{g(1)} = \dfrac{4(1)^2 + 4(1) + 1}{4(1) + 2} = \dfrac{9}{6} = \dfrac{3}{2}$

33. Evaluating: $(fh)(0) = f(0) \cdot h(0) = (2(0) + 1)\left(4(0)^2 + 4(0) + 1\right) = (1)(1) = 1$

35. Evaluating: $(f + g + h)(2) = f(2) + g(2) + h(2) = (2(2) + 1) + (4(2) + 2) + \left(4(2)^2 + 4(2) + 1\right) = 5 + 10 + 25 = 40$

37. Evaluating: $(h + fg)(3) = h(3) + f(3) \cdot g(3) = \left(4(3)^2 + 4(3) + 1\right) + (2(3) + 1) \cdot (4(3) + 2) = 49 + 7 \cdot 14 = 49 + 98 = 147$

39. **a.** Evaluating: $(f \circ g)(5) = f(g(5)) = f(5 + 4) = f(9) = 9^2 = 81$

    b. Evaluating: $(g \circ f)(5) = g(f(5)) = g\left(5^2\right) = g(25) = 25 + 4 = 29$

    c. Evaluating: $(f \circ g)(x) = f(g(x)) = f(x + 4) = (x + 4)^2$

    d. Evaluating: $(g \circ f)(x) = g(f(x)) = g\left(x^2\right) = x^2 + 4$

41. **a.** Evaluating: $(f \circ g)(0) = f(g(0)) = f(4 \cdot 0 - 1) = f(-1) = (-1)^2 + 3(-1) = 1 - 3 = -2$

    b. Evaluating: $(g \circ f)(0) = g(f(0)) = g\left(0^2 + 3 \cdot 0\right) = g(0) = 4(0) - 1 = -1$

    c. Evaluating: $(f \circ g)(x) = f(g(x)) = f(4x - 1) = (4x - 1)^2 + 3(4x - 1) = 16x^2 - 8x + 1 + 12x - 3 = 16x^2 + 4x - 2$

    d. Evaluating: $(g \circ f)(x) = g(f(x)) = g\left(x^2 + 3x\right) = 4\left(x^2 + 3x\right) - 1 = 4x^2 + 12x - 1$

43. Evaluating each composition:

$$(f \circ g)(x) = f(g(x)) = f\left(\frac{x+4}{5}\right) = 5\left(\frac{x+4}{5}\right) - 4 = x + 4 - 4 = x$$

$$(g \circ f)(x) = g(f(x)) = g(5x - 4) = \frac{5x - 4 + 4}{5} = \frac{5x}{5} = x$$

Thus $(f \circ g)(x) = (g \circ f)(x) = x$.

45. **a.** Finding the revenue: $R(x) = x(11.5 - 0.05x) = 11.5x - 0.05x^2$

    b. Finding the cost: $C(x) = 2x + 200$

    c. Finding the profit: $P(x) = R(x) - C(x) = \left(11.5x - 0.05x^2\right) - (2x + 200) = -0.05x^2 + 9.5x - 200$

    d. Finding the average cost: $\overline{C}(x) = \frac{C(x)}{x} = \frac{2x + 200}{x} = 2 + \frac{200}{x}$

47. **a.** The function is $M(x) = 220 - x$.   **b.** Evaluating: $M(24) = 220 - 24 = 196$ beats per minute

    c. The training heart rate function is: $T(M) = 62 + 0.6(M - 62) = 0.6M + 24.8$

        Finding the composition: $T(M(x)) = T(220 - x) = 0.6(220 - x) + 24.8 = 156.8 - 0.6x$

        Evaluating: $T(M(24)) = 156.8 - 0.6(24) \approx 142$ beats per minute

    d. Evaluating: $T(M(36)) = 156.8 - 0.6(36) \approx 135$ beats per minute

    e. Evaluating: $T(M(48)) = 156.8 - 0.6(48) \approx 128$ beats per minute

49. Solving the equation:

$$x = 0.38(75)$$
$$x = 28.5$$

51. Solving the equation:

$$0.15x = 75$$
$$x = \frac{75}{0.15} = 500$$

53. The first three terms are:

$$a_1 = 10(1) - 9 = 1$$
$$a_2 = 10(2) - 9 = 11$$
$$a_3 = 10(3) - 9 = 21$$

55. The first three terms are:

$$a_1 = (1)^2 - 10 = -9$$
$$a_2 = (2)^2 - 10 = -6$$
$$a_3 = (3)^2 - 10 = -1$$

3.8   Variation

1. The variation equation is $y = Kx$. Substituting $x = 2$ and $y = 10$:

$$10 = K \cdot 2$$
$$K = 5$$

So $y = 5x$. Substituting $x = 6$: $y = 5 \cdot 6 = 30$

3. The variation equation is $y = Kx$. Substituting $x = 4$ and $y = -32$:

$$-32 = K \cdot 4$$
$$K = -8$$

So $y = -8x$. Substituting $y = -40$:

$$-40 = -8x$$
$$x = 5$$

5. The variation equation is $r = \frac{K}{s}$. Substituting $s = 4$ and $r = -3$:

$$-3 = \frac{K}{4}$$
$$K = -12$$

So $r = \frac{-12}{s}$. Substituting $s = 2$: $r = \frac{-12}{2} = -6$

7.   The variation equation is $r = \dfrac{K}{s}$. Substituting $s = 3$ and $r = 8$:

$$8 = \frac{K}{3}$$
$$K = 24$$

So $r = \dfrac{24}{s}$. Substituting $r = 48$:

$$48 = \frac{24}{s}$$
$$48s = 24$$
$$s = \tfrac{1}{2}$$

9.   The variation equation is $d = Kr^2$. Substituting $r = 5$ and $d = 10$:

$$10 = K \cdot 5^2$$
$$10 = 25K$$
$$K = \tfrac{2}{5}$$

So $d = \tfrac{2}{5}r^2$. Substituting $r = 10$: $d = \tfrac{2}{5}(10)^2 = \tfrac{2}{5} \cdot 100 = 40$

11.   The variation equation is $d = Kr^2$. Substituting $r = 2$ and $d = 100$:

$$100 = K \cdot 2^2$$
$$100 = 4K$$
$$K = 25$$

So $d = 25r^2$. Substituting $r = 3$: $d = 25(3)^2 = 25 \cdot 9 = 225$

13.   The variation equation is $y = \dfrac{K}{x^2}$. Substituting $x = 3$ and $y = 45$:

$$45 = \frac{K}{3^2}$$
$$45 = \frac{K}{9}$$
$$K = 405$$

So $y = \dfrac{405}{x^2}$. Substituting $x = 5$: $y = \dfrac{405}{5^2} = \dfrac{405}{25} = \tfrac{81}{5}$

15.   The variation equation is $y = \dfrac{K}{x^2}$. Substituting $x = 3$ and $y = 18$:

$$18 = \frac{K}{3^2}$$
$$18 = \frac{K}{9}$$
$$K = 162$$

So $y = \dfrac{162}{x^2}$. Substituting $x = 2$: $y = \dfrac{162}{2^2} = \dfrac{162}{4} = 40.5$

17.   The variation equation is $z = Kxy^2$. Substituting $x = 3$, $y = 3$, and $z = 54$:

$$54 = K(3)(3)^2$$
$$54 = 27K$$
$$K = 2$$

So $z = 2xy^2$. Substituting $x = 2$ and $y = 4$: $z = 2(2)(4)^2 = 64$

19. The variation equation is $z = Kxy^2$. Substituting $x = 1$, $y = 4$, and $z = 64$:

$$64 = K(1)(4)^2$$
$$64 = 16K$$
$$K = 4$$

So $z = 4xy^2$. Substituting $z = 32$ and $y = 1$:

$$32 = 4x(1)^2$$
$$32 = 4x$$
$$x = 8$$

21. Let l represent the length and f represent the force. The variation equation is $l = Kf$. Substituting $f = 5$ and $l = 3$:

$$3 = K \cdot 5$$
$$K = \tfrac{3}{5}$$

So $l = \tfrac{3}{5}f$. Substituting $l = 10$:

$$10 = \tfrac{3}{5}f$$
$$50 = 3f$$
$$f = \tfrac{50}{3}$$

The force required is $\tfrac{50}{3}$ pounds.

23.   **a.**   The variation equation is $T = 4P$.

     b.   Graphing the equation:

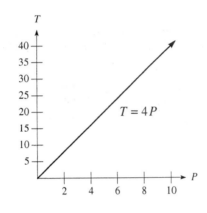

     c.   Substituting $T = 280$:

$$280 = 4P$$
$$P = 70$$

       The pressure is 70 pounds per square inch.

26. Let f represent the frequency and w represent the wavelength. The variation equation is $f = \dfrac{K}{w}$.

Substituting $w = 200$ and $f = 800$:

$$800 = \frac{K}{200}$$
$$K = 160000$$

The equation is $f = \dfrac{160000}{w}$. Substituting $w = 500$: $f = \dfrac{160000}{500} = 320$. The frequency is 320 kilocycles per second.

27.   **a.**   The variation equation is $f = \dfrac{80}{d}$.

   b.   Graphing the equation:

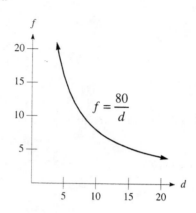

   c.   Substituting $d = 10$:

$$f = \frac{80}{10}$$
$$f = 8$$

The f-stop is 8.

29.   Let A represent the surface area, h represent the height, and r represent the radius. The variation equation is $A = Khr$.
Substituting $A = 94$, $r = 3$, and $h = 5$:

$$94 = K(3)(5)$$
$$94 = 15K$$
$$K = \tfrac{94}{15}$$

The equation is $A = \tfrac{94}{15} hr$. Substituting $r = 2$ and $h = 8$: $A = \tfrac{94}{15}(8)(2) = \tfrac{1504}{15}$. The surface area is $\tfrac{1504}{15}$ square inches

31.   Let R represent the resistance, l represent the length, and d represent the diameter. The variation equation is $R = \dfrac{Kl}{d^2}$.

Substituting $R = 10$, $l = 100$, and $d = 0.01$:

$$10 = \frac{K(100)}{(0.01)^2}$$
$$0.001 = 100K$$
$$K = 0.00001$$

The equation is $R = \dfrac{0.00001l}{d^2}$. Substituting $l = 60$ and $d = 0.02$: $R = \dfrac{0.00001(60)}{(0.02)^2} = 1.5$. The resistance is 1.5 ohms.

33.   **a.**   The variation equation is $P = 0.21\sqrt{L}$.
   b.   Graphing the equation:

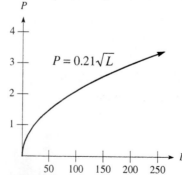

   c.   Substituting $L = 225$: $P = 0.21\sqrt{225} = 3.15$
The period is 3.15 seconds.

35. Let p represent the pitch and w represent the wavelength. The variation equation is $p = \dfrac{K}{w}$.

Substituting $p = 420$ and $w = 2.2$:
$$420 = \frac{K}{2.2}$$
$$K = 924$$

The equation is $p = \dfrac{924}{w}$. Substituting $p = 720$:
$$720 = \frac{924}{w}$$
$$720w = 924$$
$$w \approx 1.28$$

The wavelength is approximately 1.28 meters.

37. The variation equation is $F = \dfrac{Gm_1 m_2}{d^2}$.

39. Solving the inequality:
$$\left| \frac{x}{5} + 1 \right| \geq \frac{4}{5}$$

$\dfrac{x}{5} + 1 \leq -\dfrac{4}{5}$ or $\dfrac{x}{5} + 1 \geq \dfrac{4}{5}$

$x + 5 \leq -4$ $x + 5 \geq 4$

$x \leq -9$ $x \geq -1$

Graphing the solution set:

41. Since $|3 - 4t| > -5$ is always true, the solution set is all real numbers. Graphing the solution set:

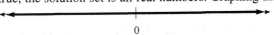

43. Solving the inequality:
$$-8 + |3y + 5| < 5$$
$$|3y + 5| < 13$$
$$-13 < 3y + 5 < 13$$
$$-18 < 3y < 8$$
$$-6 < y < \frac{8}{3}$$

Graphing the solution set:

45. Substitute $W_1 = 85$, $d_1 = 4$, and $W_2 = 120$ into the law of levers:
$$W_1 \cdot d_1 = W_2 \cdot d_2$$
$$85 \cdot 4 = 120 \cdot d_2$$
$$340 = 120 d_2$$
$$d_2 = \frac{17}{6} = 2\frac{5}{6}$$

Her brother should be $2\frac{5}{6}$ feet from the center of the balance.

Chapter 3 Review

1.   Graphing the line:

3.   Graphing the line:

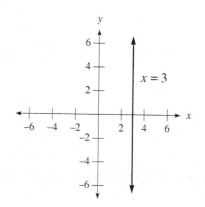

5.   Finding the slope: $m = \dfrac{2-2}{3-(-4)} = \dfrac{0}{7} = 0$

7.   Since the points are (4,0) and (0,–6), the slope is given by: $m = \dfrac{-6-0}{0-4} = \dfrac{-6}{-4} = \dfrac{3}{2}$

9.   First find the slope of l: $m = \dfrac{-2-2}{-3-5} = \dfrac{-4}{-8} = \dfrac{1}{2}$

    Since a perpendicular to l must have a slope which is the opposite reciprocal, its slope is: $\dfrac{-1}{1/2} = -2$

11.   Using the slope-intercept formula, the slope is $y = -2x$.

13.   Solving for y:
$$2x - 3y = 9$$
$$-3y = -2x + 9$$
$$y = \tfrac{2}{3}x - 3$$
    The slope is $m = \tfrac{2}{3}$ and the y-intercept is $b = -3$.

15.   Using the point-slope formula:
$$y - 1 = -\tfrac{1}{3}(x + 3)$$
$$y - 1 = -\tfrac{1}{3}x - 1$$
$$y = -\tfrac{1}{3}x$$

17.   First find the slope: $m = \dfrac{7-7}{4-(-3)} = \dfrac{0}{7} = 0$. Since the line is horizontal, its equation is $y = 7$.

19.   First find the slope by solving for y:
$$2x - y = 4$$
$$-y = -2x + 4$$
$$y = 2x - 4$$
    The parallel slope is also $m = 2$. Now using the point-slope formula:
$$y + 3 = 2(x - 2)$$
$$y + 3 = 2x - 4$$
$$y = 2x - 7$$

21.   Graphing the inequality:

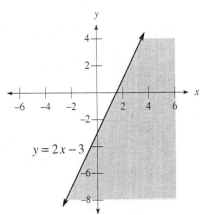

23.   The domain is $\{2,3,4\}$ and the range is $\{2,3,4\}$. This is a function.

25. $f(-3) = 0$

27.   Evaluating the function: $f(0) = 2(0)^2 - 4(0) + 1 = 0 - 0 + 1 = 1$

29.   Evaluating the function: $(f \circ g)(0) = f[g(0)] = f[3(0) + 2] = f(2) = 2(2)^2 - 4(2) + 1 = 8 - 8 + 1 = 1$

31.   The variation equation is $y = Kx$. Substituting $x = 2$ and $y = 6$:
$$6 = K \cdot 2$$
$$K = 3$$
The equation is $y = 3x$. Substituting $x = 8$: $y = 3 \cdot 8 = 24$

33.   The variation equation is $y = \dfrac{K}{x^2}$. Substituting $x = 2$ and $y = 9$:
$$9 = \frac{K}{2^2}$$
$$9 = \frac{K}{4}$$
$$K = 36$$
The equation is $y = \dfrac{36}{x^2}$. Substituting $x = 3$: $y = \dfrac{36}{3^2} = \dfrac{36}{9} = 4$

35.   The variation equation is $t = Kd$. Substituting $t = 42$ and $d = 2$:
$$42 = K \cdot 2$$
$$K = 21$$
The equation is $t = 21d$. Substituting $d = 4$: $t = 21 \cdot 4 = 84$. The tension is 84 pounds.

Chapters 1-3 Cumulative Review

1.   Simplifying: $9 - 6 \div 2 + 2 \cdot 3 = 9 - 3 + 6 = 12$ **3.**   Simplifying: $12 - 7 - 2 - (-3) = 12 - 7 - 2 + 3 = 6$

5.   Simplifying: $5(3 - 7)^2 - 2(4 - 6)^3 = 5(-4)^2 - 2(-2)^3 = 5(16) - 2(-8) = 80 + 16 = 96$

7.   Solving the equation:

$$\frac{2}{3}a - 4 = 6$$
$$\frac{2}{3}a = 10$$
$$2a = 30$$
$$a = 15$$

9.   Solving the equation:

$$-4 + 3(3x + 2) = 7$$
$$-4 + 9x + 6 = 7$$
$$9x + 2 = 7$$
$$9x = 5$$
$$x = \frac{5}{9}$$

11.   Solving the equation:

$$-4x + 9 = -3(-2x + 1) - 2$$
$$-4x + 9 = 6x - 3 - 2$$
$$-4x + 9 = 6x - 5$$
$$-10x + 9 = -5$$
$$-10x = -14$$
$$x = \frac{7}{5}$$

13.   Solving the equation:

$$|2x - 3| - 7 = 1$$
$$|2x - 3| = 8$$
$$2x - 3 = -8, 8$$
$$2x = -5, 11$$
$$x = -\frac{5}{2}, \frac{11}{2}$$

15.   Solving the inequality:

$$|3x - 1| - 1 \le 2$$
$$|3x - 1| \le 3$$
$$-3 \le 3x - 1 \le 3$$
$$-2 \le 3x \le 4$$
$$-\frac{2}{3} \le x \le \frac{4}{3}$$

17.   Solving the inequality:

$$\begin{array}{ccc} 3y - 6 \ge 3 & & 3y - 6 \le -3 \\ 3y \ge 9 & \text{or} & 3y \le 3 \\ y \ge 3 & & y \le 1 \end{array}$$

Graphing the solution set:

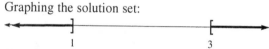

19.   Solving the inequality:

$$|3x + 4| - 7 > 3$$
$$|3x + 4| > 10$$
$$\begin{array}{ccc} 3x + 4 < -10 & \text{or} & 3x + 4 > 10 \\ 3x < -14 & & 3x > 6 \\ x < -\frac{14}{3} & & x > 2 \end{array}$$

Graphing the solution set:

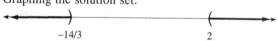

21.   Graphing the equation:

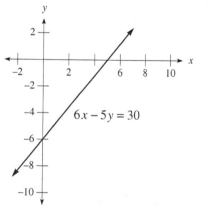

23.   Graphing the equation:

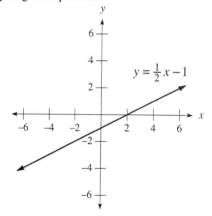

25.   Finding the slope: $m = \dfrac{3 - (-2)}{-5 - (-3)} = \dfrac{3 + 2}{-5 + 3} = -\dfrac{5}{2}$

27.   Since this line is horizontal, its slope is 0.

29. Solving for y:
$$4x - 5y = 20$$
$$-5y = -4x + 20$$
$$y = \frac{4}{5}x - 4$$
The slope is $m = \frac{4}{5}$ and the y-intercept is $b = -4$.

31. Solving for y:
$$5x + 2y = -1$$
$$2y = -5x - 1$$
$$y = -\frac{5}{2}x - \frac{1}{2}$$
Using the slope of $-\frac{5}{2}$ in the point-slope formula:
$$y + 2 = -\frac{5}{2}(x + 4)$$
$$y + 2 = -\frac{5}{2}x - 10$$
$$y = -\frac{5}{2}x - 12$$

33. Evaluating the function: $h(0) = 3(0) - 1 = -1$

35. Evaluating the function: $f[g(-2)] = f(-2-1) = f(-3) = (-3)^2 - 3(-3) = 9 + 9 = 18$

37. Evaluating: $(-3)\left(\frac{6}{15}\right) - \frac{2}{5} = -\frac{6}{5} - \frac{2}{5} = -\frac{8}{5}$

39. Reducing to lowest terms: $\frac{721}{927} = \frac{7 \cdot 103}{9 \cdot 103} = \frac{7}{9}$

41. The set is $A \cup B = \{0,1,2,3,6,7\}$.

43. Finding the percent:
$$13.5 = p \cdot 54$$
$$p = \frac{13.5}{54} = 25\%$$

45. Solving the equation:
$$0.04x = 12$$
$$x = \frac{12}{0.04} = 300$$

47. The domain is $\{-1,2,3\}$ and the range is $\{-1,3\}$. This is a function.

49. Let w represent the width and $2w + 3$ represent the length. Using the perimeter formula:
$$2(w) + 2(2w + 3) = 48$$
$$2w + 4w + 6 = 48$$
$$6w + 6 = 48$$
$$6w = 42$$
$$w = 7$$
The width is 7 feet and the length is 17 feet.

Chapter 3 Test

1. The x-intercept is 3, the y-intercept is 6, and the slope is -2.

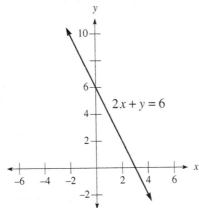

2.   The *x*-intercept is $-\frac{3}{2}$, the *y*-intercept is -3, and the slope is -2.

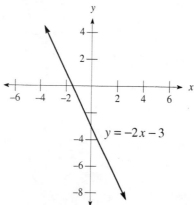

$$y = -2x - 3$$

3.   The *x*-intercept is $-\frac{8}{3}$, the *y*-intercept is 4, and the slope is $\frac{3}{2}$.

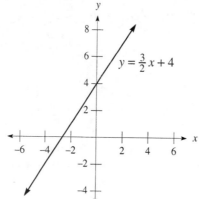

$$y = \frac{3}{2}x + 4$$

4.   The *x*-intercept is -2, there is no *y*-intercept, and there is no slope.

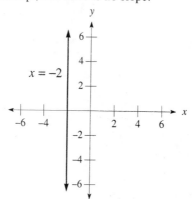

$$x = -2$$

5.   Using the point-slope formula:
$$y - 3 = 2(x + 1)$$
$$y - 3 = 2x + 2$$
$$y = 2x + 5$$

6. First find the slope: $m = \dfrac{-1-2}{4-(-3)} = \dfrac{-3}{4+3} = -\dfrac{3}{7}$. Using the point-slope formula:

$$y - 2 = -\tfrac{3}{7}(x+3)$$
$$y - 2 = -\tfrac{3}{7}x - \tfrac{9}{7}$$
$$y = -\tfrac{3}{7}x + \tfrac{5}{7}$$

7. First solve for y to find the slope:

$$2x - 5y = 10$$
$$-5y = -2x + 10$$
$$y = \tfrac{2}{5}x - 2$$

The parallel line will also have a slope of $\tfrac{2}{5}$. Now using the point-slope formula:

$$y - (-3) = \tfrac{2}{5}(x-5)$$
$$y + 3 = \tfrac{2}{5}x - 2$$
$$y = \tfrac{2}{5}x - 5$$

8. The perpendicular slope is $-\tfrac{1}{3}$. Using the point-slope formula:

$$y - (-2) = -\tfrac{1}{3}(x-(-1))$$
$$y + 2 = -\tfrac{1}{3}x - \tfrac{1}{3}$$
$$y = -\tfrac{1}{3}x - \tfrac{7}{3}$$

9. Since the line is vertical, its equation is $x = 4$.

10. Graphing the inequality:

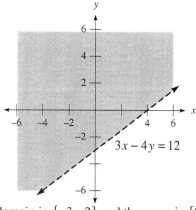

11. Graphing the inequality:

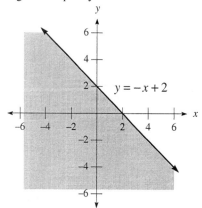

12. The domain is $\{-3, -2\}$ and the range is $\{0, 1\}$. This is not a function.

13. The domain is all real numbers and the range is $\{y \mid y \geq -9\}$. This is a function.

14. Evaluating the function: $f(3) + g(2) = [3-2] + [3 \cdot 2 + 4] = 1 + 10 = 11$

15. Evaluating the function: $h(0) + g(0) = [3 \cdot 0^2 - 2 \cdot 0 - 8] + [3 \cdot 0 + 4] = -8 + 4 = -4$

16. Evaluating the function: $(f \circ g)(2) = f[g(2)] = f(3 \cdot 2 + 4) = f(10) = 10 - 2 = 8$

17. Evaluating the function: $(g \circ f)(2) = g[f(2)] = g(2-2) = g(0) = 3 \cdot 0 + 4 = 4$

18. The restriction on the variable is $0 < x < 4$.

19. The height of the box is x, and the dimensions of the base are $8 - 2x$ by $8 - 2x$. Therefore the volume is given by $V(x) = x(8-2x)^2$.

20. $V(2) = 2(8 - 2 \cdot 2)^2 = 2(4)^2 = 32$ cubic inches

    This represents the volume of the box if a square with 2-inch sides is cut from each corner.

21. The variation equation is $y = Kx^2$. Substituting $x = 5$ and $y = 50$:

$$50 = K(5)^2$$
$$50 = 25K$$
$$K = 2$$

The equation is $y = 2x^2$. Substituting $x = 3$: $y = 2(3)^2 = 2 \cdot 9 = 18$

22. The variation equation is $z = Kxy^3$. Substituting $x = 5$, $y = 2$, and $z = 15$:

$$15 = K(5)(2)^3$$
$$15 = 40K$$
$$K = \frac{3}{8}$$

The equation is $z = \frac{3}{8}xy^3$. Substituting $x = 2$ and $y = 3$: $z = \frac{3}{8}(2)(3)^3 = \frac{3}{8} \cdot 54 = \frac{81}{4}$

23. The variation equation is $L = \frac{Kwd^2}{l}$. Substituting $l = 10$, $w = 3$, $d = 4$, and $L = 800$:

$$800 = \frac{K(3)(4)^2}{10}$$
$$8000 = 48K$$
$$K = \frac{500}{3}$$

The equation is $L = \frac{500wd^2}{3l}$. Substituting $l = 12$, $w = 3$, and $d = 4$: $L = \frac{500(3)(4)^2}{3(12)} = \frac{2000}{3}$

The beam can safely hold $\frac{2000}{3}$ pounds.

Chapter 4
Systems of Linear Equations and Inequalities

4.1   Systems of Linear Equations in Two Variables

1.   The intersection point is (4,3).

3.   The intersection point is (–5,–6).

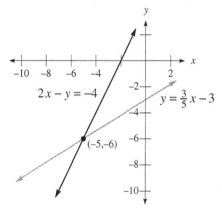

5.   The intersection point is (4,2).

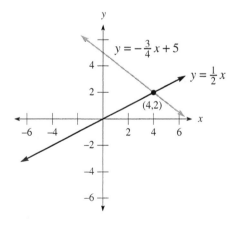

7.   The lines are parallel. There is no solution to the system.

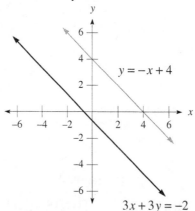

$$y = -x + 4$$

$$3x + 3y = -2$$

9.   The lines coincide. Any solution to one of the equations is a solution to the other.

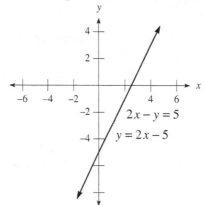

$$2x - y = 5$$

$$y = 2x - 5$$

11.   Solving the two equations:
$$x + y = 5$$
$$3x - y = 3$$
Adding yields:
$$4x = 8$$
$$x = 2$$
The solution is $(2,3)$.

13.   Multiply the first equation by -1:
$$-3x - y = 4$$
$$4x + y = 5$$
Adding yields: $x = 1$. The solution is $(1,1)$.

15.   Multiply the first equation by -2:
$$-6x + 4y = -12$$
$$6x - 4y = 12$$
Adding yields $0 = 0$, so the lines coincide. The solution is $\{(x, y) \mid 3x - 2y = 6\}$.

17.   Multiply the first equation by 3:
$$3x + 6y = 0$$
$$2x - 6y = 5$$
Adding yields:
$$5x = 5$$
$$x = 1$$
The solution is $\left(1, -\frac{1}{2}\right)$.

19. Multiply the first equation by −2:
$$-4x + 10y = -32$$
$$4x - 3y = 11$$
Adding yields:
$$7y = -21$$
$$y = -3$$
The solution is $\left(\frac{1}{2}, -3\right)$.

21. Multiply the first equation by 3 and the second equation by −2:
$$18x + 9y = -3$$
$$-18x - 10y = -2$$
Adding yields:
$$-y = -5$$
$$y = 5$$
The solution is $\left(-\frac{8}{3}, 5\right)$.

23. Multiply the first equation by 2 and the second equation by 3:
$$8x + 6y = 28$$
$$27x - 6y = 42$$
Adding yields:
$$35x = 70$$
$$x = 2$$
The solution is $(2,2)$.

25. Multiply the first equation by 2:
$$4x - 10y = 6$$
$$-4x + 10y = 3$$
Adding yields $0 = 9$, which is false (parallel lines). There is no solution ($\varnothing$).

27. To clear each equation of fractions, multiply the first equation by 12 and the second equation by 30:
$$3x - 2y = -24$$
$$-5x + 6y = 120$$
Multiply the first equation by 3:
$$9x - 6y = -72$$
$$-5x + 6y = 120$$
Adding yields $4x = 48$, so $x = 12$. Substituting into the first equation:
$$36 - 2y = -24$$
$$-2y = -60$$
$$y = 30$$
The solution is $(12,30)$.

29. To clear each equation of fractions, multiply the first equation by 6 and the second equation by 20:
$$3x + 2y = 78$$
$$8x + 5y = 200$$
Multiply the first equation by 5 and the second equation by −2:
$$15x + 10y = 390$$
$$-16x - 10y = -400$$
Adding yields:
$$-x = -10$$
$$x = 10$$
The solution is $(10,24)$.

31. Substituting into the first equation:

$$7(2y+9)-y=24$$
$$14y+63-y=24$$
$$13y=-39$$
$$y=-3$$

The solution is $(3,-3)$.

35. Substituting $x=y+4$ into the second equation:

$$2(y+4)-3y=6$$
$$2y+8-3y=6$$
$$-y+8=6$$
$$-y=-2$$
$$y=2$$

The solution is $(6,2)$.

39. Solving the first equation for y yields $y=2x-5$. Substituting into the second equation:

$$4x-2(2x-5)=10$$
$$4x-4x+10=10$$
$$10=10$$

Since this statement is true, the two lines coincide. The solution is $\{(x,y)\mid 2x-y=5\}$.

41. Substituting into the first equation:

$$\tfrac{1}{3}\left(\tfrac{3}{2}y\right)-\tfrac{1}{2}y=0$$
$$\tfrac{1}{2}y-\tfrac{1}{2}y=0$$
$$0=0$$

Since this statement is true, the two lines coincide. The solution is $\left\{(x,y)\mid x=\tfrac{3}{2}y\right\}$.

43. Multiply the first equation by 2 and the second equation by 7:

$$8x-14y=6$$
$$35x+14y=-21$$

Adding yields:

$$43x=-15$$
$$x=-\tfrac{15}{43}$$

Substituting into the original second equation:

$$5\left(-\tfrac{15}{43}\right)+2y=-3$$
$$-\tfrac{75}{43}+2y=-3$$
$$2y=-\tfrac{54}{43}$$
$$y=-\tfrac{27}{43}$$

The solution is $\left(-\tfrac{15}{43},-\tfrac{27}{43}\right)$.

33. Substituting into the first equation:
$$6x-\left(-\tfrac{3}{4}x-1\right)=10$$
$$6x+\tfrac{3}{4}x+1=10$$
$$\tfrac{27}{4}x=9$$
$$27x=36$$
$$x=\tfrac{4}{3}$$

The solution is $\left(\tfrac{4}{3},-2\right)$.

37. Substituting into the first equation:

$$4x-4=3x-2$$
$$x-4=-2$$
$$x=2$$

The solution is $(2,4)$.

45. Multiply the first equation by 3 and the second equation by 8:
$$27x - 24y = 12$$
$$16x + 24y = 48$$
Adding yields:
$$43x = 60$$
$$x = \frac{60}{43}$$
Substituting into the original second equation:
$$2\left(\frac{60}{43}\right) + 3y = 6$$
$$\frac{120}{43} + 3y = 6$$
$$3y = \frac{138}{43}$$
$$y = \frac{46}{43}$$
The solution is $\left(\frac{60}{43}, \frac{46}{43}\right)$.

47. Multiply the first equation by 2 and the second equation by 5:
$$6x - 10y = 4$$
$$35x + 10y = 5$$
Adding yields:
$$41x = 9$$
$$x = \frac{9}{41}$$
Substituting into the original second equation:
$$7\left(\frac{9}{41}\right) + 2y = 1$$
$$\frac{63}{41} + 2y = 1$$
$$2y = -\frac{22}{41}$$
$$y = -\frac{11}{41}$$
The solution is $\left(\frac{9}{41}, -\frac{11}{41}\right)$.

49. Multiply the second equation by 100:
$$x + y = 10000$$
$$6x + 5y = 56000$$
Multiply the first equation by −5:
$$-5x - 5y = -50000$$
$$6x + 5y = 56000$$
Adding yields $x = 6000$. The solution is $(6000, 4000)$.

51. Multiplying the first equation by $\frac{2}{3}$ yields the equation $4x - 6y = 2$. For the lines to coincide, the value is $c = 2$.

53. **a.** The percent of women was 12.8% and the percent of men was 18.76%.
b. Setting the two equations equal:
$$0.55x + 6.2 = 0.78x + 9.4$$
$$-0.23x = 3.2$$
$$x \approx -13.9$$
No, there was no year in which the percents were equal.

55. **a.** The range of possible values is $0 \le x \le 12$.
b. Setting the two functions equal:
$$\frac{10}{3}x + 20 = -\frac{10}{3}x + 80$$
$$\frac{20}{3}x = 60$$
$$20x = 180$$
$$x = 9$$
The solution is $(9, 50)$. In the year 1989, both inpatient and outpatient surgeries were 50%.

57. Finding the slope: $m = \dfrac{5-(-1)}{-2-(-4)} = \dfrac{5+1}{-2+4} = \dfrac{6}{2} = 3$

59. Solving for y:

$$2x - 3y = 6$$
$$-3y = -2x + 6$$
$$y = \tfrac{2}{3}x - 2$$

The slope is $m = \tfrac{2}{3}$ and the y-intercept is $b = -2$.

61. Using the point-slope formula:

$$y - 2 = \tfrac{2}{3}(x + 6)$$
$$y - 2 = \tfrac{2}{3}x + 4$$
$$y = \tfrac{2}{3}x + 6$$

63. First find the slope: $m = \dfrac{-2-0}{0-3} = \tfrac{2}{3}$. The slope-intercept form is $y = \tfrac{2}{3}x - 2$.

65. Substituting the points $(1,-2)$ and $(3,1)$ results in the two equations:

$$a - 2b = 7$$
$$3a + b = 7$$

Multiply the second equation by 2:

$$a - 2b = 7$$
$$6a + 2b = 14$$

Adding the two equations:

$$7a = 21$$
$$a = 3$$

Substituting into the second equation:

$$9 + b = 7$$
$$b = -2$$

So $a = 3$ and $b = -2$.

67. **a.** Graphing the system:

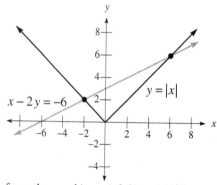

b. The solutions (by inspection from the graph) are $(-2,2)$ and $(6,6)$.

4.2 Systems of Linear Equations in Three Variables

1. Adding the first two equations and the first and third equations results in the system:
$$2x + 3z = 5$$
$$2x - 2z = 0$$
Solving the second equation yields $x = z$, now substituting:
$$2z + 3z = 5$$
$$5z = 5$$
$$z = 1$$
So $x = 1$, now substituting into the original first equation:
$$1 + y + 1 = 4$$
$$y + 2 = 4$$
$$y = 2$$
The solution is $(1,2,1)$.

3. Adding the first two equations and the first and third equations results in the system:
$$2x + 3z = 13$$
$$3x - 3z = -3$$
Adding yields:
$$5x = 10$$
$$x = 2$$
Substituting to find z:
$$2(2) + 3z = 13$$
$$4 + 3z = 13$$
$$3z = 9$$
$$z = 3$$
Substituting into the original first equation:
$$2 + y + 3 = 6$$
$$y + 5 = 6$$
$$y = 1$$
The solution is $(2,1,3)$.

5. Adding the second and third equations:
$$5x + z = 11$$
Multiplying the second equation by 2:
$$x + 2y + z = 3$$
$$4x - 2y + 4z = 12$$
Adding yields:
$$5x + 5z = 15$$
$$x + z = 3$$
So the system becomes:
$$5x + z = 11$$
$$x + z = 3$$
Multiply the second equation by -1:
$$5x + z = 11$$
$$-x - z = -3$$
Adding yields:
$$4x = 8$$
$$x = 2$$
Substituting to find z:
$$5(2) + z = 11$$
$$z + 10 = 11$$
$$z = 1$$

Substituting into the original first equation:
$$2 + 2y + 1 = 3$$
$$2y + 3 = 3$$
$$2y = 0$$
$$y = 0$$
The solution is $(2,0,1)$.

7.   Multiply the second equation by -1 and add it to the first equation:
$$2x + 3y - 2z = 4$$
$$-x - 3y + 3z = -4$$
Adding results in the equation $x + z = 0$. Multiply the second equation by 2 and add it to the third equation:
$$2x + 6y - 6z = 8$$
$$3x - 6y + z = -3$$
Adding results in the equation:
$$5x - 5z = 5$$
$$x - z = 1$$
So the system becomes:
$$x - z = 1$$
$$x + z = 0$$
Adding yields:
$$2x = 1$$
$$x = \tfrac{1}{2}$$
Substituting to find z:
$$\tfrac{1}{2} + z = 0$$
$$z = -\tfrac{1}{2}$$
Substituting into the original first equation:
$$2\left(\tfrac{1}{2}\right) + 3y - 2\left(-\tfrac{1}{2}\right) = 4$$
$$1 + 3y + 1 = 4$$
$$3y + 2 = 4$$
$$3y = 2$$
$$y = \tfrac{2}{3}$$
The solution is $\left(\tfrac{1}{2}, \tfrac{2}{3}, -\tfrac{1}{2}\right)$.

9.   Multiply the first equation by 2 and add it to the second equation:
$$-2x + 8y - 6z = 4$$
$$2x - 8y + 6z = 1$$
Adding yields $0 = 5$, which is false. There is no solution (inconsistent system).

11.   To clear the system of fractions, multiply the first equation by 2 and the second equation by 3:
$$x - 2y + 2z = 0$$
$$6x + y + 3z = 6$$
$$x + y + z = -4$$
Multiply the third equation by 2 and add it to the first equation:
$$x - 2y + 2z = 0$$
$$2x + 2y + 2z = -8$$
Adding yields the equation $3x + 4z = -8$. Multiply the third equation by -1 and add it to the second equation:
$$6x + y + 3z = 6$$
$$-x - y - z = 4$$
Adding yields the equation $5x + 2z = 10$. So the system becomes:
$$3x + 4z = -8$$
$$5x + 2z = 10$$

Multiply the second equation by –2:
$$3x + 4z = -8$$
$$-10x - 4z = -20$$
Adding yields:
$$-7x = -28$$
$$x = 4$$
Substituting to find z:
$$3(4) + 4z = -8$$
$$12 + 4z = -8$$
$$4z = -20$$
$$z = -5$$
Substituting into the original third equation:
$$4 + y - 5 = -4$$
$$y - 1 = -4$$
$$y = -3$$
The solution is $(4,-3,-5)$.

13. Multiply the first equation by –2 and add it to the third equation:
$$-4x + 2y + 6z = -2$$
$$4x - 2y - 6z = 2$$
Adding yields $0 = 0$, which is true. Since there are now less equations than unknowns, there is no unique solution (dependent system).

15. Multiply the second equation by 3 and add it to the first equation:
$$2x - y + 3z = 4$$
$$3x + 6y - 3z = -9$$
Adding yields the equation $5x + 5y = -5$, or $x + y = -1$.

Multiply the second equation by 2 and add it to the third equation:
$$2x + 4y - 2z = -6$$
$$4x + 3y + 2z = -5$$
Adding yields the equation $6x + 7y = -11$. So the system becomes:
$$6x + 7y = -11$$
$$x + y = -1$$
Multiply the second equation by –6:
$$6x + 7y = -11$$
$$-6x - 6y = 6$$
Adding yields $y = -5$. Substituting to find x:
$$6x + 7(-5) = -11$$
$$6x - 35 = -11$$
$$6x = 24$$
$$x = 4$$
Substituting into the original first equation:
$$2(4) - (-5) + 3z = 4$$
$$13 + 3z = 4$$
$$3z = -9$$
$$z = -3$$
The solution is $(4,-5,-3)$.

17. Adding the second and third equations results in the equation $x + y = 9$. Since this is the same as the first equation, there are less equations than unknowns. There is no unique solution (dependent system).

19.   Adding the second and third equations results in the equation $4x + y = 3$. So the system becomes:

$$4x + y = 3$$
$$2x + y = 2$$

Multiplying the second equation by -1:

$$4x + y = 3$$
$$-2x - y = -2$$

Adding yields:

$$2x = 1$$
$$x = \tfrac{1}{2}$$

Substituting to find y:

$$2\left(\tfrac{1}{2}\right) + y = 2$$
$$1 + y = 2$$
$$y = 1$$

Substituting into the original second equation:

$$1 + z = 3$$
$$z = 2$$

The solution is $\left(\tfrac{1}{2}, 1, 2\right)$.

21.   Multiply the third equation by 2 and adding it to the second equation:

$$6y - 4z = 1$$
$$2x + 4z = 2$$

Adding yields the equation $2x + 6y = 3$. So the system becomes:

$$2x - 3y = 0$$
$$2x + 6y = 3$$

Multiply the first equation by 2:

$$4x - 6y = 0$$
$$2x + 6y = 3$$

Adding yields:

$$6x = 3$$
$$x = \tfrac{1}{2}$$

Substituting to find y:

$$2\left(\tfrac{1}{2}\right) + 6y = 3$$
$$1 + 6y = 3$$
$$6y = 2$$
$$y = \tfrac{1}{3}$$

Substituting into the original third equation to find z:

$$\tfrac{1}{2} + 2z = 1$$
$$2z = \tfrac{1}{2}$$
$$z = \tfrac{1}{4}$$

The solution is $\left(\tfrac{1}{2}, \tfrac{1}{3}, \tfrac{1}{4}\right)$.

23. To clear each equation of fractions, multiply the first equation by 6, the second equation by 10, and the third equation by 12:

$$3x + 4y = 15$$
$$2x - 5z = -3$$
$$4y - 3z = 9$$

Multiply the third equation by -1 and add it to the first equation:

$$3x + 4y = 15$$
$$-4y + 3z = -9$$

Adding yields:

$$3x + 3z = 6$$
$$x + z = 2$$

So the system becomes:

$$x + z = 2$$
$$2x - 5z = -3$$

Multiply the first equation by 5:

$$5x + 5z = 10$$
$$2x - 5z = -3$$

Adding yields:

$$7x = 7$$
$$x = 1$$

Substituting yields $z = 1$. Substituting to find y:

$$3 + 4y = 15$$
$$4y = 12$$
$$y = 3$$

The solution is $(1,3,1)$.

25. To clear each equation of fractions, multiply the first equation by 4, the second equation by 12, and the third equation by 6:

$$2x - y + 2z = -8$$
$$3x - y - 4z = 3$$
$$x + 2y - 3z = 9$$

Multiply the first equation by -1 and add it to the first equation:

$$-2x + y - 2z = 8$$
$$3x - y - 4z = 3$$

Adding yields the equation $x - 6z = 11$. Multiply the first equation by 2 and add it to the third equation:

$$4x - 2y + 4z = -16$$
$$x + 2y - 3z = 9$$

Adding yields the equation $5x + z = -7$. So the system becomes:

$$5x + z = -7$$
$$x - 6z = 11$$

Multiply the first equation by 6:

$$30x + 6z = -42$$
$$x - 6z = 11$$

Adding yields:

$$31x = -31$$
$$x = -1$$

Substituting to find z:

$$-5 + z = -7$$
$$z = -2$$

Substituting to find y:

$$-1 + 2y + 6 = 9$$
$$2y + 5 = 9$$
$$2y = 4$$
$$y = 2$$

The solution is $(-1,2,-2)$.

27. Divide the second equation by 5 and the third equation by 10 to produce the system:
$$x - y - z = 0$$
$$x + 4y = 16$$
$$2y - z = 5$$
Multiply the third equation by -1 and add it to the first equation:
$$x - y - z = 0$$
$$-2y + z = -5$$
Adding yields the equation $x - 3y = -5$. So the system becomes:
$$x + 4y = 16$$
$$x - 3y = -5$$
Multiply the second equation by -1:
$$x + 4y = 16$$
$$-x + 3y = 5$$
Adding yields:
$$7y = 21$$
$$y = 3$$
Substituting to find x:
$$x + 12 = 16$$
$$x = 4$$
Substituting to find z:
$$6 - z = 5$$
$$z = 1$$
The currents are 4 amps, 3 amps, and 1 amp.

29. a.   Cost for A: $z = 36(3) + 0.20(350) = \$178$ Cost for B: $z = 32(3) + 0.22(350) = \$173$

    b.   Substituting $x = 5$ into the two equations:
$$z = 36(5) + 0.20y = 180 + 0.20y$$
$$z = 32(5) + 0.22y = 160 + 0.22y$$
Substituting for z:
$$160 + 0.22y = 180 + 0.20y$$
$$0.02y = 20$$
$$y = 1000$$
The mileage is 1000 miles.

31. The variation equation is $y = Kx^2$. Substituting $x = 5$ and $y = 75$:
$$75 = K \cdot 5^2$$
$$75 = 25K$$
$$K = 3$$
So $y = 3x^2$. Substituting $x = 7$: $y = 3 \cdot 7^2 = 3 \cdot 49 = 147$

33. The variation equation is $y = \dfrac{K}{x}$. Substituting $x = 25$ and $y = 10$:
$$10 = \frac{K}{25}$$
$$K = 250$$
So $y = \dfrac{250}{x}$. Substituting $y = 5$:
$$5 = \frac{250}{x}$$
$$5x = 250$$
$$x = 50$$

35. The variation equation is $z = Kxy^2$. Substituting $z = 40$, $x = 5$, and $y = 2$:
$$40 = K \cdot 5 \cdot 2^2$$
$$40 = 20K$$
$$K = 2$$
So $z = 2xy^2$. Substituting $x = 2$ and $y = 5$: $z = 2 \cdot 2 \cdot 5^2 = 100$

37. Adding the first and second, the first and third, and third and fourth equations results in the system:
$$2x + 3y + 2w = 16$$
$$2x + 3w = 14$$
$$2x - 3y - w = -8$$
Adding the first and third equations results in the system:
$$4x + w = 8$$
$$2x + 3w = 14$$
Multiply the second equation by –2:
$$4x + w = 8$$
$$-4x - 6w = -28$$
Adding yields:
$$-5w = -20$$
$$w = 4$$
Substituting to find x:
$$4x + 4 = 8$$
$$4x = 4$$
$$x = 1$$
Substituting to find y:
$$2 + 3y + 8 = 16$$
$$3y + 10 = 16$$
$$3y = 6$$
$$y = 2$$
Substituting to find z:
$$1 + 2 + z + 4 = 10$$
$$z + 7 = 10$$
$$z = 3$$
The solution is (1,2,3,4).

4.3   Introduction to Determinants

1. Evaluating the determinant: $\begin{vmatrix} 1 & 0 \\ 2 & 3 \end{vmatrix} = 1 \cdot 3 - 0 \cdot 2 = 3 - 0 = 3$

3. Evaluating the determinant: $\begin{vmatrix} 2 & 1 \\ 3 & 4 \end{vmatrix} = 2 \cdot 4 - 1 \cdot 3 = 8 - 3 = 5$

5. Evaluating the determinant: $\begin{vmatrix} 0 & 1 \\ 1 & 0 \end{vmatrix} = 0 \cdot 0 - 1 \cdot 1 = 0 - 1 = -1$

7. Evaluating the determinant: $\begin{vmatrix} -3 & 2 \\ 6 & -4 \end{vmatrix} = (-3) \cdot (-4) - 6 \cdot 2 = 12 - 12 = 0$

9. Solving the equation:
$$\begin{vmatrix} 2x & 1 \\ x & 3 \end{vmatrix} = 10$$
$$6x - x = 10$$
$$5x = 10$$
$$x = 2$$

11. Solving the equation:
$$\begin{vmatrix} 1 & 2x \\ 2 & -3x \end{vmatrix} = 21$$
$$-3x - 4x = 21$$
$$-7x = 21$$
$$x = -3$$

13. Solving the equation:
$$\begin{vmatrix} 2x & -4 \\ 2 & x \end{vmatrix} = -8x$$
$$2x^2 + 8 = -8x$$
$$2x^2 + 8x + 8 = 0$$
$$x^2 + 4x + 4 = 0$$
$$(x+2)^2 = 0$$
$$x = -2$$

15. Solving the equation:
$$\begin{vmatrix} x^2 & 3 \\ x & 1 \end{vmatrix} = 10$$
$$x^2 - 3x = 10$$
$$x^2 - 3x - 10 = 0$$
$$(x-5)(x+2) = 0$$
$$x = -2, 5$$

17. Duplicating the first two columns:
$$\begin{vmatrix} 1 & 2 & 0 \\ 0 & 2 & 1 \\ 1 & 1 & 1 \end{vmatrix}\begin{matrix} 1 & 2 \\ 0 & 2 \\ 1 & 1 \end{matrix} = 1\cdot2\cdot1 + 2\cdot1\cdot1 + 0\cdot0\cdot1 - 1\cdot2\cdot0 - 1\cdot1\cdot1 - 1\cdot0\cdot2 = 2 + 2 + 0 - 0 - 1 - 0 = 3$$

19. Duplicating the first two columns:
$$\begin{vmatrix} 1 & 2 & 3 \\ 3 & 2 & 1 \\ 1 & 1 & 1 \end{vmatrix}\begin{matrix} 1 & 2 \\ 3 & 2 \\ 1 & 1 \end{matrix} = 1\cdot2\cdot1 + 2\cdot1\cdot1 + 3\cdot3\cdot1 - 1\cdot2\cdot3 - 1\cdot1\cdot1 - 1\cdot3\cdot2 = 2 + 2 + 9 - 6 - 1 - 6 = 0$$

21. Expanding across the first row:
$$\begin{vmatrix} 0 & 1 & 2 \\ 1 & 0 & 1 \\ -1 & 2 & 0 \end{vmatrix} = 0\begin{vmatrix} 0 & 1 \\ 2 & 0 \end{vmatrix} - 1\begin{vmatrix} 1 & 1 \\ -1 & 0 \end{vmatrix} + 2\begin{vmatrix} 1 & 0 \\ -1 & 2 \end{vmatrix} = 0(0-2) - 1(0+1) + 2(2-0) = 0 - 1 + 4 = 3$$

23. Expanding across the first row:
$$\begin{vmatrix} 3 & 0 & 2 \\ 0 & -1 & -1 \\ 4 & 0 & 0 \end{vmatrix} = 3\begin{vmatrix} -1 & -1 \\ 0 & 0 \end{vmatrix} - 0\begin{vmatrix} 0 & -1 \\ 4 & 0 \end{vmatrix} + 2\begin{vmatrix} 0 & -1 \\ 4 & 0 \end{vmatrix} = 3(0-0) - 0(0+4) + 2(0+4) = 0 - 0 + 8 = 8$$

25. Expanding across the first row: $\begin{vmatrix} 2 & -1 & 0 \\ 1 & 0 & -2 \\ 0 & 1 & 2 \end{vmatrix} = 2\begin{vmatrix} 0 & -2 \\ 1 & 2 \end{vmatrix} + 1\begin{vmatrix} 1 & -2 \\ 0 & 2 \end{vmatrix} + 0\begin{vmatrix} 1 & 0 \\ 0 & 1 \end{vmatrix} = 2(0+2) + 1(2-0) + 0 = 4 + 2 = 6$

27. Expanding across the first row:
$$\begin{vmatrix} 1 & 3 & 7 \\ -2 & 6 & 4 \\ 3 & 7 & -1 \end{vmatrix} = 1\begin{vmatrix} 6 & 4 \\ 7 & -1 \end{vmatrix} - 3\begin{vmatrix} -2 & 4 \\ 3 & -1 \end{vmatrix} + 7\begin{vmatrix} -2 & 6 \\ 3 & 7 \end{vmatrix} = 1(-6-28) - 3(2-12) + 7(-14-18) = -34 + 30 - 224 = -228$$

29. The determinant equation is:
$$\begin{vmatrix} y & x \\ m & 1 \end{vmatrix} = b$$
$$y - mx = b$$
$$y = mx + b$$

31. a. Writing the determinant equation:
$$\begin{vmatrix} x & -1.7 \\ 2 & 0.3 \end{vmatrix} = y$$
$$0.3x + 3.4 = y$$
$$y = 0.3x + 3.4$$

  b. Substituting $x = 2$: $y = 0.3(2) + 3.4 = 0.6 + 3.4 = 4$ billion dollars

33. Substituting $x = 6$: $y = \begin{vmatrix} 0.1 & 6.9 \\ -2 & 6 \end{vmatrix} = 0.6 + 13.8 = 14.4$ million

35. The domain is $\{1, 3, 4\}$ and the range is $\{2, 4\}$. This is a function.

37. The domain is $\{1, 2, 3\}$ and the range is $\{1, 2, 3\}$. This is a function.

39. Since this passes the vertical line test, it is a function.

41. Since this fails the vertical line test, it is not a function.

43. Expanding across row 1:

$$\begin{vmatrix} 2 & 0 & 1 & -3 \\ -1 & 2 & 0 & 1 \\ -3 & 0 & 1 & 0 \\ 1 & 1 & 0 & 0 \end{vmatrix} = 2\begin{vmatrix} 2 & 0 & 1 \\ 0 & 1 & 0 \\ 1 & 0 & 0 \end{vmatrix} - 0 + 1\begin{vmatrix} -1 & 2 & 1 \\ -3 & 0 & 0 \\ 1 & 1 & 0 \end{vmatrix} + 3\begin{vmatrix} -1 & 2 & 0 \\ -3 & 0 & 1 \\ 1 & 1 & 0 \end{vmatrix}$$

$$= 2 \cdot 1\begin{vmatrix} 2 & 1 \\ 1 & 0 \end{vmatrix} + 1 \cdot 3\begin{vmatrix} 2 & 1 \\ 1 & 0 \end{vmatrix} + 3\left(-1\begin{vmatrix} 0 & 1 \\ 1 & 0 \end{vmatrix} - 2\begin{vmatrix} -3 & 1 \\ 1 & 0 \end{vmatrix} \right)$$

$$= 2(-1) + 3(-1) + 3(1+2)$$

$$= -2 - 3 + 9$$

$$= 4$$

45. Expanding down column 3:

$$\begin{vmatrix} 2 & 0 & 1 & -3 \\ -1 & 2 & 0 & 1 \\ -3 & 0 & 1 & 0 \\ 1 & 1 & 0 & 0 \end{vmatrix} = 1\begin{vmatrix} -1 & 2 & 1 \\ -3 & 0 & 0 \\ 1 & 1 & 0 \end{vmatrix} + 1\begin{vmatrix} 2 & 0 & -3 \\ -1 & 2 & 1 \\ 1 & 1 & 0 \end{vmatrix}$$

$$= 1 \cdot 3\begin{vmatrix} 2 & 1 \\ 1 & 0 \end{vmatrix} + 1 \cdot \left(2\begin{vmatrix} 2 & 1 \\ 1 & 0 \end{vmatrix} - 3\begin{vmatrix} -1 & 2 \\ 1 & 1 \end{vmatrix} \right)$$

$$= 3(-1) + 1(-2+9)$$

$$= -3 + 7$$

$$= 4$$

4.4   Cramer's Rule

1. First find the determinants:

$$D = \begin{vmatrix} 2 & -3 \\ 4 & -2 \end{vmatrix} = -4 + 12 = 8$$

$$D_x = \begin{vmatrix} 3 & -3 \\ 10 & -2 \end{vmatrix} = -6 + 30 = 24$$

$$D_y = \begin{vmatrix} 2 & 3 \\ 4 & 10 \end{vmatrix} = 20 - 12 = 8$$

Now use Cramer's rule:

$$x = \frac{D_x}{D} = \frac{24}{8} = 3 \qquad y = \frac{D_y}{D} = \frac{8}{8} = 1$$

The solution is (3,1).

3. First find the determinants:

$$D = \begin{vmatrix} 5 & -2 \\ -10 & 4 \end{vmatrix} = 20 - 20 = 0$$

$$D_x = \begin{vmatrix} 4 & -2 \\ 1 & 4 \end{vmatrix} = 16 + 2 = 18$$

$$D_y = \begin{vmatrix} 5 & 4 \\ -10 & 1 \end{vmatrix} = 5 + 40 = 45$$

Since $D = 0$ and other determinants are nonzero, there is no solution, or $\varnothing$.

5. First find the determinants:

$$D = \begin{vmatrix} 4 & -7 \\ 5 & 2 \end{vmatrix} = 8 + 35 = 43$$

$$D_x = \begin{vmatrix} 3 & -7 \\ -3 & 2 \end{vmatrix} = 6 - 21 = -15$$

$$D_y = \begin{vmatrix} 4 & 3 \\ 5 & -3 \end{vmatrix} = -12 - 15 = -27$$

Now use Cramer's rule:

$$x = \frac{D_x}{D} = -\frac{15}{43} \qquad y = \frac{D_y}{D} = -\frac{27}{43}$$

The solution is $\left(-\frac{15}{43}, -\frac{27}{43}\right)$.

7. First find the determinants:

$$D = \begin{vmatrix} 9 & -8 \\ 2 & 3 \end{vmatrix} = 27 + 16 = 43$$

$$D_x = \begin{vmatrix} 4 & -8 \\ 6 & 3 \end{vmatrix} = 12 + 48 = 60$$

$$D_y = \begin{vmatrix} 9 & 4 \\ 2 & 6 \end{vmatrix} = 54 - 8 = 46$$

Now use Cramer's rule:

$$x = \frac{D_x}{D} = \frac{60}{43} \qquad y = \frac{D_y}{D} = \frac{46}{43}$$

The solution is $\left(\frac{60}{43}, \frac{46}{43}\right)$.

9. First find the determinants:

$$D = \begin{vmatrix} 1 & 1 & 1 \\ 1 & -1 & -1 \\ 2 & 2 & -1 \end{vmatrix} = 1\begin{vmatrix} -1 & -1 \\ 2 & -1 \end{vmatrix} - 1\begin{vmatrix} 1 & -1 \\ 2 & -1 \end{vmatrix} + 1\begin{vmatrix} 1 & -1 \\ 2 & 2 \end{vmatrix} = 3 - 1 + 4 = 6$$

$$D_x = \begin{vmatrix} 4 & 1 & 1 \\ 2 & -1 & -1 \\ 2 & 2 & -1 \end{vmatrix} = 4\begin{vmatrix} -1 & -1 \\ 2 & -1 \end{vmatrix} - 1\begin{vmatrix} 2 & -1 \\ 2 & -1 \end{vmatrix} + 1\begin{vmatrix} 2 & -1 \\ 2 & 2 \end{vmatrix} = 12 - 0 + 6 = 18$$

$$D_y = \begin{vmatrix} 1 & 4 & 1 \\ 1 & 2 & -1 \\ 2 & 2 & -1 \end{vmatrix} = 1\begin{vmatrix} 2 & -1 \\ 2 & -1 \end{vmatrix} - 4\begin{vmatrix} 1 & -1 \\ 2 & -1 \end{vmatrix} + 1\begin{vmatrix} 1 & 2 \\ 2 & 2 \end{vmatrix} = 0 - 4 - 2 = -6$$

$$D_z = \begin{vmatrix} 1 & 1 & 4 \\ 1 & -1 & 2 \\ 2 & 2 & 2 \end{vmatrix} = 1\begin{vmatrix} -1 & 2 \\ 2 & 2 \end{vmatrix} - 1\begin{vmatrix} 1 & 2 \\ 2 & 2 \end{vmatrix} + 4\begin{vmatrix} 1 & -1 \\ 2 & 2 \end{vmatrix} = -6 + 2 + 16 = 12$$

Now use Cramer's rule:

$$x = \frac{D_x}{D} = \frac{18}{6} = 3 \qquad y = \frac{D_y}{D} = \frac{-6}{6} = -1 \qquad z = \frac{D_z}{D} = \frac{12}{6} = 2$$

The solution is $(3, -1, 2)$.

11. First find the determinants:

$$D = \begin{vmatrix} 1 & 1 & -1 \\ -1 & 1 & 1 \\ 1 & 1 & 1 \end{vmatrix} = 1\begin{vmatrix} 1 & 1 \\ 1 & 1 \end{vmatrix} - 1\begin{vmatrix} -1 & 1 \\ 1 & 1 \end{vmatrix} - 1\begin{vmatrix} -1 & 1 \\ 1 & 1 \end{vmatrix} = 0 + 2 + 2 = 4$$

$$D_x = \begin{vmatrix} 2 & 1 & -1 \\ 3 & 1 & 1 \\ 4 & 1 & 1 \end{vmatrix} = 2\begin{vmatrix} 1 & 1 \\ 1 & 1 \end{vmatrix} - 1\begin{vmatrix} 3 & 1 \\ 4 & 1 \end{vmatrix} - 1\begin{vmatrix} 3 & 1 \\ 4 & 1 \end{vmatrix} = 0 + 1 + 1 = 2$$

$$D_y = \begin{vmatrix} 1 & 2 & -1 \\ -1 & 3 & 1 \\ 1 & 4 & 1 \end{vmatrix} = 1\begin{vmatrix} 3 & 1 \\ 4 & 1 \end{vmatrix} - 2\begin{vmatrix} -1 & 1 \\ 1 & 1 \end{vmatrix} - 1\begin{vmatrix} -1 & 3 \\ 1 & 4 \end{vmatrix} = -1 + 4 + 7 = 10$$

$$D_z = \begin{vmatrix} 1 & 1 & 2 \\ -1 & 1 & 3 \\ 1 & 1 & 4 \end{vmatrix} = 1\begin{vmatrix} 1 & 3 \\ 1 & 4 \end{vmatrix} - 1\begin{vmatrix} -1 & 3 \\ 1 & 4 \end{vmatrix} + 2\begin{vmatrix} -1 & 1 \\ 1 & 1 \end{vmatrix} = 1 + 7 - 4 = 4$$

Now use Cramer's rule:

$$x = \frac{D_x}{D} = \frac{2}{4} = \frac{1}{2} \qquad\qquad y = \frac{D_y}{D} = \frac{10}{4} = \frac{5}{2} \qquad\qquad z = \frac{D_z}{D} = \frac{4}{4} = 1$$

The solution is $\left(\frac{1}{2}, \frac{5}{2}, 1\right)$.

13. First find the determinants:

$$D = \begin{vmatrix} 3 & -1 & 2 \\ 6 & -2 & 4 \\ 1 & -5 & 2 \end{vmatrix} = 3\begin{vmatrix} -2 & 4 \\ -5 & 2 \end{vmatrix} + 1\begin{vmatrix} 6 & 4 \\ 1 & 2 \end{vmatrix} + 2\begin{vmatrix} 6 & -2 \\ 1 & -5 \end{vmatrix} = 48 + 8 - 56 = 0$$

$$D_x = \begin{vmatrix} 4 & -1 & 2 \\ 8 & -2 & 4 \\ 1 & -5 & 2 \end{vmatrix} = 4\begin{vmatrix} -2 & 4 \\ -5 & 2 \end{vmatrix} + 1\begin{vmatrix} 8 & 4 \\ 1 & 2 \end{vmatrix} + 2\begin{vmatrix} 8 & -2 \\ 1 & -5 \end{vmatrix} = 64 + 12 - 76 = 0$$

$$D_y = \begin{vmatrix} 3 & 4 & 2 \\ 6 & 8 & 4 \\ 1 & 1 & 2 \end{vmatrix} = 3\begin{vmatrix} 8 & 4 \\ 1 & 2 \end{vmatrix} - 4\begin{vmatrix} 6 & 4 \\ 1 & 2 \end{vmatrix} + 2\begin{vmatrix} 6 & 8 \\ 1 & 1 \end{vmatrix} = 36 - 32 - 4 = 0$$

$$D_z = \begin{vmatrix} 3 & -1 & 4 \\ 6 & -2 & 8 \\ 1 & -5 & 1 \end{vmatrix} = 3\begin{vmatrix} -2 & 8 \\ -5 & 1 \end{vmatrix} + 1\begin{vmatrix} 6 & 8 \\ 1 & 1 \end{vmatrix} + 4\begin{vmatrix} 6 & -2 \\ 1 & -5 \end{vmatrix} = 114 - 2 - 112 = 0$$

Since $D = 0$ and the other determinants are also 0, there is no unique solution (dependent).

15. First find the determinants:

$$D = \begin{vmatrix} 2 & -1 & 3 \\ 1 & -5 & -2 \\ -4 & -2 & 1 \end{vmatrix} = 2\begin{vmatrix} -5 & -2 \\ -2 & 1 \end{vmatrix} + 1\begin{vmatrix} 1 & -2 \\ -4 & 1 \end{vmatrix} + 3\begin{vmatrix} 1 & -5 \\ -4 & -2 \end{vmatrix} = -18 - 7 - 66 = -91$$

$$D_x = \begin{vmatrix} 4 & -1 & 3 \\ 1 & -5 & -2 \\ 3 & -2 & 1 \end{vmatrix} = 4\begin{vmatrix} -5 & -2 \\ -2 & 1 \end{vmatrix} + 1\begin{vmatrix} 1 & -2 \\ 3 & 1 \end{vmatrix} + 3\begin{vmatrix} 1 & -5 \\ 3 & -2 \end{vmatrix} = -36 + 7 + 39 = 10$$

$$D_y = \begin{vmatrix} 2 & 4 & 3 \\ 1 & 1 & -2 \\ -4 & 3 & 1 \end{vmatrix} = 2\begin{vmatrix} 1 & -2 \\ 3 & 1 \end{vmatrix} - 4\begin{vmatrix} 1 & -2 \\ -4 & 1 \end{vmatrix} + 3\begin{vmatrix} 1 & 1 \\ -4 & 3 \end{vmatrix} = 14 + 28 + 21 = 63$$

$$D_z = \begin{vmatrix} 2 & -1 & 4 \\ 1 & -5 & 1 \\ -4 & -2 & 3 \end{vmatrix} = 2\begin{vmatrix} -5 & 1 \\ -2 & 3 \end{vmatrix} + 1\begin{vmatrix} 1 & 1 \\ -4 & 3 \end{vmatrix} + 4\begin{vmatrix} 1 & -5 \\ -4 & -2 \end{vmatrix} = -26 + 7 - 88 = -107$$

Now use Cramer's rule:

$$x = \frac{D_x}{D} = -\frac{10}{91} \qquad y = \frac{D_y}{D} = -\frac{63}{91} = -\frac{9}{13} \qquad z = \frac{D_z}{D} = \frac{-107}{-91} = \frac{107}{91}$$

The solution is $\left(-\frac{10}{91}, -\frac{9}{13}, \frac{107}{91}\right)$.

17. First find the determinants:

$$D = \begin{vmatrix} -1 & -7 & 0 \\ 1 & 0 & 3 \\ 0 & 2 & 1 \end{vmatrix} = -1\begin{vmatrix} 0 & 3 \\ 2 & 1 \end{vmatrix} + 7\begin{vmatrix} 1 & 3 \\ 0 & 1 \end{vmatrix} + 0\begin{vmatrix} 1 & 0 \\ 0 & 2 \end{vmatrix} = 6 + 7 + 0 = 13$$

$$D_x = \begin{vmatrix} 1 & -7 & 0 \\ 11 & 0 & 3 \\ 0 & 2 & 1 \end{vmatrix} = 1\begin{vmatrix} 0 & 3 \\ 2 & 1 \end{vmatrix} + 7\begin{vmatrix} 11 & 3 \\ 0 & 1 \end{vmatrix} + 0\begin{vmatrix} 11 & 0 \\ 0 & 2 \end{vmatrix} = -6 + 77 + 0 = 71$$

$$D_y = \begin{vmatrix} -1 & 1 & 0 \\ 1 & 11 & 3 \\ 0 & 0 & 1 \end{vmatrix} = -1\begin{vmatrix} 11 & 3 \\ 0 & 1 \end{vmatrix} - 1\begin{vmatrix} 1 & 3 \\ 0 & 1 \end{vmatrix} + 0\begin{vmatrix} 1 & 11 \\ 0 & 0 \end{vmatrix} = -11 - 1 + 0 = -12$$

$$D_z = \begin{vmatrix} -1 & -7 & 1 \\ 1 & 0 & 11 \\ 0 & 2 & 0 \end{vmatrix} = -1\begin{vmatrix} 0 & 11 \\ 2 & 0 \end{vmatrix} + 7\begin{vmatrix} 1 & 11 \\ 0 & 0 \end{vmatrix} + 1\begin{vmatrix} 1 & 0 \\ 0 & 2 \end{vmatrix} = 22 + 0 + 2 = 24$$

Now use Cramer's rule:

$$x = \frac{D_x}{D} = \frac{71}{13} \qquad\qquad y = \frac{D_y}{D} = -\frac{12}{13} \qquad\qquad z = \frac{D_z}{D} = \frac{24}{13}$$

The solution is $\left(\frac{71}{13}, -\frac{12}{13}, \frac{24}{13}\right)$.

19. First find the determinants:

$$D = \begin{vmatrix} 1 & -1 & 0 \\ 3 & 0 & 1 \\ 0 & 1 & -2 \end{vmatrix} = 1\begin{vmatrix} 0 & 1 \\ 1 & -2 \end{vmatrix} + 1\begin{vmatrix} 3 & 1 \\ 0 & -2 \end{vmatrix} + 0\begin{vmatrix} 3 & 0 \\ 0 & 1 \end{vmatrix} = -1 - 6 + 0 = -7$$

$$D_x = \begin{vmatrix} 2 & -1 & 0 \\ 11 & 0 & 1 \\ -3 & 1 & -2 \end{vmatrix} = 2\begin{vmatrix} 0 & 1 \\ 1 & -2 \end{vmatrix} + 1\begin{vmatrix} 11 & 1 \\ -3 & -2 \end{vmatrix} + 0\begin{vmatrix} 11 & 0 \\ -3 & 1 \end{vmatrix} = -2 - 19 + 0 = -21$$

$$D_y = \begin{vmatrix} 1 & 2 & 0 \\ 3 & 11 & 1 \\ 0 & -3 & -2 \end{vmatrix} = 1\begin{vmatrix} 11 & 1 \\ -3 & -2 \end{vmatrix} - 2\begin{vmatrix} 3 & 1 \\ 0 & -2 \end{vmatrix} + 0\begin{vmatrix} 3 & 11 \\ 0 & -3 \end{vmatrix} = -19 + 12 + 0 = -7$$

$$D_z = \begin{vmatrix} 1 & -1 & 2 \\ 3 & 0 & 11 \\ 0 & 1 & -3 \end{vmatrix} = 1\begin{vmatrix} 0 & 11 \\ 1 & -3 \end{vmatrix} + 1\begin{vmatrix} 3 & 11 \\ 0 & -3 \end{vmatrix} + 2\begin{vmatrix} 3 & 0 \\ 0 & 1 \end{vmatrix} = -11 - 9 + 6 = -14$$

Now use Cramer's rule:

$$x = \frac{D_x}{D} = \frac{-21}{-7} = 3 \qquad\qquad y = \frac{D_y}{D} = \frac{-7}{-7} = 1 \qquad\qquad z = \frac{D_z}{D} = \frac{-14}{-7} = 2$$

The solution is $(3, 1, 2)$.

21. First rewrite the system as:
$$-10x + y = 100$$
$$-12x + y = 0$$
Now find the determinants:
$$D = \begin{vmatrix} -10 & 1 \\ -12 & 1 \end{vmatrix} = -10 + 12 = 2$$
$$D_x = \begin{vmatrix} 100 & 1 \\ 0 & 1 \end{vmatrix} = 100 - 0 = 100$$
$$D_y = \begin{vmatrix} -10 & 100 \\ -12 & 0 \end{vmatrix} = 0 + 1200 = 1200$$
Now using Cramer's rule:
$$x = \frac{D_x}{D} = \frac{100}{2} = 50 \qquad\qquad y = \frac{D_y}{D} = \frac{1200}{2} = 600$$
The company must sell 50 items per week to break even.

23. First find the determinants:
$$D = \begin{vmatrix} -164.2 & 1 \\ 1 & 0 \end{vmatrix} = 0 - 1 = -1$$
$$D_x = \begin{vmatrix} 719 & 1 \\ 5 & 0 \end{vmatrix} = 0 - 5 = -5$$
$$D_H = \begin{vmatrix} -164.2 & 719 \\ 1 & 5 \end{vmatrix} = -821 - 719 = -1540$$
Now using Cramer's rule:
$$x = \frac{D_x}{D} = \frac{-5}{-1} = 5 \qquad\qquad H = \frac{D_H}{D} = \frac{-1540}{-1} = 1540$$
There were 1,540 heart transplants in the year 1990.

25. Evaluating the function: $f(0) = \frac{1}{2}(0) + 3 = 3$

27. Evaluating the function: $g(2) = 2^2 - 4 = 0$

29. Evaluating the function: $f(-4) = \frac{1}{2}(-4) + 3 = 1$

31. Evaluating the function: $f[g(2)] = f(0) = \frac{1}{2}(0) + 3 = 3$

33. First find the determinants:
$$D = \begin{vmatrix} a & b \\ b & a \end{vmatrix} = a^2 - b^2 = (a+b)(a-b)$$
$$D_x = \begin{vmatrix} -1 & b \\ 1 & a \end{vmatrix} = -a + b = -(a-b)$$
$$D_y = \begin{vmatrix} a & -1 \\ b & 1 \end{vmatrix} = a + b$$
Now using Cramer's rule:
$$x = \frac{D_x}{D} = \frac{-(a+b)}{(a+b)(a-b)} = \frac{1}{b-a} \qquad\qquad y = \frac{D_y}{D} = \frac{a+b}{(a+b)(a-b)} = \frac{1}{a-b}$$
The solution is $\left(\dfrac{1}{b-a}, \dfrac{1}{a-b} \right)$.

35.   First find the determinants:

$$D = \begin{vmatrix} a^2 & b \\ b^2 & a \end{vmatrix} = a^3 - b^3 = (a-b)\left(a^2 + ab + b^2\right)$$

$$D_x = \begin{vmatrix} 1 & b \\ 1 & a \end{vmatrix} = a - b$$

$$D_y = \begin{vmatrix} a^2 & 1 \\ b^2 & 1 \end{vmatrix} = a^2 - b^2 = (a-b)(a+b)$$

Now using Cramer's rule:

$$x = \frac{D_x}{D} = \frac{a-b}{(a-b)\left(a^2 + ab + b^2\right)} = \frac{1}{a^2 + ab + b^2} \qquad y = \frac{D_y}{D} = \frac{(a-b)(a+b)}{(a-b)\left(a^2 + ab + b^2\right)} = \frac{a+b}{a^2 + ab + b^2}$$

The solution is $\left(\dfrac{1}{a^2 + ab + b^2}, \dfrac{a+b}{a^2 + ab + b^2} \right)$.

37.   The system is:
$$x + 2y = 1$$
$$3x + 4y = 0$$

4.5   Applications

1.   Let x and y represent the two numbers. The system of equations is:
$$y = 2x + 3$$
$$x + y = 18$$
Substituting into the second equation:
$$x + 2x + 3 = 18$$
$$3x = 15$$
$$x = 5$$
$$y = 2(5) + 3 = 13$$
The two numbers are 5 and 13.

3.   Let x and y represent the two numbers. The system of equations is:
$$y - x = 6$$
$$2x = 4 + y$$
The second equation is $y = 2x - 4$. Substituting into the first equation:
$$2x - 4 - x = 6$$
$$x = 10$$
$$y = 2(10) - 4 = 16$$
The two numbers are 10 and 16.

5.   Let x, y, and z represent the three numbers. The system of equations is:
$$x + y + z = 8$$
$$2x = z - 2$$
$$x + z = 5$$
The third equation is $z = 5 - x$. Substituting into the second equation:
$$2x = 5 - x - 2$$
$$3x = 3$$
$$x = 1$$
$$z = 5 - 1 = 4$$
Substituting into the first equation:
$$1 + y + 4 = 8$$
$$y = 3$$
The three numbers are 1, 3, and 4.

7. Let a represent the number of adult tickets and c represent the number of children's tickets. The system of equations is:

$a + c = 925$

$2a + c = 1150$

Multiply the first equation by -1:

$-a - c = -925$

$2a + c = 1150$

Adding yields:

$a = 225$

$c = 700$

There were 225 adult tickets and 700 children's tickets sold.

9. Let x represent the amount invested at 6% and y represent the amount invested at 7%. The system of equations is:

$x + y = 20000$

$0.06x + 0.07y = 1280$

Multiplying the first equation by -0.06:

$-0.06x - 0.06y = -1200$

$0.06x + 0.07y = 1280$

Adding yields:

$0.01y = 80$

$y = 8000$

$x = 12000$

Mr. Jones invested \$12,000 at 6% and \$8,000 at 7%.

11. Let x represent the amount invested at 6% and $2x$ represent the amount invested at 7.5%. The equation is:

$0.075(2x) + 0.06(x) = 840$

$0.21x = 840$

$x = 4000$

$2x = 8000$

Susan invested \$4,000 at 6% and \$8,000 at 7.5%.

13. Let x, y and z represent the amounts invested in the three accounts. The system of equations is:

$x + y + z = 2200$

$z = 3x$

$0.06x + 0.08y + 0.09z = 178$

Substituting into the first equation:

$x + y + 3x = 2200$

$4x + y = 2200$

Substituting into the third equation:

$0.06x + 0.08y + 0.09(3x) = 178$

$0.33x + 0.08y = 178$

The system of equations becomes:

$4x + y = 2200$

$0.33x + 0.08y = 178$

Multiply the first equation by -0.08:

$-0.32x - 0.08y = -176$

$0.33x + 0.08y = 178$

Adding yields:

$0.01x = 2$

$x = 200$

$z = 3(200) = 600$

$y = 2200 - 4(200) = 1400$

He invested \$200 at 6%, \$1,400 at 8%, and \$600 at 9%.

15. Let x represent the amount of 20% alcohol and y represent the amount of 50% alcohol. The system of equations is:

$$x + y = 9$$
$$0.20x + 0.50y = 0.30(9)$$

Multiplying the first equation by -0.2:

$$-0.20x - 0.20y = -1.8$$
$$0.20x + 0.50y = 2.7$$

Adding yields:

$$0.30y = 0.9$$
$$y = 3$$
$$x = 6$$

The mixture contains 3 gallons of 50% alcohol and 6 gallons of 20% alcohol.

17. Let x represent the amount of 20% disinfectant and y represent the amount of 14% disinfectant. The system of equations is:

$$x + y = 15$$
$$0.20x + 0.14y = 0.16(15)$$

Multiplying the first equation by -0.14:

$$-0.14x - 0.14y = -2.1$$
$$0.20x + 0.14y = 2.4$$

Adding yields:

$$0.06x = 0.3$$
$$x = 5$$
$$y = 10$$

The mixture contains 5 gallons of 20% disinfectant and 10 gallons of 14% disinfectant.

19. Let x represent the amount 40% copper alloy and y represent the amount of 60% copper alloy. The system of equations is:

$$x + y = 50$$
$$0.4x + 0.6y = 0.55(50)$$

Multiplying the first equation by -0.4:

$$-0.4x - 0.4y = -20$$
$$0.4x + 0.6y = 27.5$$

Adding yields:

$$0.2y = 7.5$$
$$y = 37.5$$
$$x = 12.5$$

The mixture contains 12.5 pounds of 40% copper alloy and 37.5 pounds of 60% copper alloy.

21. Let b represent the rate of the boat and c represent the rate of the current. The system of equations is:

$$2(b + c) = 24$$
$$3(b - c) = 18$$

The system of equations simplifies to:

$$b + c = 12$$
$$b - c = 6$$

Adding yields:

$$2b = 18$$
$$b = 9$$
$$c = 3$$

The rate of the boat is 9 mph and the rate of the current is 3 mph.

23. Let b represent the rate of the boat and c represent the rate of the current. The system of equations is:
$$2(b+c) = 20$$
$$6(b-c) = 12$$
The system of equations simplifies to:
$$b+c = 10$$
$$b-c = 2$$
Adding yields:
$$2b = 12$$
$$b = 6$$
$$c = 4$$
The rate of the boat is 6 mph and the rate of the current is 4 mph.

25. Let n represent the number of nickels and d represent the number of dimes. The system of equations is:
$$n + d = 20$$
$$0.05n + 0.10d = 1.40$$
Multiplying the first equation by -0.05:
$$-0.05n - 0.05d = -1$$
$$0.05n + 0.10d = 1.40$$
Adding yields:
$$0.05d = 0.40$$
$$d = 8$$
$$n = 12$$
Bob has 12 nickels and 8 dimes.

27. Let n, d, and q represent the number of nickels, dimes, and quarters. The system of equations is:
$$n + d + q = 9$$
$$0.05n + 0.10d + 0.25q = 1.20$$
$$d = n$$
Substituting into the first equation:
$$n + n + q = 9$$
$$2n + q = 9$$
Substituting into the second equation:
$$0.05n + 0.10n + 0.25q = 1.20$$
$$0.15n + 0.25q = 1.20$$
The system of equations becomes:
$$2n + q = 9$$
$$0.15n + 0.25q = 1.20$$
Multiplying the first equation by -0.25:
$$-0.50n - 0.25q = -2.25$$
$$0.15n + 0.25q = 1.20$$
Adding yields:
$$-0.35n = -1.05$$
$$n = 3$$
$$d = 3$$
$$q = 9 - 2(3) = 3$$
The collection contains 3 nickels, 3 dimes, and 3 quarters.

29.   Let n, d, and q represent the number of nickels, dimes, and quarters. The system of equations is:
$$n + d + q = 140$$
$$0.05n + 0.10d + 0.25q = 10.00$$
$$d = 2q$$
Substituting into the first equation:
$$n + 2q + q = 140$$
$$n + 3q = 140$$
Substituting into the second equation:
$$0.05n + 0.10(2q) + 0.25q = 10.00$$
$$0.05n + 0.45q = 10.00$$
The system of equations becomes:
$$n + 3q = 140$$
$$0.05n + 0.45q = 10.00$$
Multiplying the first equation by -0.05:
$$-0.05n - 0.15q = -7$$
$$0.05n + 0.45q = 10$$
Adding yields:
$$0.30q = 3$$
$$q = 10$$
$$d = 2(10) = 20$$
$$n = 140 - 3(10) = 110$$
There are 110 nickels in the collection.

31.   Let $x = mp + b$ represent the relationship. Using the points $(2,300)$ and $(1.5,400)$ results in the system:
$$300 = 2m + b$$
$$400 = 1.5m + b$$
Multiplying the second equation by -1:
$$300 = 2m + b$$
$$-400 = -1.5m - b$$
Adding yields:
$$-100 = 0.5m$$
$$m = -200$$
$$b = 300 - 2(-200) = 700$$
The equation is $x = -200p + 700$. Substituting $p = 3$: $x = -200(3) + 700 = 100$ items

33.   Let $C = mx + b$ represent the relationship. Using the points $(5,25.60)$ and $(7,27.10)$ results in the system:
$$5m + b = 25.60$$
$$7m + b = 27.10$$
Multiplying the first equation by -1:
$$-5m - b = -25.60$$
$$7m + b = 27.10$$
Adding yields:
$$2m = 1.5$$
$$m = 0.75$$
$$b = 25.60 - 5(0.75) = 21.85$$
The equation is $C = 0.75x + 21.85$. Substituting $x = 12$: $C = 0.75(12) + 21.85 = \$30.85$

35.   The system of equations is:
$$a + b + c = 128$$
$$9a + 3b + c = 128$$
$$25a + 5b + c = 0$$
Multiply the first equation by –1 and add it to the second equation:
$$-a - b - c = -128$$
$$9a + 3b + c = 128$$
Adding yields:
$$8a + 2b = 0$$
$$4a + b = 0$$
Multiply the first equation by –1 and add it to the third equation:
$$-a - b - c = -128$$
$$25a + 5b + c = 0$$
Adding yields:
$$24a + 4b = -128$$
$$6a + b = -32$$
The system simplifies to:
$$4a + b = 0$$
$$6a + b = -32$$
Multiplying the first equation by –1:
$$-4a - b = 0$$
$$6a + b = -32$$
Adding yields:
$$2a = -32$$
$$a = -16$$
Substituting to find b:
$$4(-16) + b = 0$$
$$b = 64$$
Substituting to find c:
$$-16 + 64 + c = 128$$
$$c = 80$$
The equation for the height is $h = -16t^2 + 64t + 80$.

37.   Let f, m and s represent the full price, member, and student tickets sold. The system of equations is:
$$f + m + s = 1050$$
$$5f + 4m + 2.5s = 4035$$
$$s = f + 35$$
Substituting into the first equation:
$$f + m + f + 35 = 1050$$
$$2f + m = 1015$$
Substituting into the second equation:
$$5f + 4m + 2.5(f + 35) = 4035$$
$$7.5f + 4m = 3947.5$$
So the system of equations becomes:
$$2f + m = 1015$$
$$7.5f + 4m = 3947.5$$
Multiplying the first equation by –4:
$$-8f - 4m = -4060$$
$$7.5f + 4m = 3947.5$$

Adding yields:
$$-0.5f = -112.5$$
$$f = 225$$
$$s = 225 + 35 = 260$$
$$m = 1015 - 2(225) = 565$$

There were 225 full price tickets, 565 member tickets, and 260 student tickets sold.

39. Graphing the inequality:

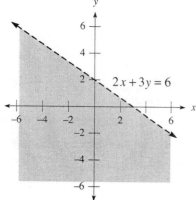

41. Graphing the inequality:

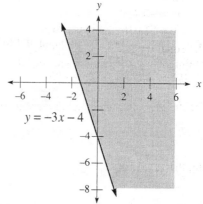

43. Graphing the inequality:

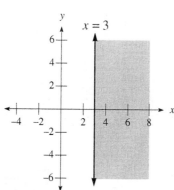

45. Graphing the data:

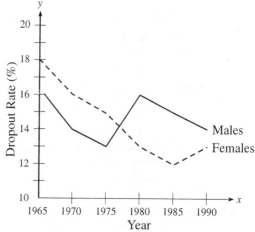

47. **a.** Using the points $(1975,13)$ and $(1980,16)$ in the model $M = mx + b$ results in the system:
$$1975m + b = 13$$
$$1980m + b = 16$$
Multiply the first equation by -1:
$$-1975m - b = -13$$
$$1980m + b = 16$$
Adding yields:
$$5m = 3$$
$$m = 0.6$$
$$b = 16 - 1980(0.6) = -1172$$
The equation is $M = 0.6x - 1172$.

b. Using the points (1975,15) and (1980,13) in the model $F = mx + b$ results in the system:
$$1975m + b = 15$$
$$1980m + b = 13$$
Multiply the first equation by –1:
$$-1975m - b = -15$$
$$1980m + b = 13$$
Adding yields:
$$5m = -2$$
$$m = -0.4$$
$$b = 13 - 1980(-0.4) = 805$$

The equation is $F = -0.4x + 805$.

c. Finding where the two rates are equal:
$$0.6x - 1172 = -0.4x + 805$$
$$x = 1977$$
The dropout rates are equal in the year 1977.

4.6 Systems of Linear Inequalities

1.   Graphing the solution set:

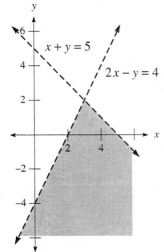

3.   Graphing the solution set:

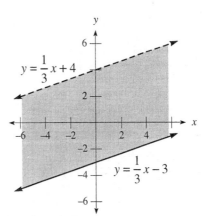

5.   Graphing the solution set:

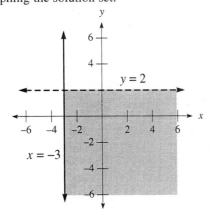

7.   Graphing the solution set:

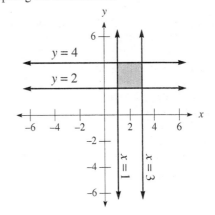

9.   Graphing the solution set:

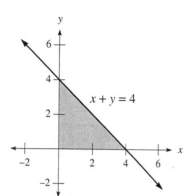

11.   Graphing the solution set:

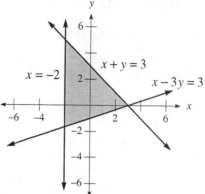

13.   Graphing the solution set:

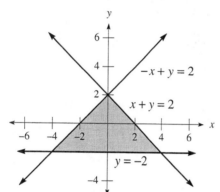

15.   Graphing the solution set:

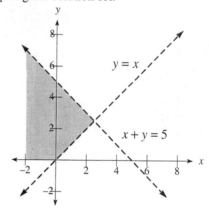

17.   Graphing the solution set:

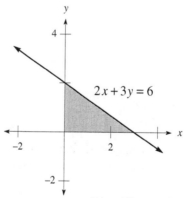

19.   The system of inequalities is:
$$x + y \le 4$$
$$-x + y < 4$$

21.   The system of inequalities is:
$$x + y \ge 4$$
$$-x + y < 4$$

23.   **a.**   The system of inequalities is:

$$0.55x + 0.65y \le 40$$
$$x \ge 2y$$
$$x > 15$$
$$y \ge 0$$

Graphing the solution set:

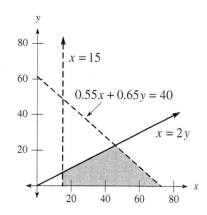

b.   Substitute $x = 20$:

$$2y \le 20$$
$$y \le 10$$

The most he can purchase is 10 65-cent stamps.

25.   Graphing the line:

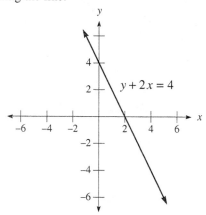

27.   Graphing the line:

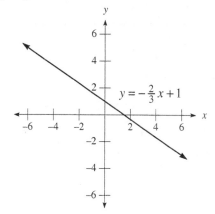

Chapter 4 Review

1.   Adding the two equations yields:

$$3x = 18$$
$$x = 6$$

Substituting to find y:

$$6 + y = 4$$
$$y = -2$$

The solution is $(6, -2)$.

3.   Multiply the second equation by 2:

$$2x - 4y = 5$$
$$-2x + 4y = 6$$

Adding yields $0 = 11$, which is false. There is no solution (parallel lines).

5.   Multiply the second equation by -2:
$$6x - 5y = -5$$
$$-6x - 2y = -2$$
Adding yields:
$$-7y = -7$$
$$y = 1$$
Substituting to find x:
$$3x + 1 = 1$$
$$3x = 0$$
$$x = 0$$
The solution is $(0,1)$.

7.   Multiply the first equation by 4 and the second equation by 3:
$$12x - 28y = 8$$
$$-12x + 18y = -18$$
Adding yields:
$$-10y = -10$$
$$y = 1$$
Substitute to find x:
$$3x - 7 = 2$$
$$3x = 9$$
$$x = 3$$
The solution is $(3,1)$.

9.   Multiply the first equation by 3 and the second equation by 4:
$$-21x + 12y = -3$$
$$20x - 12y = 0$$
Adding yields $x = 3$. Substitute to find y:
$$-21 + 4y = -1$$
$$4y = 20$$
$$y = 5$$
The solution is $(3,5)$.

11.   To clear each equation of fractions, multiply the first equation by 6 and the second equation by 6:
$$4x - y = 0$$
$$8x + 5y = 84$$
Multiply the first equation by -2:
$$-8x + 2y = 0$$
$$8x + 5y = 84$$
Adding yields:
$$7y = 84$$
$$y = 12$$
Substitute to find x:
$$4x - 12 = 0$$
$$4x = 12$$
$$x = 3$$
The solution is $(3,12)$.

13.   Substitute into the first equation:
$$x + x - 1 = 2$$
$$2x - 1 = 2$$
$$2x = 3$$
$$x = \tfrac{3}{2}$$
$$y = \tfrac{3}{2} - 1 = \tfrac{1}{2}$$
The solution is $\left(\tfrac{3}{2}, \tfrac{1}{2}\right)$.

15. Write the first equation as $y = 4 - x$. Substitute into the second equation:
$$2x + 5(4 - x) = 2$$
$$2x + 20 - 5x = 2$$
$$-3x = -18$$
$$x = 6$$
$$y = 4 - 6 = -2$$
The solution is $(6, -2)$.

17. Substitute into the first equation:
$$3(-3y + 4) + 7y = 6$$
$$-9y + 12 + 7y = 6$$
$$-2y = -6$$
$$y = 3$$
$$x = -9 + 4 = -5$$
The solution is $(-5, 3)$.

19. Adding the first and second equations yields:
$$2x - 2z = -2$$
$$x - z = -1$$
Adding the second and third equations yields $2x - 5z = -14$. So the system becomes:
$$x - z = -1$$
$$2x - 5z = -14$$
Multiply the first equation by -2:
$$-2x + 2z = 2$$
$$2x - 5z = -14$$
Adding yields:
$$-3z = -12$$
$$z = 4$$
Substitute to find x:
$$x - 4 = -1$$
$$x = 3$$
Substitute to find y:
$$3 + y + 4 = 6$$
$$y = -1$$
The solution is $(3, -1, 4)$.

21. Multiply the second equation by 2 and add it to the first equation:
$$5x + 8y - 4z = -7$$
$$14x + 8y + 4z = -4$$
Adding yields the equation $19x + 16y = -11$. Multiply the first equation by 2 and add it to the third equation:
$$10x + 16y - 8z = -14$$
$$3x - 2y + 8z = 8$$
Adding yields the equation $13x + 14y = -6$. So the system of equations becomes:
$$19x + 16y = -11$$
$$13x + 14y = -6$$
Multiply the first equation by 7 and the second equation by -8:
$$133x + 112y = -77$$
$$-104x - 112y = 48$$
Adding yields:
$$29x = -29$$
$$x = -1$$

Substituting to find y:
$$-13 + 14y = -6$$
$$14y = 7$$
$$y = \tfrac{1}{2}$$

Substituting to find z:
$$7(-1) + 4\left(\tfrac{1}{2}\right) + 2z = -2$$
$$-5 + 2z = -2$$
$$2z = 3$$
$$z = \tfrac{3}{2}$$

The solution is $\left(-1, \tfrac{1}{2}, \tfrac{3}{2}\right)$.

23. Multiply the second equation by 2 and add it to the third equation:
$$-6x + 8y - 2z = 4$$
$$6x - 8y + 2z = -4$$
Adding yields $0 = 0$. Since this is a true statement, there is no unique solution (dependent system).

25. Multiply the third equation by 2 and add it to the second equation:
$$3x - 2z = -2$$
$$10y + 2z = -2$$
Adding yields the equation $3x + 10y = -4$. So the system becomes:
$$2x - y = 5$$
$$3x + 10y = -4$$
Multiplying the first equation by 10:
$$20x - 10y = 50$$
$$3x + 10y = -4$$
Adding yields:
$$23x = 46$$
$$x = 2$$
Substituting to find y:
$$4 - y = 5$$
$$y = -1$$
Substituting to find z:
$$-5 + z = -1$$
$$z = 4$$
The solution is $(2, -1, 4)$.

27. Evaluating the determinant: $\begin{vmatrix} 2 & 3 \\ -5 & 4 \end{vmatrix} = 2 \cdot 4 - 3(-5) = 8 + 15 = 23$

29. Evaluating the determinant: $\begin{vmatrix} 1 & 0 \\ -7 & -3 \end{vmatrix} = 1(-3) - 0(-7) = -3 - 0 = -3$

31. Evaluating the determinant: $\begin{vmatrix} 3 & -1 & 0 \\ 0 & 2 & -4 \\ 6 & 0 & 2 \end{vmatrix} = 3\begin{vmatrix} 2 & -4 \\ 0 & 2 \end{vmatrix} + 1\begin{vmatrix} 0 & -4 \\ 6 & 2 \end{vmatrix} = 3(4 - 0) + 1(0 + 24) = 12 + 24 = 36$

33. Solving for x:
$$\begin{vmatrix} 2 & 3x \\ -1 & 2x \end{vmatrix} = 4$$
$$4x + 3x = 4$$
$$7x = 4$$
$$x = \tfrac{4}{7}$$

35. First find the determinants:

$$D = \begin{vmatrix} 3 & -5 \\ 7 & -2 \end{vmatrix} = -6 + 35 = 29$$

$$D_x = \begin{vmatrix} 4 & -5 \\ 3 & -2 \end{vmatrix} = -8 + 15 = 7$$

$$D_y = \begin{vmatrix} 3 & 4 \\ 7 & 3 \end{vmatrix} = 9 - 28 = -19$$

Now using Cramer's rule:

$$x = \frac{D_x}{D} = \frac{7}{29} \qquad\qquad y = \frac{D_y}{D} = -\frac{19}{29}$$

The solution is $\left(\frac{7}{29}, -\frac{19}{29}\right)$.

37. First find the determinants:

$$D = \begin{vmatrix} 3 & -6 \\ 2 & -4 \end{vmatrix} = -12 + 12 = 0$$

$$D_x = \begin{vmatrix} 9 & -6 \\ 6 & -4 \end{vmatrix} = -36 + 36 = 0$$

$$D_y = \begin{vmatrix} 3 & 9 \\ 2 & 6 \end{vmatrix} = 18 - 18 = 0$$

Since all of the determinants are zero, there is no unique solution (lines coincide).

39. First find the determinants:

$$D = \begin{vmatrix} -6 & 3 \\ 5 & -8 \end{vmatrix} = 48 - 15 = 33$$

$$D_x = \begin{vmatrix} 7 & 3 \\ -2 & -8 \end{vmatrix} = -56 + 6 = -50$$

$$D_y = \begin{vmatrix} -6 & 7 \\ 5 & -2 \end{vmatrix} = 12 - 35 = -23$$

Now using Cramer's rule:

$$x = \frac{D_x}{D} = -\frac{50}{33} \qquad\qquad y = \frac{D_y}{D} = -\frac{23}{33}$$

The solution is $\left(-\frac{50}{33}, -\frac{23}{33}\right)$.

41. First find the determinants:

$$D = \begin{vmatrix} 4 & -5 & 0 \\ 2 & 0 & 3 \\ 0 & 3 & -1 \end{vmatrix} = 4\begin{vmatrix} 0 & 3 \\ 3 & -1 \end{vmatrix} + 5\begin{vmatrix} 2 & 3 \\ 0 & -1 \end{vmatrix} + 0\begin{vmatrix} 2 & 0 \\ 0 & 3 \end{vmatrix} = 4(0-9) + 5(-2-0) + 0 = -36 - 10 = -46$$

$$D_x = \begin{vmatrix} -3 & -5 & 0 \\ 4 & 0 & 3 \\ 8 & 3 & -1 \end{vmatrix} = -3\begin{vmatrix} 0 & 3 \\ 3 & -1 \end{vmatrix} + 5\begin{vmatrix} 4 & 3 \\ 8 & -1 \end{vmatrix} + 0\begin{vmatrix} 4 & 0 \\ 8 & 3 \end{vmatrix} = -3(0-9) + 5(-4-24) + 0 = 27 - 140 = -113$$

$$D_y = \begin{vmatrix} 4 & -3 & 0 \\ 2 & 4 & 3 \\ 0 & 8 & -1 \end{vmatrix} = 4\begin{vmatrix} 4 & 3 \\ 8 & -1 \end{vmatrix} + 3\begin{vmatrix} 2 & 3 \\ 0 & -1 \end{vmatrix} + 0\begin{vmatrix} 2 & 4 \\ 0 & 8 \end{vmatrix} = 4(-4-24) + 3(-2-0) + 0 = -112 - 6 = -118$$

$$D_z = \begin{vmatrix} 4 & -5 & -3 \\ 2 & 0 & 4 \\ 0 & 3 & 8 \end{vmatrix} = 4\begin{vmatrix} 0 & 4 \\ 3 & 8 \end{vmatrix} + 5\begin{vmatrix} 2 & 4 \\ 0 & 8 \end{vmatrix} - 3\begin{vmatrix} 2 & 0 \\ 0 & 3 \end{vmatrix} = 4(0-12) + 5(16-0) - 3(6-0) = -48 + 80 - 18 = 14$$

Now using Cramer's rule:
$$x = \frac{D_x}{D} = \frac{-113}{-46} = \frac{113}{46} \qquad y = \frac{D_y}{D} = \frac{-118}{-46} = \frac{59}{23} \qquad z = \frac{D_z}{D} = \frac{14}{-46} = -\frac{7}{23}$$
The solution is $\left(\frac{113}{46}, \frac{59}{23}, -\frac{7}{23}\right)$.

43.   Let a represent the adult tickets and c represent the children's tickets. The system of equations is:
$$a + c = 127$$
$$2a + 1.5c = 214$$
Multiply the first equation by –2:
$$-2a - 2c = -254$$
$$2a + 1.5c = 214$$
Adding yields:
$$-0.5c = -40$$
$$c = 80$$
$$a = 47$$
There were 47 adult tickets and 80 children's tickets sold.

45.   Let x represent the amount invested at 12% and y represent the amount invested at 15%. The system of equations is:
$$x + y = 12000$$
$$0.12x + 0.15y = 1650$$
Multiplying the first equation by –0.12:
$$-0.12x - 0.12y = -1440$$
$$0.12x + 0.15y = 1650$$
Adding yields:
$$0.03y = 210$$
$$y = 7000$$
$$x = 5000$$
Ms. Jones invested \$5,000 at 12% and \$7,000 at 15%.

47.   Graphing the solution set:

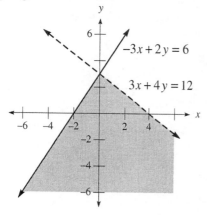

49.   Graphing the solution set:

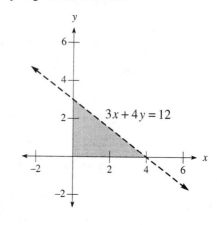

Chapters 1-4 Cumulative Review

1.   Simplifying: $2^4 - 4(5 - 12 + 4) = 2^4 - 4(-3) = 16 + 12 = 28$

3.   Simplifying: $18 - 7 - 5 - (-3) = 18 - 7 - 5 + 3 = 9$ **5.**   Simplifying: $\frac{3}{5} \cdot \frac{8}{9} \cdot 10 = \frac{240}{45} = \frac{16}{3}$

7.   Solving the equation:

$$-\frac{4}{7}a - 3 = 9$$
$$-\frac{4}{7}a = 12$$
$$a = -21$$

9.   Solving the equation:

$$\frac{3}{5}(2x - 3) + \frac{4}{5} = 5$$
$$\frac{6}{5}x - \frac{9}{5} + \frac{4}{5} = 5$$
$$\frac{6}{5}x - 1 = 5$$
$$\frac{6}{5}x = 6$$
$$x = 5$$

11.   Solving the equation:

$$|2y + 3| = |2y - 7|$$

$\quad 2y + 3 = 2y - 7 \qquad$ or $\qquad 2y + 3 = -2y + 7$

$\qquad\quad 3 = -7 \qquad\qquad\qquad\qquad 4y = 4$

$\qquad\quad y = \text{impossible} \qquad\qquad\quad y = 1$

13.   Solving the inequality:

$$-2(4x - 4) \geq -4(3x + 1)$$
$$-8x + 8 \geq -12x - 4$$
$$4x \geq -12$$
$$x \geq -3$$

Graphing the solution set:

15.   Graphing the line:

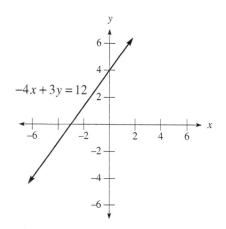

$-4x + 3y = 12$

17.   The slope is 0.

19.   Multiply the first equation by 2:

$$6x + 2y = 4$$
$$-6x + 2y = -4$$

Adding yields $4y = 0$, so $y = 0$. Substituting into the first equation:

$$6x = 4$$
$$x = \frac{2}{3}$$

The solution is $\left(\frac{2}{3}, 0\right)$.

21.   Multiply the first equation by 3 and the second equation by -2:
$$12x - 24y = 18$$
$$-12x + 24y = -12$$
Adding yields $0 = 6$, which is false. There is no solution, or $\varnothing$. The system is inconsistent (lines are parallel).

23.   First clear each equation of fractions by multiplying the first equation by 15 and the second equation by 12:
$$-3x + 20y = 14$$
$$4x - 3y = 5$$
Multiplying the first equation by 4 and the second equation by 3:
$$-12x + 80y = 56$$
$$12x - 9y = 15$$
Adding yields $71y = 71$, so $y = 1$. Substituting into the second equation:
$$4x - 3 = 5$$
$$4x = 8$$
$$x = 2$$
The solution is $(2,1)$.

25.   Evaluating the determinant: $\begin{vmatrix} -5 & -2 \\ 0 & -6 \end{vmatrix} = (-5)(-6) - (-2)(0) = 30 - 0 = 30$

27.   Expanding along the second row: $\begin{vmatrix} 3 & 5 & -1 \\ 0 & 4 & 2 \\ -6 & 0 & 7 \end{vmatrix} = -0\begin{vmatrix} 5 & -1 \\ 0 & 7 \end{vmatrix} + 4\begin{vmatrix} 3 & -1 \\ -6 & 7 \end{vmatrix} - 2\begin{vmatrix} 3 & 5 \\ -6 & 0 \end{vmatrix} = 0 + 4(15) - 2(30) = 0$

29.   First find the determinants:
$$D = \begin{vmatrix} 3 & -5 \\ 2 & 7 \end{vmatrix} = 21 + 10 = 31$$
$$D_x = \begin{vmatrix} 1 & -5 \\ -1 & 7 \end{vmatrix} = 7 - 5 = 2$$
$$D_y = \begin{vmatrix} 3 & 1 \\ 2 & -1 \end{vmatrix} = -3 - 2 = -5$$
Now using Cramer's rule:
$$x = \frac{D_x}{D} = \frac{2}{31} \qquad\qquad y = \frac{D_y}{D} = -\frac{5}{31}$$
The solution is $\left(\frac{2}{31}, -\frac{5}{31}\right)$.

31.   Solving the inequality:
$$3 - 2t < 7$$
$$-2t < 4$$
$$t > -2$$
The solution is the interval $(-2, \infty)$.

33.   The pattern is to add 4, so the next number is $3 + 4 = 7$. This is an arithmetic sequence.

35.   The irrational numbers are: $-\sqrt{2}, \sqrt{3}$ 37.   The opposite is $-\frac{7}{3}$ and the reciprocal is $\frac{3}{7}$.

39.   Yes, the graph passes the vertical line test, so it is the graph of a function.

41.   Finding the function value: $f[g(-4)] = f(-4 + 5) = f(1) = 2(1)^2 - 1 + 3 = 2 - 1 + 3 = 4$

43.   Using the slope formula:
$$\frac{y - (-8)}{1 - (-2)} = 3$$
$$\frac{y + 8}{3} = 3$$
$$y + 8 = 9$$
$$y = 1$$

45. Using the slope-intercept formula, the equation is $y = -\frac{3}{4}x + 2$.

47. The variation equation is $z = Kxy^3$. Substituting $z = -108$, $x = 4$, and $y = 3$:
$$-108 = K(4)(3)^3$$
$$-108 = 108K$$
$$K = -1$$
So the equation is $z = -xy^3$. Substituting $z = 8$ and $y = -2$:
$$8 = -x(-2)^3$$
$$8 = 8x$$
$$x = 1$$

49. Let x represent the amount of 30% HCl and y represent the amount of 70% HCl solution. The system is:
$$x + y = 15$$
$$0.3x + 0.7y = 0.5(15)$$
Multiply the first equation by -0.3:
$$-0.3x - 0.3y = -4.5$$
$$0.3x + 0.7y = 7.5$$
Adding yields:
$$0.4y = 3$$
$$y = 7.5$$
$$x = 7.5$$
The mixture should be made with 7.5 ounces of 30% HCl and 7.5 ounces of 70% HCl.

Chapter 4 Test

1. Multiply the second equation by 5:
$$2x - 5y = -8$$
$$15x + 5y = 25$$
Adding yields:
$$17x = 17$$
$$x = 1$$
Substituting to find y:
$$3 + y = 5$$
$$y = 2$$
The solution is $(1, 2)$.

2. Multiply the first equation by 5 and the second equation by 4:
$$20x - 35y = -10$$
$$-20x + 24y = -12$$
Adding yields:
$$-11y = -22$$
$$y = 2$$
Substituting to find x:
$$4x - 14 = -2$$
$$4x = 12$$
$$x = 3$$
The solution is $(3, 2)$.

3.   To clear each equation of fractions, multiply the first equation by 6 and the second equation by 20:
$$2x - y = 18$$
$$-4x + 5y = 0$$
Multiply the first equation by 2:
$$4x - 2y = 36$$
$$-4x + 5y = 0$$
Adding yields:
$$3y = 36$$
$$y = 12$$
Substituting to find x:
$$2x - 12 = 18$$
$$2x = 30$$
$$x = 15$$
The solution is $(15,12)$.

4.   Substituting into the first equation:
$$2x - 5(3x + 8) = 14$$
$$2x - 15x - 40 = 14$$
$$-13x = 54$$
$$x = -\frac{54}{13}$$
Substituting to find y: $y = 3\left(-\frac{54}{13}\right) + 8 = -\frac{162}{13} + \frac{104}{13} = -\frac{58}{13}$. The solution is $\left(-\frac{54}{13}, -\frac{58}{13}\right)$.

5.   The first equation is equivalent to $y = 2x$. Substituting into the second equation:
$$x + 2(2x) = 5$$
$$5x = 5$$
$$x = 1$$
$$y = 2$$
The solution is $(1,2)$.

6.   Adding the first and third equations:
$$5x = 15$$
$$x = 3$$
Adding the first and second equations:
$$3x - 2z = 7$$
$$9 - 2z = 7$$
$$-2z = -2$$
$$z = 1$$
Substituting to find y:
$$3 + y - 3 = -2$$
$$y = -2$$
The solution is $(3,-2,1)$.

7.   Evaluating the determinant: $\begin{vmatrix} 3 & -5 \\ -4 & 2 \end{vmatrix} = 6 - 20 = -14$

8.   Evaluating the determinant: $\begin{vmatrix} 1 & 0 & -3 \\ 2 & 1 & 0 \\ 0 & 5 & 4 \end{vmatrix} = 1\begin{vmatrix} 1 & 0 \\ 5 & 4 \end{vmatrix} - 0\begin{vmatrix} 2 & 0 \\ 0 & 4 \end{vmatrix} - 3\begin{vmatrix} 2 & 1 \\ 0 & 5 \end{vmatrix} = 1(4 - 0) - 0 - 3(10 - 0) = 4 - 30 = -26$

9. First find the determinants:

$$D = \begin{vmatrix} 5 & -4 \\ -2 & 1 \end{vmatrix} = 5 - 8 = -3$$

$$D_x = \begin{vmatrix} 2 & -4 \\ 3 & 1 \end{vmatrix} = 2 + 12 = 14$$

$$D_y = \begin{vmatrix} 5 & 2 \\ -2 & 3 \end{vmatrix} = 15 + 4 = 19$$

Now using Cramer's rule:

$$x = \frac{D_x}{D} = -\frac{14}{3} \qquad\qquad y = \frac{D_y}{D} = -\frac{19}{3}$$

The solution is $\left(-\frac{14}{3}, -\frac{19}{3}\right)$.

10. First find the determinants:

$$D = \begin{vmatrix} 2 & 4 \\ -4 & -8 \end{vmatrix} = -16 + 16 = 0$$

$$D_x = \begin{vmatrix} 3 & 4 \\ -6 & -8 \end{vmatrix} = -24 + 24 = 0$$

$$D_y = \begin{vmatrix} 2 & 3 \\ -4 & -6 \end{vmatrix} = -12 + 12 = 0$$

Since all three determinants are equal to 0, there is no unique solution (lines coincide). The solution is any point on the line, or $\{(x, y) \mid 2x + 4y = 3\}$.

11. First find the determinants:

$$D = \begin{vmatrix} 2 & -1 & 3 \\ 1 & -4 & -1 \\ 3 & -2 & 1 \end{vmatrix} = 2\begin{vmatrix} -4 & -1 \\ -2 & 1 \end{vmatrix} + 1\begin{vmatrix} 1 & -1 \\ 3 & 1 \end{vmatrix} + 3\begin{vmatrix} 1 & -4 \\ 3 & -2 \end{vmatrix} = 2(-4-2) + 1(1+3) + 3(-2+12) = -12 + 4 + 30 = 22$$

$$D_x = \begin{vmatrix} 2 & -1 & 3 \\ 6 & -4 & -1 \\ 4 & -2 & 1 \end{vmatrix} = 2\begin{vmatrix} -4 & -1 \\ -2 & 1 \end{vmatrix} + 1\begin{vmatrix} 6 & -1 \\ 4 & 1 \end{vmatrix} + 3\begin{vmatrix} 6 & -4 \\ 4 & -2 \end{vmatrix} = 2(-4-2) + 1(6+4) + 3(-12+16) = -12 + 10 + 12 = 10$$

$$D_y = \begin{vmatrix} 2 & 2 & 3 \\ 1 & 6 & -1 \\ 3 & 4 & 1 \end{vmatrix} = 2\begin{vmatrix} 6 & -1 \\ 4 & 1 \end{vmatrix} - 2\begin{vmatrix} 1 & -1 \\ 3 & 1 \end{vmatrix} + 3\begin{vmatrix} 1 & 6 \\ 3 & 4 \end{vmatrix} = 2(6+4) - 2(1+3) + 3(4-18) = 20 - 8 - 42 = -30$$

$$D_z = \begin{vmatrix} 2 & -1 & 2 \\ 1 & -4 & 6 \\ 3 & -2 & 4 \end{vmatrix} = 2\begin{vmatrix} -4 & 6 \\ -2 & 4 \end{vmatrix} + 1\begin{vmatrix} 1 & 6 \\ 3 & 4 \end{vmatrix} + 2\begin{vmatrix} 1 & -4 \\ 3 & -2 \end{vmatrix} = 2(-16+12) + 1(4-18) + 2(-2+12) = -8 - 14 + 20 = -2$$

Now using Cramer's rule:

$$x = \frac{D_x}{D} = \frac{10}{22} = \frac{5}{11} \qquad y = \frac{D_y}{D} = \frac{-30}{22} = -\frac{15}{11} \qquad z = \frac{D_z}{D} = \frac{-2}{22} = -\frac{1}{11}$$

The solution is $\left(\frac{5}{11}, -\frac{15}{11}, -\frac{1}{11}\right)$. Thus $y = -\frac{15}{11}$.

12. Let x and $2x - 1$ represent the two numbers. The equation is:

$$x + 2x - 1 = 14$$
$$3x = 15$$
$$x = 5$$
$$2x - 1 = 9$$

The two numbers are 5 and 9.

13.   Let x and $2x$ represent the two investments. The equation is:
$$0.05(x) + 0.06(2x) = 680$$
$$0.17x = 680$$
$$x = 4000$$
$$2x = 8000$$
John invested \$4000 at 5% and \$8000 at 6%.

14.   Let a represent the adult ticket sales and c represent the children's ticket sales. The system of equations is:
$$a + c = 750$$
$$2a + 1c = 1090$$
Multiplying the first equation by –1:
$$-a - c = -750$$
$$2a + c = 1090$$
Adding yields $a = 340$ and $c = 410$. There were 340 adult tickets and 410 children's tickets sold.

15.   Let b and c represent the rate of the boat and current. The system of equations is:
$$2(b + c) = 20$$
$$3(b - c) = 18$$
Simplifying the system:
$$b + c = 10$$
$$b - c = 6$$
Adding yields:
$$2b = 16$$
$$b = 8$$
$$c = 2$$
The boat's rate is 8 mph and the current's rate is 2 mph.

16.   Let n, d, and q represent the number of nickels, dimes, and quarters. The system of equations is:
$$n + d + q = 15$$
$$0.05n + 0.10d + 0.25q = 1.10$$
$$n = 4d - 1$$
Substituting into the first equation:
$$4d - 1 + d + q = 15$$
$$5d + q = 16$$
Substituting into the second equation:
$$0.05(4d - 1) + 0.10d + 0.25q = 1.10$$
$$0.3d + 0.25q = 1.15$$
The system of equations becomes:
$$5d + q = 16$$
$$0.3d + 0.25q = 1.15$$
Multiply the first equation by –0.25:
$$-1.25d - 0.25q = -4$$
$$0.3d + 0.25q = 1.15$$
Adding yields:
$$-0.95d = -2.85$$
$$d = 3$$
Substituting to find q:
$$15 + q = 16$$
$$q = 1$$
Substituting to find n:
$$n + 3 + 1 = 15$$
$$n = 11$$
The collection contains 11 nickels, 3 dimes, and 1 quarter.

17. Graphing the solution set:

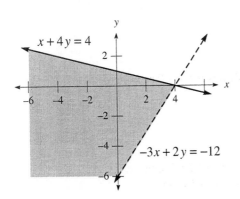

18. Graphing the solution set:

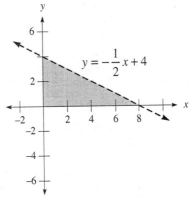

Chapter 5
Exponents and Polynomials

5.1 Properties of Exponents

1. Evaluating: $4^2 = 4 \cdot 4 = 16$

3. Evaluating: $-4^2 = -4 \cdot 4 = -16$

5. Evaluating: $-0.3^3 = -0.3 \cdot 0.3 \cdot 0.3 = -0.027$

7. Evaluating: $2^5 = 2 \cdot 2 \cdot 2 \cdot 2 \cdot 2 = 32$

9. Evaluating: $\left(\frac{1}{2}\right)^3 = \frac{1}{2} \cdot \frac{1}{2} \cdot \frac{1}{2} = \frac{1}{8}$

11. Evaluating: $\left(-\frac{5}{6}\right)^2 = \left(-\frac{5}{6}\right) \cdot \left(-\frac{5}{6}\right) = \frac{25}{36}$

13. Using properties of exponents: $x^5 \cdot x^4 = x^{5+4} = x^9$

15. Using properties of exponents: $\left(2^3\right)^2 = 2^{3 \cdot 2} = 2^6 = 64$

17. Using properties of exponents: $\left(-\frac{2}{3}x^2\right)^3 = \left(-\frac{2}{3}x^2\right)\left(-\frac{2}{3}x^2\right)\left(-\frac{2}{3}x^2\right) = -\frac{8}{27}x^6$

19. Using properties of exponents: $-3a^2\left(2a^4\right) = -6a^{2+4} = -6a^6$

21. Writing with positive exponents: $3^{-2} = \frac{1}{3^2} = \frac{1}{9}$

23. Writing with positive exponents: $(-2)^{-5} = \frac{1}{(-2)^5} = -\frac{1}{32}$

25. Writing with positive exponents: $\left(\frac{3}{4}\right)^{-2} = \left(\frac{4}{3}\right)^2 = \frac{16}{9}$

27. Writing with positive exponents: $\left(\frac{1}{3}\right)^{-2} + \left(\frac{1}{2}\right)^{-3} = 3^2 + 2^3 = 9 + 8 = 17$

29. Simplifying: $x^{-4}x^7 = x^{-4+7} = x^3$

31. Simplifying: $\left(a^2b^{-5}\right)^3 = a^6b^{-15} = \frac{a^6}{b^{15}}$

33. Simplifying: $\left(5y^4\right)^{-3}\left(2y^{-2}\right)^3 = 5^{-3}y^{-12}2^3y^{-6} = \frac{2^3}{5^3}y^{-18} = \frac{8}{125y^{18}}$

35. Simplifying: $\left(\frac{1}{2}x^3\right)\left(\frac{2}{3}x^4\right)\left(\frac{3}{5}x^{-7}\right) = \frac{1}{2} \cdot \frac{2}{3} \cdot \frac{3}{5}x^0 = \frac{1}{5}$

37. Simplifying: $\left(4a^5b^2\right)\left(2b^{-5}c^2\right)\left(3a^7c^4\right) = 24a^{5+7}b^{2-5}c^{2+4} = 24a^{12}b^{-3}c^6 = \frac{24a^{12}c^6}{b^3}$

39. Simplifying: $\left(2x^2y^{-5}\right)^3\left(3x^{-4}y^2\right)^{-4} = 2^3x^6y^{-15} \cdot 3^{-4}x^{16}y^{-8} = \frac{2^3}{3^4}x^{22}y^{-23} = \frac{8x^{22}}{81y^{23}}$

41. Simplifying: $\dfrac{x^{-1}}{x^9} = x^{-1-9} = x^{-10} = \dfrac{1}{x^{10}}$

43. Simplifying: $\dfrac{a^4}{a^{-6}} = a^{4-(-6)} = a^{4+6} = a^{10}$

45. Simplifying: $\dfrac{t^{-10}}{t^{-4}} = t^{-10-(-4)} = t^{-10+4} = t^{-6} = \dfrac{1}{t^6}$

47. Simplifying: $\left(\dfrac{x^5}{x^3}\right)^6 = \left(x^{5-3}\right)^6 = \left(x^2\right)^6 = x^{12}$

49. Simplifying: $\dfrac{\left(x^5\right)^6}{\left(x^3\right)^4} = \dfrac{x^{30}}{x^{12}} = x^{30-12} = x^{18}$

51. Simplifying: $\dfrac{\left(x^{-2}\right)^3\left(x^3\right)^{-2}}{x^{10}} = \dfrac{x^{-6}x^{-6}}{x^{10}} = \dfrac{x^{-12}}{x^{10}} = x^{-12-10} = x^{-22} = \dfrac{1}{x^{22}}$

53. Simplifying: $\dfrac{5a^8b^3}{20a^5b^{-4}} = \tfrac{5}{20}a^{8-5}b^{3-(-4)} = \tfrac{1}{4}a^3b^7 = \dfrac{a^3b^7}{4}$

55. Simplifying: $\dfrac{\left(3x^{-2}y^8\right)^4}{\left(9x^4y^{-3}\right)^2} = \dfrac{81x^{-8}y^{32}}{81x^8y^{-6}} = x^{-8-8}y^{32+6} = x^{-16}y^{38} = \dfrac{y^{38}}{x^{16}}$

57. Simplifying: $\left(\dfrac{8x^2y}{4x^4y^{-3}}\right)^4 = \left(2x^{2-4}y^{1+3}\right)^4 = \left(2x^{-2}y^4\right)^4 = 16x^{-8}y^{16} = \dfrac{16y^{16}}{x^8}$

59. Simplifying: $\left(\dfrac{x^{-5}y^2}{x^{-3}y^5}\right)^{-2} = \left(x^{-5+3}y^{2-5}\right)^{-2} = \left(x^{-2}y^{-3}\right)^{-2} = x^4y^6$

61. Writing in scientific notation: $378{,}000 = 3.78 \times 10^5$

63. Writing in scientific notation: $4{,}900 = 4.9 \times 10^3$

65. Writing in scientific notation: $0.00037 = 3.7 \times 10^{-4}$

67. Writing in scientific notation: $0.00495 = 4.95 \times 10^{-3}$

69. Writing in expanded form: $5.34 \times 10^3 = 5{,}340$

71. Writing in expanded form: $7.8 \times 10^6 = 7{,}800{,}000$

73. Writing in expanded form: $3.44 \times 10^{-3} = 0.00344$

75. Writing in expanded form: $4.9 \times 10^{-1} = 0.49$

77. Simplifying: $\left(4 \times 10^{10}\right)\left(2 \times 10^{-6}\right) = 8 \times 10^{10-6} = 8 \times 10^4$

79. Simplifying: $\dfrac{8 \times 10^{14}}{4 \times 10^5} = 2 \times 10^{14-5} = 2 \times 10^9$

81. Simplifying: $\dfrac{\left(5 \times 10^6\right)\left(4 \times 10^{-8}\right)}{8 \times 10^4} = \dfrac{20 \times 10^{-2}}{8 \times 10^4} = \dfrac{20}{8} \times 10^{-2-4} = 2.5 \times 10^{-6}$

83. **a.** Writing in scientific notation: $\$4.22 \times 10^{11}$

    b. Dividing: $\dfrac{\$4.22 \times 10^{11}}{6 \times 10^7} \approx \$7.033 \times 10^3$. The average credit card debt was $\$7{,}033$ per household.

85. **a.** For Andrew, the cost is: $\$2.65 \times 10^{10}$. For Elena, the cost is: $\$1.25 \times 10^9$.

    b. Subtracting: $\$2.65 \times 10^{10} - \$1.25 \times 10^9 = \$2.65 \times 10^{10} - \$0.125 \times 10^{10} = \$2.525 \times 10^{10}$

    Andrew was $\$2.525 \times 10^{10}$ more costly than Elena.

    c. Dividing: $\dfrac{\$2.65 \times 10^{10}}{\$1.25 \times 10^9} = 21.2$. Andrew was 21.2 times costlier than Elena.

87. Multiplying to find the distance: $\left(1.7 \times 10^6 \text{ light-years}\right)\left(5.9 \times 10^{12} \text{ miles / light-year}\right) \approx 1.003 \times 10^{19}$ miles

89. Writing in scientific notation: $630{,}000{,}000 = 6.3 \times 10^8$ seconds

91. Multiply the second equation by –2:
$$4x + 3y = 10$$
$$-4x - 2y = -8$$
Adding yields $y = 2$. Substituting into the first equation:
$$4x + 3(2) = 10$$
$$4x + 6 = 10$$
$$4x = 4$$
$$x = 1$$
The solution is (1,2).

93. Multiply the second equation by –5:
$$4x + 5y = 5$$
$$-6x - 5y = -10$$
Adding yields:
$$-2x = -5$$
$$x = \tfrac{5}{2}$$
Substituting into the first equation:
$$4\left(\tfrac{5}{2}\right) + 5y = 5$$
$$10 + 5y = 5$$
$$5y = -5$$
$$y = -1$$
The solution is $\left(\tfrac{5}{2}, -1\right)$.

95. Substituting into the first equation:
$$x + (x + 3) = 3$$
$$2x + 3 = 3$$
$$2x = 0$$
$$x = 0$$
The solution is (0,3).

97. Substituting into the first equation:
$$2x - 3(3x - 5) = -6$$
$$2x - 9x + 15 = -6$$
$$-7x + 15 = -6$$
$$-7x = -21$$
$$x = 3$$
The solution is (3,4).

99. Simplifying: $x^{m+2} \cdot x^{-2m} \cdot x^{m-5} = x^{m+2-2m+m-5} = x^{-3} = \dfrac{1}{x^3}$

101. Simplifying: $\left(y^m\right)^2 \left(y^{-3}\right)^m \left(y^{m+3}\right) = y^{2m} \cdot y^{-3m} \cdot y^{m+3} = y^{2m-3m+m+3} = y^3$

103. Simplifying: $\dfrac{x^{n+2}}{x^{n-3}} = x^{n+2-(n-3)} = x^{n+2-n+3} = x^5$

5.2   Polynomials, Sums, and Differences

1. This is a trinomial. The degree is 2 and the leading coefficient is 5.
3. This is a binomial. The degree is 1 and the leading coefficient is 3.
5. This is a trinomial. The degree is 2 and the leading coefficient is 8.
7. This is a polynomial. The degree is 3 and the leading coefficient is 4.
9. This is a monomial. The degree is 0 and the leading coefficient is $-\tfrac{3}{4}$.
11. This is a trinomial. The degree is 3 and the leading coefficient is 6.
13. Simplifying: $(4x + 2) + (3x - 1) = 7x + 1$ **15.** Simplifying: $2x^2 - 3x + 10x - 15 = 2x^2 + 7x - 15$
17. Simplifying: $12a^2 + 8ab - 15ab - 10b^2 = 12a^2 - 7ab - 10b^2$
19. Simplifying: $\left(5x^2 - 6x + 1\right) - \left(4x^2 + 7x - 2\right) = 5x^2 - 6x + 1 - 4x^2 - 7x + 2 = x^2 - 13x + 3$
21. Simplifying: $\left(\tfrac{1}{2}x^2 - \tfrac{1}{3}x - \tfrac{1}{6}\right) - \left(\tfrac{1}{4}x^2 + \tfrac{7}{12}x\right) + \left(\tfrac{1}{3}x - \tfrac{1}{12}\right) = \tfrac{1}{2}x^2 - \tfrac{1}{3}x - \tfrac{1}{6} - \tfrac{1}{4}x^2 - \tfrac{7}{12}x + \tfrac{1}{3}x - \tfrac{1}{12} = \tfrac{1}{4}x^2 - \tfrac{7}{12}x - \tfrac{1}{4}$
23. Simplifying: $\left(y^3 - 2y^2 - 3y + 4\right) - \left(2y^3 - y^2 + y - 3\right) = y^3 - 2y^2 - 3y + 4 - 2y^3 + y^2 - y + 3 = -y^3 - y^2 - 4y + 7$

25. Simplifying:
$$\left(5x^3 - 4x^2\right) - (3x+4) + \left(5x^2 - 7\right) - \left(3x^3 + 6\right) = 5x^3 - 4x^2 - 3x - 4 + 5x^2 - 7 - 3x^3 - 6 = 2x^3 + x^2 - 3x - 17$$

27. Simplifying:
$$\left(\tfrac{4}{7}x^2 - \tfrac{1}{7}xy + \tfrac{1}{14}y^2\right) - \left(\tfrac{1}{2}x^2 - \tfrac{2}{7}xy - \tfrac{9}{14}y^2\right) = \tfrac{4}{7}x^2 - \tfrac{1}{7}xy + \tfrac{1}{14}y^2 - \tfrac{1}{2}x^2 + \tfrac{2}{7}xy + \tfrac{9}{14}y^2$$
$$= \tfrac{8}{14}x^2 - \tfrac{7}{14}x^2 - \tfrac{1}{7}xy + \tfrac{2}{7}xy + \tfrac{1}{14}y^2 + \tfrac{9}{14}y^2$$
$$= \tfrac{1}{14}x^2 + \tfrac{1}{7}xy + \tfrac{5}{7}y^2$$

29. Simplifying:
$$\left(3a^3 + 2a^2 b + ab^2 - b^3\right) - \left(6a^3 - 4a^2 b + 6ab^2 - b^3\right) = 3a^3 + 2a^2 b + ab^2 - b^3 - 6a^3 + 4a^2 b - 6ab^2 + b^3$$
$$= -3a^3 + 6a^2 b - 5ab^2$$

31. Subtracting: $\left(2x^2 - 7x\right) - \left(2x^2 - 4x\right) = 2x^2 - 7x - 2x^2 + 4x = -3x$

33. Adding: $\left(x^2 - 6xy + y^2\right) + \left(2x^2 - 6xy - y^2\right) = x^2 - 6xy + y^2 + 2x^2 - 6xy - y^2 = 3x^2 - 12xy$

35. Subtracting: $\left(9x^5 - 4x^3 - 6\right) - \left(-8x^5 - 4x^3 + 6\right) = 9x^5 - 4x^3 - 6 + 8x^5 + 4x^3 - 6 = 17x^5 - 12$

37. Adding:
$$\left(11a^2 + 3ab + 2b^2\right) + \left(9a^2 - 2ab + b^2\right) + \left(-6a^2 - 3ab + 5b^2\right)$$
$$= 11a^2 + 3ab + 2b^2 + 9a^2 - 2ab + b^2 - 6a^2 - 3ab + 5b^2$$
$$= 14a^2 - 2ab + 8b^2$$

39. Simplifying: $-[2 - (4 - x)] = -(2 - 4 + x) = -(-2 + x) = 2 - x$

41. Simplifying: $-5[-(x - 3) - (x + 2)] = -5(-x + 3 - x - 2) = -5(-2x + 1) = 10x - 5$

43. Simplifying: $4x - 5[3 - (x - 4)] = 4x - 5(3 - x + 4) = 4x - 5(7 - x) = 4x - 35 + 5x = 9x - 35$

45. Simplifying:
$$-(3x - 4y) - [(4x + 2y) - (3x + 7y)] = -(3x - 4y) - (4x + 2y - 3x - 7y)$$
$$= -(3x - 4y) - (x - 5y)$$
$$= -3x + 4y - x + 5y$$
$$= -4x + 9y$$

47. Simplifying:
$$4a - \{3a + 2[a - 5(a + 1) + 4]\} = 4a - [3a + 2(a - 5a - 5 + 4)]$$
$$= 4a - [3a + 2(-4a - 1)]$$
$$= 4a - (3a - 8a - 2)$$
$$= 4a - (-5a - 2)$$
$$= 4a + 5a + 2$$
$$= 9a + 2$$

49. Evaluating when $x = 2$: $2(2)^2 - 3(2) - 4 = 2(4) - 3(2) - 4 = 8 - 6 - 4 = -2$

51. Evaluating when $x = 12$: $P(12) = \tfrac{3}{2}(12)^2 - \tfrac{3}{4}(12) + 1 = \tfrac{3}{2}(144) - \tfrac{3}{4}(12) + 1 = 216 - 9 + 1 = 208$

53. Evaluating when $x = -2$: $Q(-2) = (-2)^3 - (-2)^2 + (-2) - 1 = -8 - 4 - 2 - 1 = -15$

55. Evaluating when $a = 3$:
$$(a + 4)^2 = (3 + 4)^2 = 7^2 = 49$$
$$a^2 + 16 = 3^2 + 16 = 9 + 16 = 25$$
$$a^2 + 8a + 16 = 3^2 + 8(3) + 16 = 9 + 24 + 16 = 49$$

57. Substituting $t = 3$: $h = -16(3)^2 + 128(3) = 240$ feet   Substituting $t = 5$: $h = -16(5)^2 + 128(5) = 240$ feet

59. The weekly profit is given by:
$$P = R - C = \left(100x - 0.5x^2\right) - (60x + 300) = 100x - 0.5x^2 - 60x - 300 = -0.5x^2 + 40x - 300$$

Substituting $x = 60$: $P = -0.5(60)^2 + 40(60) - 300 = \$300$

61. The weekly profit is given by:

$$P = R - C = \left(10x - 0.002x^2\right) - (800 + 6.5x) = 10x - 0.002x^2 - 800 - 6.5x = -0.002x^2 + 3.5x - 800$$

Substituting $x = 1000$: $P = -0.002(1000)^2 + 3.5(1000) - 800 = \$700$

63. Simplifying: $4x^3\left(5x^2\right) = 20x^{3+2} = 20x^5$ 65. Simplifying: $2a^2b\left(ab^2\right) = 2a^{2+1}b^{1+2} = 2a^3b^3$

67. **a.** From the graph: $f(-3) = 5$ **b.** From the graph: $f(0) = -4$
     c. From the graph: $f(1) = -3$ **d.** From the graph: $g(-1) = 3$
     e. From the graph: $g(0) = 4$ **f.** From the graph: $g(2) = 0$
     g. From the graph: $f[g(2)] = f(0) = -4$ **h.** From the graph: $g[f(2)] = g(0) = 4$

5.3 Multiplication of Polynomials

1. Multiplying: $2x\left(6x^2 - 5x + 4\right) = 2x \cdot 6x^2 - 2x \cdot 5x + 2x \cdot 4 = 12x^3 - 10x^2 + 8x$

3. Multiplying: $-3a^2\left(a^3 - 6a^2 + 7\right) = -3a^2 \cdot a^3 - \left(-3a^2\right) \cdot 6a^2 + \left(-3a^2\right) \cdot 7 = -3a^5 + 18a^4 - 21a^2$

5. Multiplying: $2a^2b\left(a^3 - ab + b^3\right) = 2a^2b \cdot a^3 - 2a^2b \cdot ab + 2a^2b \cdot b^3 = 2a^5b - 2a^3b^2 + 2a^2b^4$

7. Multiplying using the vertical format:

$$\begin{array}{rrr} x & -5 & \\ x & +3 & \\ \hline x^2 & -5x & \\ & +3x & -15 \\ \hline x^2 & -2x & -15 \end{array}$$

The product is $x^2 - 2x - 15$.

9. Multiplying using the vertical format:

$$\begin{array}{rrr} 2x^2 & -3 & \\ 3x^2 & -5 & \\ \hline 6x^4 & -9x^2 & \\ & -10x^2 & +15 \\ \hline 6x^4 & -19x^2 & +15 \end{array}$$

The product is $6x^4 - 19x^2 + 15$.

11. Multiplying using the vertical format:

$$\begin{array}{rrrr} x^2 & +6x & +5 & \\ & x & +3 & \\ \hline x^3 & +6x^2 & +5x & \\ & +3x^2 & +18x & +15 \\ \hline x^3 & +9x^2 & +23x & +15 \end{array}$$

The product is $x^3 + 9x^2 + 23x + 15$.

13. Multiplying using the vertical format:

$$\begin{array}{rrrr} a^2 & +ab & +b^2 & \\ & a & -b & \\ \hline a^3 & +a^2b & +ab^2 & \\ & -a^2b & -ab^2 & -b^3 \\ \hline a^3 & & & -b^3 \end{array}$$

The product is $a^3 - b^3$.

15. Multiplying using the vertical format:

$$\begin{array}{rrrr} 4x^2 & -2xy & +y^2 & \\ & 2x & +y & \\ \hline 8x^3 & -4x^2y & +2xy^2 & \\ & +4x^2y & -2xy^2 & +y^3 \\ \hline 8x^3 & & & +y^3 \end{array}$$

The product is $8x^3 + y^3$.

17. Multiplying using the vertical format:

$$\begin{array}{rrrr} a^2 & +ab & +b^2 & \\ & 2a & -3b & \\ \hline 2a^3 & +2a^2b & +2ab^2 & \\ & -3a^2b & -3ab^2 & -3b^3 \\ \hline 2a^3 & -a^2b & -ab^2 & -3b^3 \end{array}$$

The product is $2a^3 - a^2b - ab^2 - 3b^3$.

19. Multiplying using FOIL: $(x - 2)(x + 3) = x^2 + 3x - 2x - 6 = x^2 + x - 6$

21. Multiplying using FOIL: $(2a + 3)(3a + 2) = 6a^2 + 4a + 9a + 6 = 6a^2 + 13a + 6$

23. Multiplying using FOIL: $(5 - 3t)(4 + 2t) = 20 + 10t - 12t - 6t^2 = 20 - 2t - 6t^2$

25. Multiplying using FOIL: $\left(x^3 + 3\right)\left(x^3 - 5\right) = x^6 - 5x^3 + 3x^3 - 15 = x^6 - 2x^3 - 15$

27. Multiplying using FOIL: $(5x - 6y)(4x + 3y) = 20x^2 + 15xy - 24xy - 18y^2 = 20x^2 - 9xy - 18y^2$

29. Multiplying using FOIL: $\left(3t + \frac{1}{3}\right)\left(6t - \frac{2}{3}\right) = 18t^2 - 2t + 2t - \frac{2}{9} = 18t^2 - \frac{2}{9}$

31.   Finding the product: $(5x + 2y)^2 = (5x)^2 + 2(5x)(2y) + (2y)^2 = 25x^2 + 20xy + 4y^2$

33.   Finding the product: $\left(5 - 3t^3\right)^2 = (5)^2 - 2(5)\left(3t^3\right) + \left(3t^3\right)^2 = 25 - 30t^3 + 9t^6$

35.   Finding the product: $(2a + 3b)(2a - 3b) = (2a)^2 - (3b)^2 = 4a^2 - 9b^2$

37.   Finding the product: $\left(3r^2 + 7s\right)\left(3r^2 - 7s\right) = \left(3r^2\right)^2 - (7s)^2 = 9r^4 - 49s^2$

39.   Finding the product: $\left(\frac{1}{3}x - \frac{2}{5}\right)\left(\frac{1}{3}x + \frac{2}{5}\right) = \left(\frac{1}{3}x\right)^2 - \left(\frac{2}{5}\right)^2 = \frac{1}{9}x^2 - \frac{4}{25}$

41.   Expanding and simplifying: **43.**   Expanding and simplifying:

$$(x-2)^3 = (x-2)(x-2)^2$$
$$= (x-2)\left(x^2 - 4x + 4\right)$$
$$= x^3 - 4x^2 + 4x - 2x^2 + 8x - 8$$
$$= x^3 - 6x^2 + 12x - 8$$

$$\left(x - \tfrac{1}{2}\right)^3 = \left(x - \tfrac{1}{2}\right)\left(x - \tfrac{1}{2}\right)^2$$
$$= \left(x - \tfrac{1}{2}\right)\left(x^2 - x + \tfrac{1}{4}\right)$$
$$= x^3 - x^2 + \tfrac{1}{4}x - \tfrac{1}{2}x^2 + \tfrac{1}{2}x - \tfrac{1}{8}$$
$$= x^3 - \tfrac{3}{2}x^2 + \tfrac{3}{4}x - \tfrac{1}{8}$$

45.   Expanding and simplifying:

$$3(x-1)(x-2)(x-3) = 3(x-1)\left(x^2 - 5x + 6\right)$$
$$= 3\left(x^3 - 5x^2 + 6x - x^2 + 5x - 6\right)$$
$$= 3\left(x^3 - 6x^2 + 11x - 6\right)$$
$$= 3x^3 - 18x^2 + 33x - 18$$

47.   Expanding and simplifying: $\left(b^2 + 8\right)\left(a^2 + 1\right) = a^2 b^2 + b^2 + 8a^2 + 8$

49.   Expanding and simplifying:

$$(x+1)^2 + (x+2)^2 + (x+3)^2 = x^2 + 2x + 1 + x^2 + 4x + 4 + x^2 + 6x + 9 = 3x^2 + 12x + 14$$

51.   Expanding and simplifying:

$$(2x+3)^2 - (2x-3)^2 = \left(4x^2 + 12x + 9\right) - \left(4x^2 - 12x + 9\right) = 4x^2 + 12x + 9 - 4x^2 + 12x - 9 = 24x$$

53.   Multiplying using FOIL: $\left[(x+y) - 4\right]\left[(x+y) + 5\right] = (x+y)^2 + (x+y) - 20 = x^2 + 2xy + y^2 + x + y - 20$

55.   Evaluating when $a = 2$ and $b = 3$:

$$a^4 - b^4 = 2^4 - 3^4 = 16 - 81 = -65$$
$$(a-b)^4 = (2-3)^4 = (-1)^4 = 1$$
$$\left(a^2 + b^2\right)(a+b)(a-b) = \left(2^2 + 3^2\right)(2+3)(2-3) = (13)(5)(-1) = -65$$

57.   Since $R = xp$, substitute $x = 900 - 300p$ to obtain: $R(p) = (900 - 300p)p = 900p - 300p^2$

To find $R(x)$, we first solve for p:

$$x = 900 - 300p$$
$$300p = 900 - x$$
$$p = 3 - \frac{x}{300}$$

Now substitute to obtain: $R(x) = x\left(3 - \dfrac{x}{300}\right) = 3x - \dfrac{x^2}{300}$

Substituting $p = \$1.60$: $R(1.60) = 900(1.60) - 300(1.60)^2 = \$672$. The revenue is $672.

59. Since $R = xp$, substitute $x = 350 - 10p$ to obtain: $R(p) = (350 - 10p)p = 350p - 10p^2$

To find $R(x)$, we first solve for p:
$$x = 350 - 10p$$
$$10p = 350 - x$$
$$p = 35 - \frac{x}{10}$$

Now substitute to obtain: $R(x) = x\left(35 - \frac{x}{10}\right) = 35x - \frac{x^2}{10}$

Substituting $x = 65$: $R(65) = 35(65) - \frac{(65)^2}{10} = \$1,852.50$. The revenue is $\$1,852.50$.

61. Since $R(x) = 35x - \frac{x^2}{10}$ and $C(x) = 5x + 500$, then:

$$P(x) = R(x) - C(x) = 35x - \frac{x^2}{10} - (5x + 500) = -\frac{x^2}{10} + 30x - 500$$

$$P(60) = -\frac{(60)^2}{10} + 30(60) - 500 = \$940$$

63. Expanding the formula:
$$A = 100(1 + r)^4$$
$$= 100(1 + r)^2(1 + r)^2$$
$$= 100\left(1 + 2r + r^2\right)\left(1 + 2r + r^2\right)$$
$$= 100\left(1 + 2r + r^2 + 2r + 4r^2 + 2r^3 + r^2 + 2r^3 + r^4\right)$$
$$= 100\left(1 + 4r + 6r^2 + 4r^3 + r^4\right)$$
$$= 100 + 400r + 600r^2 + 400r^3 + 100r^4$$

65. Multiply the first equation by -1 and add it to the second equation:
$$-x - y - z = -6$$
$$2x - y + z = 3$$

Adding yields the equation $x - 2y = -3$. Multiply the first equation by 3 and add it to the third equation:
$$3x + 3y + 3z = 18$$
$$x + 2y - 3z = -4$$

Adding yields the equation $4x + 5y = 14$. So the system becomes:
$$x - 2y = -3$$
$$4x + 5y = 14$$

Multiply the first equation by -4 and add it to the second equation:
$$-4x + 8y = 12$$
$$4x + 5y = 14$$

Adding yields:
$$13y = 26$$
$$y = 2$$

Substituting to find x:
$$4x + 5(2) = 14$$
$$4x + 10 = 14$$
$$4x = 4$$
$$x = 1$$

Substituting into the original first equation:
$$1 + 2 + z = 6$$
$$z + 3 = 6$$
$$z = 3$$

The solution is $(1, 2, 3)$.

67.   Multiply the third equation by -1 and add it to the first equation:
$$3x + 4y = 15$$
$$-4y + 3z = -9$$
Adding yields the equation $3x + 3z = 6$, or $x + z = 2$. So the system becomes:
$$2x - 5z = -3$$
$$x + z = 2$$
Multiply the second equation by -2:
$$2x - 5z = -3$$
$$-2x - 2z = -4$$
Adding yields:
$$-7z = -7$$
$$z = 1$$
Substituting to find x:
$$2x - 5(1) = -3$$
$$2x - 5 = -3$$
$$2x = 2$$
$$x = 1$$
Substituting into the original first equation:
$$3(1) + 4y = 15$$
$$4y + 3 = 15$$
$$4y = 12$$
$$y = 3$$
The solution is $(1,3,1)$.

69.   Drawing the figure:

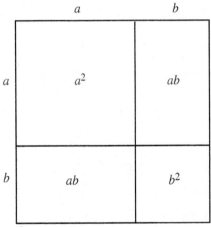

Adding up the individual areas: $(a + b)^2 = a^2 + ab + ab + b^2 = a^2 + 2ab + b^2$

5.4 The Greatest Common Factor and Factoring by Grouping

1. Factoring the expression: $10x^3 - 15x^2 = 5x^2(2x - 3)$

3. Factoring the expression: $9y^6 + 18y^3 = 9y^3(y^3 + 2)$

5. Factoring the expression: $9a^2b - 6ab^2 = 3ab(3a - 2b)$

7. Factoring the expression: $21xy^4 + 7x^2y^2 = 7xy^2(3y^2 + x)$

9. Factoring the expression: $3a^2 - 21a + 30 = 3(a^2 - 7a + 10)$

11. Factoring the expression: $4x^3 - 16x^2 - 20x = 4x(x^2 - 4x - 5)$

13. Factoring the expression: $10x^4y^2 + 20x^3y^3 - 30x^2y^4 = 10x^2y^2(x^2 + 2xy - 3y^2)$

15. Factoring the expression: $-x^2y + xy^2 - x^2y^2 = xy(-x + y - xy)$

17. Factoring the expression: $4x^3y^2z - 8x^2y^2z^2 + 6xy^2z^3 = 2xy^2z(2x^2 - 4xz + 3z^2)$

19. Factoring the expression: $20a^2b^2c^2 - 30ab^2c + 25a^2bc^2 = 5abc(4abc - 6b + 5ac)$

21. Factoring the expression: $5x(a - 2b) - 3y(a - 2b) = (a - 2b)(5x - 3y)$

23. Factoring the expression: $3x^2(x + y)^2 - 6y^2(x + y)^2 = 3(x + y)^2(x^2 - 2y^2)$

25. Factoring the expression: $2x^2(x + 5) + 7x(x + 5) + 6(x + 5) = (x + 5)(2x^2 + 7x + 6)$

27. Factoring by grouping: $3xy + 3y + 2ax + 2a = 3y(x + 1) + 2a(x + 1) = (x + 1)(3y + 2a)$

29. Factoring by grouping: $x^2y + x + 3xy + 3 = x(xy + 1) + 3(xy + 1) = (x + 3)(xy + 1)$

31. Factoring by grouping: $3xy^2 - 6y^2 + 4x - 8 = 3y^2(x - 2) + 4(x - 2) = (x - 2)(3y^2 + 4)$

33. Factoring by grouping: $x^2 - ax - bx + ab = x(x - a) - b(x - a) = (x - a)(x - b)$

35. Factoring by grouping: $ab + 5a - b - 5 = a(b + 5) - 1(b + 5) = (b + 5)(a - 1)$

37. Factoring by grouping: $a^4b^2 + a^4 - 5b^2 - 5 = a^4(b^2 + 1) - 5(b^2 + 1) = (b^2 + 1)(a^4 - 5)$

39. Factoring by grouping: $x^3 + 3x^2 - 4x - 12 = x^2(x + 3) - 4(x + 3) = (x + 3)(x^2 - 4)$

41. Factoring by grouping: $x^3 + 2x^2 - 25x - 50 = x^2(x + 2) - 25(x + 2) = (x + 2)(x^2 - 25)$

43. Factoring by grouping: $2x^3 + 3x^2 - 8x - 12 = x^2(2x + 3) - 4(2x + 3) = (2x + 3)(x^2 - 4)$

45. Factoring by grouping: $4x^3 + 12x^2 - 9x - 27 = 4x^2(x + 3) - 9(x + 3) = (x + 3)(4x^2 - 9)$

47. It will be $3 \cdot 2 = 6$.

49. Using factoring by grouping:
$$P + Pr + (P + Pr)r = (P + Pr) + (P + Pr)r = (P + Pr)(1 + r) = P(1 + r)(1 + r) = P(1 + r)^2$$

51. Factoring the revenue: $R(x) = 11.5x - 0.05x^2 = x(11.5 - 0.05x)$. So the price is $p = 11.5 - 0.05x$.
    Substituting $x = 125$: $p = 11.5 - 0.05(125) = \$5.25$. The price is $5.25.

53. Factoring the revenue: $R(x) = 35x - 0.1x^2 = x(35 - 0.1x)$. So the price is $p = 35 - 0.1x$.
    Substituting $x = 65$: $p = 35 - 0.1(65) = \$28.50$. The price is $28.50.

55. Multiplying using FOIL: $(x + 2)(x + 3) = x^2 + 3x + 2x + 6 = x^2 + 5x + 6$

57. Multiplying using FOIL: $(2y + 5)(3y - 7) = 6y^2 - 14y + 15y - 35 = 6y^2 + y - 35$

59. Multiplying using FOIL: $(4 - 3a)(5 - a) = 20 - 4a - 15a + 3a^2 = 20 - 19a + 3a^2$

61. Evaluating the determinant: $\begin{vmatrix} 3 & 5 \\ -6 & 2 \end{vmatrix} = 3(2) - 5(-6) = 6 + 30 = 36$

63.   Expanding along the first column:
$$\begin{vmatrix} 1 & -2 & 3 \\ 0 & 4 & -1 \\ 2 & -4 & 6 \end{vmatrix} = 1\begin{vmatrix} 4 & -1 \\ -4 & 6 \end{vmatrix} - 0\begin{vmatrix} -2 & 3 \\ -4 & 6 \end{vmatrix} + 2\begin{vmatrix} -2 & 3 \\ 4 & -1 \end{vmatrix} = 1(24-4)-0+2(2-12) = 20-20 = 0$$

5.5   Factoring Trinomials

1.   Factoring the trinomial: $x^2 + 7x + 12 = (x+3)(x+4)$ **3.**   Factoring the trinomial: $x^2 - x - 12 = (x+3)(x-4)$

5.   Factoring the trinomial: $y^2 + y - 6 = (y+3)(y-2)$ **7.**   Factoring the trinomial: $16 - 6x - x^2 = (2-x)(8+x)$

9.   Factoring the trinomial: $12 + 8x + x^2 = (2+x)(6+x)$

11.   Factoring the trinomial: $3a^2 - 21a + 30 = 3(a^2 - 7a + 10) = 3(a-5)(a-2)$

13.   Factoring the trinomial: $4x^3 - 16x^2 - 20x = 4x(x^2 - 4x - 5) = 4x(x-5)(x+1)$

15.   Factoring the trinomial: $x^2 + 3xy + 2y^2 = (x+2y)(x+y)$

17.   Factoring the trinomial: $a^2 + 3ab - 18b^2 = (a+6b)(a-3b)$

19.   Factoring the trinomial: $x^2 - 2xa - 48a^2 = (x-8a)(x+6a)$

21.   Factoring the trinomial: $x^2 - 12xb + 36b^2 = (x-6b)^2$

23.   Factoring the trinomial: $3x^2 - 6xy - 9y^2 = 3(x^2 - 2xy - 3y^2) = 3(x-3y)(x+y)$

25.   Factoring the trinomial: $2a^5 + 4a^4b + 4a^3b^2 = 2a^3(a^2 + 2ab + 2b^2)$

27.   Factoring the trinomial: $10x^4y^2 + 20x^3y^3 - 30x^2y^4 = 10x^2y^2(x^2 + 2xy - 3y^2) = 10x^2y^2(x+3y)(x-y)$

29.   Factoring completely: $2x^2 + 7x - 15 = (2x-3)(x+5)$ **31.**   Factoring completely: $2x^2 + x - 15 = (2x-5)(x+3)$

33.   Factoring completely: $2x^2 - 13x + 15 = (2x-3)(x-5)$

35.   Factoring completely: $2x^2 - 11x + 15 = (2x-5)(x-3)$

37.   The trinomial $2x^2 + 7x + 15$ does not factor (prime).   **39.**   Factoring completely: $2 + 7a + 6a^2 = (2+3a)(1+2a)$

41.   Factoring completely: $60y^2 - 15y - 45 = 15(4y^2 - y - 3) = 15(4y+3)(y-1)$

43.   Factoring completely: $6x^4 - x^3 - 2x^2 = x^2(6x^2 - x - 2) = x^2(3x-2)(2x+1)$

45.   Factoring completely: $40r^3 - 120r^2 + 90r = 10r(4r^2 - 12r + 9) = 10r(2r-3)^2$

47.   Factoring completely: $4x^2 - 11xy - 3y^2 = (4x+y)(x-3y)$

49.   Factoring completely: $10x^2 - 3xa - 18a^2 = (2x-3a)(5x+6a)$

51.   Factoring completely: $18a^2 + 3ab - 28b^2 = (3a+4b)(6a-7b)$

53.   Factoring completely: $600 + 800t - 800t^2 = 200(3 + 4t - 4t^2) = 200(1+2t)(3-2t)$

55.   Factoring completely: $9y^4 + 9y^3 - 10y^2 = y^2(9y^2 + 9y - 10) = y^2(3y-2)(3y+5)$

57.   Factoring completely: $24a^2 - 2a^3 - 12a^4 = 2a^2(12 - a - 6a^2) = 2a^2(3+2a)(4-3a)$

59.   Factoring completely: $8x^4y^2 - 2x^3y^3 - 6x^2y^4 = 2x^2y^2(4x^2 - xy - 3y^2) = 2x^2y^2(4x+3y)(x-y)$

61.   Factoring completely: $300x^4 + 1000x^2 + 300 = 100(3x^4 + 10x^2 + 3) = 100(3x^2 + 1)(x^2 + 3)$

63.   Factoring completely: $20a^4 + 37a^2 + 15 = (5a^2 + 3)(4a^2 + 5)$

65.   Factoring completely: $9 + 3r^2 - 12r^4 = 3(3 + r^2 - 4r^4) = 3(3 + 4r^2)(1 - r^2)$

67. Factoring completely: $2x^2(x+5)+7x(x+5)+6(x+5)=(x+5)(2x^2+7x+6)=(x+5)(2x+3)(x+2)$

69. Factoring completely: $x^2(2x+3)+7x(2x+3)+10(2x+3)=(2x+3)(x^2+7x+10)=(2x+3)(x+5)(x+2)$

71. Multiplying out, the polynomial is: $(3x+5y)(3x-5y)=9x^2-25y^2$

73. The polynomial factors as $a^2+260a+2500=(a+10)(a+250)$, so the other factor is $a+250$.

75. Factoring the right side: $y=4x^2+18x-10=2(2x^2+9x-5)=2(2x-1)(x+5)$

Evaluating when $x=\frac{1}{2}$: $y=2(2\cdot\frac{1}{2}-1)(\frac{1}{2}+5)=2(0)(\frac{11}{2})=0$

Evaluating when $x=-5$: $y=2(2\cdot(-5)-1)(-5+5)=2(-11)(0)=0$

Evaluating when $x=2$: $y=2(2\cdot2-1)(2+5)=2(3)(7)=42$

77. Factoring the right side: $h(t)=96+80t-16t^2=16(6+5t-t^2)=16(6-t)(1+t)$

Evaluating when $t=6$: $h(6)=16(6-6)(1+6)=16(0)(7)=0$ feet

Evaluating when $t=3$: $h(3)=16(6-3)(1+3)=16(3)(4)=192$ feet

79. Multiplying: $(2x-3)(2x+3)=(2x)^2-(3)^2=4x^2-9$

81. Multiplying: $(2x-3)^2=(2x-3)(2x-3)=4x^2-12x+9$

83. Multiplying: $(2x-3)(4x^2+6x+9)=8x^3+12x^2+18x-12x^2-18x-27=8x^3-27$

85. Factoring completely: $8x^6+26x^3y^2+15y^4=(2x^3+5y^2)(4x^3+3y^2)$

87. Factoring completely: $3x^2+295x-500=(3x-5)(x+100)$

89. Factoring completely: $\frac{1}{8}x^2+x+2=(\frac{1}{4}x+1)(\frac{1}{2}x+2)$

91. Factoring completely: $2x^2+1.5x+0.25=(2x+0.5)(x+0.5)$

5.6   Special Factoring

1. Factoring the trinomial: $x^2-6x+9=(x-3)^2$

3. Factoring the trinomial: $a^2-12a+36=(a-6)^2$

5. Factoring the trinomial: $25-10t+t^2=(5-t)^2$

7. Factoring the trinomial: $\frac{1}{9}x^2+2x+9=(\frac{1}{3}x+3)^2$

9. Factoring the trinomial: $4y^4-12y^2+9=(2y^2-3)^2$

11. Factoring the trinomial: $16a^2+40ab+25b^2=(4a+5b)^2$

13. Factoring the trinomial: $\frac{1}{25}+\frac{1}{10}t^2+\frac{1}{16}t^4=(\frac{1}{5}+\frac{1}{4}t^2)^2$

15. Factoring the trinomial: $16x^2-48x+36=4(4x^2-12x+9)=4(2x-3)^2$

17. Factoring the trinomial: $75a^3+30a^2+3a=3a(25a^2+10a+1)=3a(5a+1)^2$

19. Factoring the trinomial: $(x+2)^2+6(x+2)+9=(x+2+3)^2=(x+5)^2$

21. Factoring: $x^2-9=(x+3)(x-3)$

23. Factoring: $49x^2-64y^2=(7x+8y)(7x-8y)$

25. Factoring: $4a^2-\frac{1}{4}=(2a+\frac{1}{2})(2a-\frac{1}{2})$

27. Factoring: $x^2-\frac{9}{25}=(x+\frac{3}{5})(x-\frac{3}{5})$

29. Factoring: $9x^2-16y^2=(3x+4y)(3x-4y)$

31. Factoring: $250-10t^2=10(25-t^2)=10(5+t)(5-t)$

33. Factoring: $x^4-81=(x^2+9)(x^2-9)=(x^2+9)(x+3)(x-3)$

35. Factoring: $9x^6-1=(3x^3+1)(3x^3-1)$

37. Factoring: $16a^4-81=(4a^2+9)(4a^2-9)=(4a^2+9)(2a+3)(2a-3)$

39. Factoring: $\frac{1}{81} - \frac{y^4}{16} = \left(\frac{1}{9} + \frac{y^2}{4}\right)\left(\frac{1}{9} - \frac{y^2}{4}\right) = \left(\frac{1}{9} + \frac{y^2}{4}\right)\left(\frac{1}{3} + \frac{y}{2}\right)\left(\frac{1}{3} - \frac{y}{2}\right)$

41. Factoring: $x^6 - y^6 = \left(x^3 + y^3\right)\left(x^3 - y^3\right) = (x+y)\left(x^2 - xy + y^2\right)(x-y)\left(x^2 + xy + y^2\right)$

43. Factoring: $2a^7 - 128a = 2a\left(a^6 - 64\right) = 2a\left(a^3 + 8\right)\left(a^3 - 8\right) = 2a(a+2)\left(a^2 - 2a + 4\right)(a-2)\left(a^2 + 2a + 4\right)$

45. Factoring: $(x-2)^2 - 9 = (x-2+3)(x-2-3) = (x+1)(x-5)$

47. Factoring: $(y+4)^2 - 16 = (y+4+4)(y+4-4) = y(y+8)$

49. Factoring: $x^2 - 10x + 25 - y^2 = (x-5)^2 - y^2 = (x-5+y)(x-5-y)$

51. Factoring: $a^2 + 8a + 16 - b^2 = (a+4)^2 - b^2 = (a+4+b)(a+4-b)$

53. Factoring: $x^2 + 2xy + y^2 - a^2 = (x+y)^2 - a^2 = (x+y+a)(x+y-a)$

55. Factoring: $x^3 + 3x^2 - 4x - 12 = x^2(x+3) - 4(x+3) = (x+3)\left(x^2 - 4\right) = (x+3)(x+2)(x-2)$

57. Factoring: $x^3 + 2x^2 - 25x - 50 = x^2(x+2) - 25(x+2) = (x+2)\left(x^2 - 25\right) = (x+2)(x+5)(x-5)$

59. Factoring: $2x^3 + 3x^2 - 8x - 12 = x^2(2x+3) - 4(2x+3) = (2x+3)\left(x^2 - 4\right) = (2x+3)(x+2)(x-2)$

61. Factoring: $4x^3 + 12x^2 - 9x - 27 = 4x^2(x+3) - 9(x+3) = (x+3)\left(4x^2 - 9\right) = (x+3)(2x+3)(2x-3)$

63. Factoring: $x^3 - y^3 = (x-y)\left(x^2 + xy + y^2\right)$ 　　　65. Factoring: $a^3 + 8 = (a+2)\left(a^2 - 2a + 4\right)$

67. Factoring: $27 + x^3 = (3+x)\left(9 - 3x + x^2\right)$ 　　　69. Factoring: $y^3 - 1 = (y-1)\left(y^2 + y + 1\right)$

71. Factoring: $10r^3 - 1250 = 10\left(r^3 - 125\right) = 10(r-5)\left(r^2 + 5r + 25\right)$

73. Factoring: $64 + 27a^3 = (4+3a)\left(16 - 12a + 9a^2\right)$ 　　75. Factoring: $8x^3 - 27y^3 = (2x-3y)\left(4x^2 + 6xy + 9y^2\right)$

77. Factoring: $t^3 + \frac{1}{27} = \left(t + \frac{1}{3}\right)\left(t^2 - \frac{1}{3}t + \frac{1}{9}\right)$ 　　79. Factoring: $27x^3 - \frac{1}{27} = \left(3x - \frac{1}{3}\right)\left(9x^2 + x + \frac{1}{9}\right)$

81. Factoring: $64a^3 + 125b^3 = (4a+5b)\left(16a^2 - 20ab + 25b^2\right)$

83. Since $9x^2 + 30x + 25 = (3x+5)^2$ and $9x^2 - 30x + 25 = (3x-5)^2$, two values of b are $b = 30$ and $b = -30$.

85. Evaluating with $r = 0.12$: $A = 100\left(1 + 0.12 + \frac{0.12^2}{4}\right) = \$112.36$

87. Evaluating the determinants:

$$D = \begin{vmatrix} 4 & -7 \\ 5 & 2 \end{vmatrix} = 4(2) - (-7)(5) = 8 + 35 = 43$$

$$D_x = \begin{vmatrix} 3 & -7 \\ -3 & 2 \end{vmatrix} = 3(2) - (-7)(-3) = 6 - 21 = -15$$

$$D_y = \begin{vmatrix} 4 & 3 \\ 5 & -3 \end{vmatrix} = 4(-3) - 3(5) = -12 - 15 = -27$$

Using Cramer's rule: $x = \dfrac{D_x}{D} = -\dfrac{15}{43}$, $y = \dfrac{D_y}{D} = -\dfrac{27}{43}$

The solution is $\left(-\frac{15}{43}, -\frac{27}{43}\right)$.

89.   Evaluating the determinants:

$$D = \begin{vmatrix} 3 & 4 & 0 \\ 2 & 0 & -5 \\ 0 & 4 & -3 \end{vmatrix} = 3\begin{vmatrix} 0 & -5 \\ 4 & -3 \end{vmatrix} - 4\begin{vmatrix} 2 & -5 \\ 0 & -3 \end{vmatrix} = 3(0+20) - 4(-6-0) = 60+24 = 84$$

$$D_x = \begin{vmatrix} 15 & 4 & 0 \\ -3 & 0 & -5 \\ 9 & 4 & -3 \end{vmatrix} = 15\begin{vmatrix} 0 & -5 \\ 4 & -3 \end{vmatrix} - 4\begin{vmatrix} -3 & -5 \\ 9 & -3 \end{vmatrix} = 15(0+20) - 4(9+45) = 300-216 = 84$$

$$D_y = \begin{vmatrix} 3 & 15 & 0 \\ 2 & -3 & -5 \\ 0 & 9 & -3 \end{vmatrix} = 3\begin{vmatrix} -3 & -5 \\ 9 & -3 \end{vmatrix} - 15\begin{vmatrix} 2 & -5 \\ 0 & -3 \end{vmatrix} = 3(9+45) - 15(-6-0) = 162+90 = 252$$

$$D_z = \begin{vmatrix} 3 & 4 & 15 \\ 2 & 0 & -3 \\ 0 & 4 & 9 \end{vmatrix} = 3\begin{vmatrix} 0 & -3 \\ 4 & 9 \end{vmatrix} - 2\begin{vmatrix} 4 & 15 \\ 4 & 9 \end{vmatrix} = 3(0+12) - 2(36-60) = 36+48 = 84$$

Using Cramer's rule: $x = \dfrac{D_x}{D} = \dfrac{84}{84} = 1, y = \dfrac{D_y}{D} = \dfrac{252}{84} = 3, z = \dfrac{D_z}{D} = \dfrac{84}{84} = 1$. The solution is $(1,3,1)$.

5.7   Factoring: A General Review

1.   Factoring: $x^2 - 81 = (x+9)(x-9)$ **3.**   Factoring: $x^2 + 2x - 15 = (x-3)(x+5)$

5.   Factoring: $x^2(x+2) + 6x(x+2) + 9(x+2) = (x+2)\left(x^2 + 6x + 9\right) = (x+2)(x+3)^2$

7.   Factoring: $x^2 y^2 + 2y^2 + x^2 + 2 = y^2\left(x^2 + 2\right) + 1\left(x^2 + 2\right) = \left(x^2 + 2\right)\left(y^2 + 1\right)$

9.   Factoring: $2a^3 b + 6a^2 b + 2ab = 2ab\left(a^2 + 3a + 1\right)$ **11.**   The polynomial $x^2 + x + 1$ does not factor (prime).

13.   Factoring: $12a^2 - 75 = 3\left(4a^2 - 25\right) = 3(2a+5)(2a-5)$

15.   Factoring: $9x^2 - 12xy + 4y^2 = (3x-2y)^2$ **17.**   Factoring: $25 - 10t + t^2 = (5-t)^2$

19.   Factoring: $4x^3 + 16xy^2 = 4x\left(x^2 + 4y^2\right)$

21.   Factoring: $2y^3 + 20y^2 + 50y = 2y\left(y^2 + 10y + 25\right) = 2y(y+5)^2$

23.   Factoring: $a^7 + 8a^4 b^3 = a^4\left(a^3 + 8b^3\right) = a^4(a+2b)\left(a^2 - 2ab + 4b^2\right)$

25.   Factoring: $t^2 + 6t + 9 - x^2 = (t+3)^2 - x^2 = (t+3+x)(t+3-x)$

27.   Factoring: $x^3 + 5x^2 - 9x - 45 = x^2(x+5) - 9(x+5) = (x+5)\left(x^2 - 9\right) = (x+5)(x+3)(x-3)$

29.   Factoring: $5a^2 + 10ab + 5b^2 = 5\left(a^2 + 2ab + b^2\right) = 5(a+b)^2$

31.   The polynomial $x^2 + 49$ does not factor (prime).

33.   Factoring: $3x^2 + 15xy + 18y^2 = 3\left(x^2 + 5xy + 6y^2\right) = 3(x+2y)(x+3y)$

35.   Factoring: $9a^2 + 2a + \frac{1}{9} = \left(3a + \frac{1}{3}\right)^2$

37.   Factoring: $x^2(x-3) - 14x(x-3) + 49(x-3) = (x-3)\left(x^2 - 14x + 49\right) = (x-3)(x-7)^2$

39.   Factoring: $x^2 - 64 = (x+8)(x-8)$ **41.**   Factoring: $8 - 14x - 15x^2 = (2-5x)(4+3x)$

43.   Factoring: $49a^7 - 9a^5 = a^5\left(49a^2 - 9\right) = a^5(7a+3)(7a-3)$

45.   Factoring: $r^2 - \frac{1}{25} = \left(r + \frac{1}{5}\right)\left(r - \frac{1}{5}\right)$ **47.**   The polynomial $49x^2 + 9y^2$ does not factor (prime).

49.   Factoring: $100x^2 - 100x - 600 = 100\left(x^2 - x - 6\right) = 100(x-3)(x+2)$

51. Factoring: $25a^3 + 20a^2 + 3a = a\left(25a^2 + 20a + 3\right) = a(5a+3)(5a+1)$

53. Factoring: $3x^4 - 14x^2 - 5 = \left(3x^2 + 1\right)\left(x^2 - 5\right)$

55. Factoring: $24a^5b - 3a^2b = 3a^2b\left(8a^3 - 1\right) = 3a^2b(2a-1)\left(4a^2 + 2a + 1\right)$

57. Factoring: $64 - r^3 = (4-r)\left(16 + 4r + r^2\right)$

59. Factoring: $20x^4 - 45x^2 = 5x^2\left(4x^2 - 9\right) = 5x^2(2x+3)(2x-3)$

61. Factoring: $400t^2 - 900 = 100\left(4t^2 - 9\right) = 100(2t+3)(2t-3)$

63. Factoring: $16x^5 - 44x^4 + 30x^3 = 2x^3\left(8x^2 - 22x + 15\right) = 2x^3(4x-5)(2x-3)$

65. Factoring: $y^6 - 1 = \left(y^3 + 1\right)\left(y^3 - 1\right) = (y+1)\left(y^2 - y + 1\right)(y-1)\left(y^2 + y + 1\right)$

67. Factoring: $50 - 2a^2 = 2\left(25 - a^2\right) = 2(5+a)(5-a)$

69. Factoring: $12x^4y^2 + 36x^3y^3 + 27x^2y^4 = 3x^2y^2\left(4x^2 + 12xy + 9y^2\right) = 3x^2y^2(2x+3y)^2$

71. Factoring: $x^2 - 4x + 4 - y^2 = (x-2)^2 - y^2 = (x-2+y)(x-2-y)$

73. Let g and d represent the number of geese and ducks. The system of equations is:
$$g + d = 108$$
$$1.4g + 0.6d = 112.80$$
Substituting $d = 108 - g$ into the second equation:
$$1.4g + 0.6(108 - g) = 112.80$$
$$1.4g + 64.8 - 0.6g = 112.80$$
$$0.8g = 48$$
$$g = 60$$
$$d = 48$$
He bought 60 geese and 48 ducks.

75. Let o represent the number of oranges and a represent the number of apples. The system of equations is:
$$\frac{o}{3}(0.10) + \frac{a}{12}(0.15) = 6.80$$
$$\frac{5o}{3}(0.10) + \frac{a}{48}(0.15) = 25.45$$
Clearing each equation of fractions:

$$12 \cdot \frac{o}{3}(0.10) + 12 \cdot \frac{a}{12}(0.15) = 12 \cdot 6.80 \qquad 48 \cdot \frac{5o}{3}(0.10) + 48 \cdot \frac{a}{48}(0.15) = 48 \cdot 25.45$$
$$0.4o + 0.15a = 81.6 \qquad\qquad\qquad 8o + 0.15a = 1221.6$$

So the system becomes:
$$0.4o + 0.15a = 81.6$$
$$8o + 0.15a = 1221.6$$
Multiply the first equation by –20:
$$-8o - 3a = -1632$$
$$8o + 0.15a = 1221.6$$
Adding yields:
$$-2.85a = -410.4$$
$$a = 144$$
Substituting to find o:
$$8o + 0.15(144) = 1221.6$$
$$8o + 21.6 = 1221.6$$
$$8a = 1200$$
$$a = 150$$
So 150 oranges and 144 apples were bought.

5.8   Solving Equations by Factoring

1.   Solving the equation:

$$x^2 - 5x - 6 = 0$$
$$(x+1)(x-6) = 0$$
$$x = -1, 6$$

3.   Solving the equation:
$$x^3 - 5x^2 + 6x = 0$$
$$x(x^2 - 5x + 6) = 0$$
$$x(x-2)(x-3) = 0$$
$$x = 0, 2, 3$$

5.   Solving the equation:

$$3y^2 + 11y - 4 = 0$$
$$(3y-1)(y+4) = 0$$
$$y = -4, \tfrac{1}{3}$$

7.   Solving the equation:
$$60x^2 - 130x + 60 = 0$$
$$10(6x^2 - 13x + 6) = 0$$
$$10(3x-2)(2x-3) = 0$$
$$x = \tfrac{2}{3}, \tfrac{3}{2}$$

9.   Solving the equation:
$$\tfrac{1}{10}t^2 - \tfrac{5}{2} = 0$$
$$10\left(\tfrac{1}{10}t^2 - \tfrac{5}{2}\right) = 10(0)$$
$$t^2 - 25 = 0$$
$$(t+5)(t-5) = 0$$
$$t = -5, 5$$

11.   Solving the equation:

$$100x^4 = 400x^3 + 2100x^2$$
$$100x^4 - 400x^3 - 2100x^2 = 0$$
$$100x^2(x^2 - 4x - 21) = 0$$
$$100x^2(x-7)(x+3) = 0$$
$$x = -3, 0, 7$$

13.   Solving the equation:
$$\tfrac{1}{5}y^2 - 2 = -\tfrac{3}{10}y$$
$$10\left(\tfrac{1}{5}y^2 - 2\right) = 10\left(-\tfrac{3}{10}y\right)$$
$$2y^2 - 20 = -3y$$
$$2y^2 + 3y - 20 = 0$$
$$(y+4)(2y-5) = 0$$
$$y = -4, \tfrac{5}{2}$$

15.   Solving the equation:

$$9x^2 - 12x = 0$$
$$3x(3x-4) = 0$$
$$x = 0, \tfrac{4}{3}$$

17.   Solving the equation:
$$0.02r + 0.01 = 0.15r^2$$
$$2r + 1 = 15r^2$$
$$15r^2 - 2r - 1 = 0$$
$$(5r+1)(3r-1) = 0$$
$$r = -\tfrac{1}{5}, \tfrac{1}{3}$$

19.   Solving the equation:
$$9a^3 = 16a$$
$$9a^3 - 16a = 0$$
$$a(9a^2 - 16) = 0$$
$$a(3a+4)(3a-4) = 0$$
$$a = -\tfrac{4}{3}, 0, \tfrac{4}{3}$$

21.   Solving the equation:

$$-100x = 10x^2$$
$$0 = 10x^2 + 100x$$
$$0 = 10x(x+10)$$
$$x = -10, 0$$

23.   Solving the equation:
$$(x+6)(x-2) = -7$$
$$x^2 + 4x - 12 = -7$$
$$x^2 + 4x - 5 = 0$$
$$(x+5)(x-1) = 0$$
$$x = -5, 1$$

25.   Solving the equation:
$$(y-4)(y+1) = -6$$
$$y^2 - 3y - 4 = -6$$
$$y^2 - 3y + 2 = 0$$
$$(y-2)(y-1) = 0$$
$$y = 1, 2$$

27.   Solving the equation:
$$(x+1)^2 = 3x + 7$$
$$x^2 + 2x + 1 = 3x + 7$$
$$x^2 - x - 6 = 0$$
$$(x+2)(x-3) = 0$$
$$x = -2, 3$$

29. Solving the equation:
$$(2r+3)(2r-1) = -(3r+1)$$
$$4r^2 + 4r - 3 = -3r - 1$$
$$4r^2 + 7r - 2 = 0$$
$$(r+2)(4r-1) = 0$$
$$r = -2, \tfrac{1}{4}$$

31. Solving the equation:
$$x^3 + 3x^2 - 4x - 12 = 0$$
$$x^2(x+3) - 4(x+3) = 0$$
$$(x+3)(x^2 - 4) = 0$$
$$(x+3)(x+2)(x-2) = 0$$
$$x = -3, -2, 2$$

33. Solving the equation:
$$x^3 + 2x^2 - 25x - 50 = 0$$
$$x^2(x+2) - 25(x+2) = 0$$
$$(x+2)(x^2 - 25) = 0$$
$$(x+2)(x+5)(x-5) = 0$$
$$x = -5, -2, 5$$

35. Solving the equation:
$$2x^3 + 3x^2 - 8x - 12 = 0$$
$$x^2(2x+3) - 4(2x+3) = 0$$
$$(2x+3)(x^2 - 4) = 0$$
$$(2x+3)(x+2)(x-2) = 0$$
$$x = -2, -\tfrac{3}{2}, 2$$

37. Solving the equation:
$$4x^3 + 12x^2 - 9x - 27 = 0$$
$$4x^2(x+3) - 9(x+3) = 0$$
$$(x+3)(4x^2 - 9) = 0$$
$$(x+3)(2x+3)(2x-3) = 0$$
$$x = -3, -\tfrac{3}{2}, \tfrac{3}{2}$$

39. Let x and $x + 2$ represent the two odd integers. The equation is:
$$x^2 + (x+2)^2 = 34$$
$$x^2 + x^2 + 4x + 4 = 34$$
$$2x^2 + 4x - 30 = 0$$
$$x^2 + 2x - 15 = 0$$
$$(x+5)(x-3) = 0$$
$$x = -5, 3$$
$$x + 2 = -3, 5$$
The integers are either –5 and –3, or 3 and 5.

41. Let x and $x + 1$ represent the two integers. The equation is:
$$(x + x + 1)^2 = 81$$
$$(2x+1)^2 = 81$$
$$4x^2 + 4x + 1 = 81$$
$$4x^2 + 4x - 80 = 0$$
$$x^2 + x - 20 = 0$$
$$(x+5)(x-4) = 0$$
$$x = -5, 4$$
$$x + 1 = -4, 5$$
The integers are either –5 and –4, or 4 and 5.

43. Let h represent the height the ladder makes with the building. Using the Pythagorean theorem:
$$7^2 + h^2 = 25^2$$
$$49 + h^2 = 625$$
$$h^2 = 576$$
$$h = 24$$
The ladder reaches a height of 24 feet along the building.

45.   Let x, $x + 2$, and $x + 4$ represent the three sides. Using the Pythagorean theorem:

$$x^2 + (x+2)^2 = (x+4)^2$$
$$x^2 + x^2 + 4x + 4 = x^2 + 8x + 16$$
$$2x^2 + 4x + 4 = x^2 + 8x + 16$$
$$x^2 - 4x - 12 = 0$$
$$(x-6)(x+2) = 0$$
$$x = 6 \quad (x = -2 \text{ is impossible})$$

The lengths of the three sides are 6, 8, and 10.

47.   Let w represent the width and $3w + 2$ represent the length. Using the area formula:

$$w(3w+2) = 16$$
$$3w^2 + 2w = 16$$
$$3w^2 + 2w - 16 = 0$$
$$(3w+8)(w-2) = 0$$
$$w = 2 \quad (w = -\tfrac{8}{3} \text{ is impossible})$$

The dimensions are 2 feet by 8 feet.

49.   Let h represent the height and $4h + 2$ represent the base. Using the area formula:

$$\tfrac{1}{2}(4h+2)(h) = 36$$
$$4h^2 + 2h = 72$$
$$4h^2 + 2h - 72 = 0$$
$$2(2h^2 + h - 36) = 0$$
$$2(2h+9)(h-4) = 0$$
$$h = 4 \quad (h = -\tfrac{9}{2} \text{ is impossible})$$

The base is 18 inches and the height is 4 inches.

51.   Setting $h = 0$ in the equation:

$$32t - 16t^2 = 0$$
$$16t(2-t) = 0$$
$$t = 0, 2$$

The object is on the ground at 0 and 2 seconds.

53.   Substituting $v = 48$ and $h = 32$:

$$h = vt - 16t^2$$
$$32 = 48t - 16t^2$$
$$0 = -16t^2 + 48t - 32$$
$$0 = -16(t^2 - 3t + 2)$$
$$0 = -16(t-1)(t-2)$$
$$t = 1, 2$$

It will reach a height of 32 feet after 1 sec and 2 sec.

55.   Substituting $v = 24$ and $h = 0$:

$$h = vt - 16t^2$$
$$0 = 24t - 16t^2$$
$$0 = -16t^2 + 24t$$
$$0 = -8t(2t - 3)$$
$$t = 0, \tfrac{3}{2}$$

It will be on the ground after 0 sec and $\tfrac{3}{2}$ sec.

57.   Substituting $h = 192$:

$$192 = 96 + 80t - 16t^2$$
$$0 = -16t^2 + 80t - 96$$
$$0 = -16(t^2 - 5t + 6)$$
$$0 = -16(t-2)(t-3)$$
$$t = 2, 3$$

The bullet will be 192 feet in the air after 2 sec and 3 sec.

59. Substituting $R = \$3,200$:
$$R = xp$$
$$3200 = (1200 - 100p)p$$
$$3200 = 1200p - 100p^2$$
$$0 = -100p^2 + 1200p - 3200$$
$$0 = -100(p^2 - 12p + 32)$$
$$0 = -100(p - 4)(p - 8)$$
$$p = 4, 8$$
The cartridges should be sold for either $4 or $8.

63. Graphing the solution set:

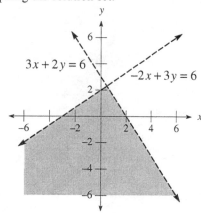

61. Substituting $R = \$7,000$:
$$R = xp$$
$$7000 = (1700 - 100p)p$$
$$7000 = 1700p - 100p^2$$
$$0 = -100p^2 + 1700p - 7000$$
$$0 = -100(p^2 - 17p + 70)$$
$$0 = -100(p - 7)(p - 10)$$
$$p = 7, 10$$
The calculators should be sold for either $7 or $10.

65. Graphing the solution set:

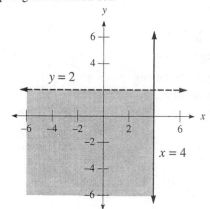

Chapter 5 Review

1. Simplifying: $x^3 \cdot x^7 = x^{3+7} = x^{10}$

3. Simplifying: $\left(2x^3 y\right)^2 \left(-2x^4 y^2\right)^3 = 4x^6 y^2 \cdot \left(-8x^{12} y^6\right) = -32x^{18} y^8$

5. Writing with positive exponents: $\left(\frac{2}{3}\right)^{-2} = \left(\frac{3}{2}\right)^2 = \frac{9}{4}$ **7.** Writing in scientific notation: $34,500,000 = 3.45 \times 10^7$

9. Writing in expanded form: $4.45 \times 10^4 = 44,500$ **11.** Simplifying: $\dfrac{a^{-4}}{a^5} = a^{-4-5} = a^{-9} = \dfrac{1}{a^9}$

13. Simplifying: $\dfrac{x^n x^{3n}}{x^{4n-2}} = \dfrac{x^{4n}}{x^{4n-2}} = x^{4n-(4n-2)} = x^{4n-4n+2} = x^2$

15. Simplifying: $\dfrac{(600,000)(0.000008)}{(4,000)(3,000,000)} = \dfrac{\left(6 \times 10^5\right)\left(8 \times 10^{-6}\right)}{\left(4 \times 10^3\right)\left(3 \times 10^6\right)} = \dfrac{48 \times 10^{-1}}{12 \times 10^9} = \dfrac{48}{12} \times 10^{-1-9} = 4 \times 10^{-10}$

17. Simplifying: $\left(x^3 - x\right) - \left(x^2 + x\right) + \left(x^3 - 3\right) - \left(x^2 + 1\right) = x^3 - x - x^2 - x + x^3 - 3 - x^2 - 1 = 2x^3 - 2x^2 - 2x - 4$

19. Simplifying: $-3\left[2x - 4(3x+1)\right] = -3(2x - 12x - 4) = -3(-10x - 4) = 30x + 12$

21. Multiplying: $3x\left(4x^2 - 2x + 1\right) = 12x^3 - 6x^2 + 3x$

23. Multiplying: $(6 - y)(3 - y) = 18 - 6y - 3y + y^2 = 18 - 9y + y^2$

25. Multiplying: $2t(t+1)(t-3) = 2t\left(t^2 - 2t - 3\right) = 2t^3 - 4t^2 - 6t$

27. Multiplying: $(2x - 3)\left(4x^2 + 6x + 9\right) = 8x^3 + 12x^2 + 18x - 12x^2 - 18x - 27 = 8x^3 - 27$

29. Multiplying: $(3x + 5)^2 = (3x)^2 + 2(3x)(5) + 5^2 = 9x^2 + 30x + 25$

31. Multiplying: $\left(x - \frac{1}{3}\right)\left(x + \frac{1}{3}\right) = (x)^2 - \left(\frac{1}{3}\right)^2 = x^2 - \frac{1}{9}$

33. Multiplying: $(x-1)^3 = (x-1)(x-1)^2 = (x-1)(x^2 - 2x + 1) = x^3 - 2x^2 + x - x^2 + 2x - 1 = x^3 - 3x^2 + 3x - 1$

35. Factoring: $6x^4 y - 9xy^4 + 18x^3 y^3 = 3xy(2x^3 - 3y^3 + 6x^2 y^2)$

37. Factoring: $8x^2 + 10 - 4x^2 y - 5y = 2(4x^2 + 5) - y(4x^2 + 5) = (4x^2 + 5)(2 - y)$

39. Factoring: $x^2 - 5x + 6 = (x - 2)(x - 3)$

41. Factoring: $20a^2 - 41ab + 20b^2 = (5a - 4b)(4a - 5b)$

43. Factoring: $24x^2 y - 6xy - 45y = 3y(8x^2 - 2x - 15) = 3y(4x + 5)(2x - 3)$

45. Factoring: $3a^4 + 18a^2 + 27 = 3(a^4 + 6a^2 + 9) = 3(a^2 + 3)^2$

47. Factoring: $5x^3 + 30x^2 y + 45xy^2 = 5x(x^2 + 6xy + 9y^2) = 5x(x + 3y)^2$

49. Factoring: $x^2 - 10x + 25 - y^2 = (x - 5)^2 - y^2 = (x - 5 + y)(x - 5 - y)$

51. Factoring: $x^3 + 4x^2 - 9x - 36 = x^2(x + 4) - 9(x + 4) = (x + 4)(x^2 - 9) = (x + 4)(x + 3)(x - 3)$

53. Solving the equation:
$$\tfrac{5}{6}y^2 = \tfrac{1}{4}y + \tfrac{1}{3}$$
$$12\left(\tfrac{5}{6}y^2\right) = 12\left(\tfrac{1}{4}y + \tfrac{1}{3}\right)$$
$$10y^2 = 3y + 4$$
$$10y^2 - 3y - 4 = 0$$
$$(5y - 4)(2y + 1) = 0$$
$$y = -\tfrac{1}{2}, \tfrac{4}{5}$$

55. Solving the equation:
$$5x^2 = -10x$$
$$5x^2 + 10x = 0$$
$$5x(x + 2) = 0$$
$$x = -2, 0$$

57. Solving the equation:
$$x^3 + 4x^2 - 9x - 36 = 0$$
$$x^2(x + 4) - 9(x + 4) = 0$$
$$(x + 4)(x^2 - 9) = 0$$
$$(x + 4)(x + 3)(x - 3) = 0$$
$$x = -4, -3, 3$$

59. Let x and $x + 1$ represent the two integers. The equation is:
$$x^2 + (x + 1)^2 = 41$$
$$x^2 + x^2 + 2x + 1 = 41$$
$$2x^2 + 2x - 40 = 0$$
$$x^2 + x - 20 = 0$$
$$(x + 5)(x - 4) = 0$$
$$x = -5, 4$$
$$x + 1 = -4, 5$$
The integers are either –5 and –4, or 4 and 5.

61. Let x, $x + 2$, and $x + 4$ represent the three sides. Using the Pythagorean theorem:
$$x^2 + (x + 2)^2 = (x + 4)^2$$
$$x^2 + x^2 + 4x + 4 = x^2 + 8x + 16$$
$$2x^2 + 4x + 4 = x^2 + 8x + 16$$
$$x^2 - 4x - 12 = 0$$
$$(x - 6)(x + 2) = 0$$
$$x = 6 \quad (x = -2 \text{ is impossible})$$
The lengths of the three sides are 6, 8, and 10.

Chapters 1-5 Cumulative Review

1. Simplifying: $15 - 12 \div 4 - 3 \cdot 2 = 15 - 3 - 6 = 6$ 3. Simplifying: $\left(\frac{2}{5}\right)^{-2} = \left(\frac{5}{2}\right)^{2} = \frac{25}{4}$

5. Simplifying: $5 - 3\left[2x - 4(x-2)\right] = 5 - 3(2x - 4x + 8) = 5 - 3(-2x + 8) = 5 + 6x - 24 = 6x - 19$

7. Simplifying: $(2x+3)\left(x^2 - 4x + 2\right) = 2x^3 - 8x^2 + 4x + 3x^2 - 12x + 6 = 2x^3 - 5x^2 - 8x + 6$

9. Solving the equation:
$$-6 + 2(2x + 3) = 0$$
$$-6 + 4x + 6 = 0$$
$$4x = 0$$
$$x = 0$$

11. Solving the equation:
$$|2x - 3| + 7 = 1$$
$$|2x - 3| = -6$$
Since this statement is false, there is no solution, or $\varnothing$.

13. Multiply the first equation by 3 and the second equation by –2:
$$24x + 18y = 12$$
$$-24x - 18y = -16$$
Adding yields $0 = -4$. Since this statement is false, there is no solution (parallel lines).

15. Multiply the second equation by –2 and add it to the third equation:
$$-8y + 2z = 18$$
$$3x - 2z = -6$$
Adding yields the equation $3x - 8y = 12$. The system of equations becomes:
$$2x + y = 8$$
$$3x - 8y = 12$$
Multiply the first equation by 8:
$$16x + 8y = 64$$
$$3x - 8y = 12$$
Adding yields:
$$19x = 76$$
$$x = 4$$
Substituting to find y:
$$8 + y = 8$$
$$y = 0$$
Substituting to find z:
$$0 - z = -9$$
$$z = 9$$
The solution is (4,0,9).

17. Solving the inequality:
$$|2x - 7| \le 3$$
$$-3 \le 2x - 7 \le 3$$
$$4 \le 2x \le 10$$
$$2 \le x \le 5$$
The solution set is $[2,5]$. Graphing the solution set:

19. Solving the inequality:
$$-3t \ge 12$$
$$t \le -4$$
The solution set is $(-\infty, -4]$.

21. The first five terms are: $\frac{1}{2}, \frac{4}{5}, 1, \frac{8}{7}, \frac{5}{4}$ 23. Finding the value: $-6 \cdot \frac{7}{54} - \frac{2}{9} = -\frac{7}{9} - \frac{2}{9} = -1$

25. The rational numbers are: $-1, 0, 2.35, 4$

27. The pattern is to add the previous two numbers, so the next two terms are $3 + 5 = 8$ and $5 + 8 = 13$.

29. Simplifying: $\left(5x^{-3}y^2z^{-2}\right)\left(6x^{-5}y^4z^{-3}\right) = 30x^{-8}y^6z^{-5} = \dfrac{30y^6}{x^8z^5}$

31. Simplifying: $\dfrac{\left(7\times10^{-5}\right)\left(21\times10^{-6}\right)}{3\times10^{-12}} = \dfrac{147\times10^{-11}}{3\times10^{-12}} = 49\times10 = 4.9\times10^2$

33. Solving the inequality:
$$5 - x \geq 0$$
$$-x \geq -5$$
$$x \leq 5$$

35. Graphing the linear inequality:

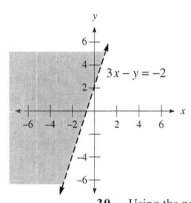

$3x - y = -2$

37. Solving for y:
$$3x - 5y = 15$$
$$-5y = -3x + 15$$
$$y = \frac{3}{5}x - 3$$
So $m = \frac{3}{5}, b = -3$.

39. Using the point-slope formula:
$$y - 8 = \tfrac{7}{3}(x - 6)$$
$$y - 8 = \tfrac{7}{3}x - 14$$
$$y = \tfrac{7}{3}x - 6$$

41. Factoring completely: $16y^2 + 2y + \frac{1}{16} = \left(4y + \frac{1}{4}\right)^2$

43. Factoring completely: $x^3 - 8 = (x - 2)\left(x^2 + 2x + 4\right)$

45. Multiply the second equation by -3:
$$7x + 9y = 2$$
$$15x - 9y = -3$$
Adding yields:
$$22x = -1$$
$$x = -\tfrac{1}{22}$$
Substituting into the first equation:
$$7\left(-\tfrac{1}{22}\right) + 9y = 2$$
$$-\tfrac{7}{22} + 9y = 2$$
$$9y = \tfrac{51}{22}$$
$$y = \tfrac{17}{66}$$
The solution is $\left(-\tfrac{1}{22}, \tfrac{17}{66}\right)$.

47. Evaluating the determinant: $\begin{vmatrix} 1 & 8 \\ -2 & 7 \end{vmatrix} = 7 + 16 = 23$

49. Let b represent the rate of the boat and c the rate of the current. The system of equations is:

$$3(b+c) = 36$$
$$9(b-c) = 36$$

Simplifying the system:

$$b + c = 12$$
$$b - c = 4$$

Adding yields:

$$2b = 16$$
$$b = 8$$
$$c = 4$$

The boat's rate is 8 mph and the current's rate is 4 mph.

Chapter 5 Test

1. Using properties of exponents: $x^4 \cdot x^7 \cdot x^{-3} = x^{4+7-3} = x^8$

2. Using properties of exponents: $2^{-5} = \dfrac{1}{2^5} = \dfrac{1}{32}$ **3.** Using properties of exponents: $\left(\frac{3}{4}\right)^{-2} = \left(\frac{4}{3}\right)^2 = \frac{16}{9}$

4. Using properties of exponents: $\left(2x^2 y\right)^3 \left(2x^3 y^4\right)^2 = \left(8x^6 y^3\right)\left(4x^6 y^8\right) = 32x^{12} y^{11}$

5. Using properties of exponents: $\dfrac{a^{-5}}{a^{-7}} = a^{-5+7} = a^2$ **6.** Using properties of exponents: $\dfrac{x^{n+1}}{x^{n-5}} = x^{n+1-n+5} = x^6$

7. Using properties of exponents:

$$\frac{\left(2ab^3\right)^{-2}\left(a^4 b^{-3}\right)}{\left(a^{-4} b^3\right)^4 \left(2a^{-2} b^2\right)^{-3}} = \frac{2^{-2} a^{-2} b^{-6} \cdot a^4 b^{-3}}{a^{-16} b^{12} \cdot 2^{-3} a^6 b^{-6}} = \frac{2^{-2} a^2 b^{-9}}{2^{-3} a^{-10} b^6} = 2^{-2+3} a^{2+10} b^{-9-6} = 2a^{12} b^{-15} = \frac{2a^{12}}{b^{15}}$$

8. Writing in scientific notation: $6,530,000 = 6.53 \times 10^6$ **9.** Writing in scientific notation: $0.00087 = 8.7 \times 10^{-4}$

10. Performing the operations: $\left(2.9 \times 10^{12}\right)\left(3 \times 10^{-5}\right) = 8.7 \times 10^7$

11. Performing the operations: $\dfrac{\left(6 \times 10^{-4}\right)\left(4 \times 10^9\right)}{8 \times 10^{-3}} = \dfrac{24 \times 10^5}{8 \times 10^{-3}} = \frac{24}{8} \times 10^{5+3} = 3 \times 10^8$

12. Simplifying: $\left(\frac{3}{4}x^3 - x^2 - \frac{3}{2}\right) - \left(\frac{1}{4}x^2 + 2x - \frac{1}{2}\right) = \frac{3}{4}x^3 - x^2 - \frac{3}{2} - \frac{1}{4}x^2 - 2x + \frac{1}{2} = \frac{3}{4}x^3 - \frac{5}{4}x^2 - 2x - 1$

13. Simplifying: $3 - 4[2x - 3(x+6)] = 3 - 4(2x - 3x - 18) = 3 - 8x + 12x + 72 = 4x + 75$

14. Multiplying: $(3y - 7)(2y + 5) = 6y^2 + 15y - 14y - 35 = 6y^2 + y - 35$

15. Multiplying: $(2x - 5)\left(x^2 + 4x - 3\right) = 2x^3 + 8x^2 - 6x - 5x^2 - 20x + 15 = 2x^3 + 3x^2 - 26x + 15$

16. Multiplying: $\left(8 - 3t^3\right)^2 = (8)^2 - 2(8)\left(3t^3\right) + \left(3t^3\right)^2 = 64 - 48t^3 + 9t^6$

17. Multiplying: $(1 - 6y)(1 + 6y) = (1)^2 - (6y)^2 = 1 - 36y^2$

18. Multiplying: $2x(x - 3)(2x + 5) = 2x\left(2x^2 - x - 15\right) = 4x^3 - 2x^2 - 30x$

19. Multiplying: $\left(5t^2 - \frac{1}{2}\right)\left(2t^2 + \frac{1}{5}\right) = 10t^4 - t^2 + t^2 - \frac{1}{10} = 10t^4 - \frac{1}{10}$

20. Factoring: $x^2 + x - 12 = (x + 4)(x - 3)$

21. Factoring: $12x^4 + 26x^2 - 10 = 2\left(6x^4 + 13x^2 - 5\right) = 2\left(3x^2 - 1\right)\left(2x^2 + 5\right)$

22. Factoring: $16a^4 - 81y^4 = \left(4a^2 + 9y^2\right)\left(4a^2 - 9y^2\right) = \left(4a^2 + 9y^2\right)(2a + 3y)(2a - 3y)$

23. Factoring: $7ax^2 - 14ay - b^2 x^2 + 2b^2 y = 7a\left(x^2 - 2y\right) - b^2\left(x^2 - 2y\right) = \left(x^2 - 2y\right)\left(7a - b^2\right)$

24. Factoring: $t^3 + \frac{1}{8} = t^3 + \left(\frac{1}{2}\right)^3 = \left(t + \frac{1}{2}\right)\left(t^2 - \frac{1}{2}t + \frac{1}{4}\right)$

25. Factoring: $4a^5b - 24a^4b^2 - 64a^3b^3 = 4a^3b(a^2 - 6ab - 16b^2) = 4a^3b(a - 8b)(a + 2b)$

26. Factoring: $x^2 - 10x + 25 - b^2 = (x - 5)^2 - b^2 = (x - 5 + b)(x - 5 - b)$

27. Factoring: $81 - x^4 = (9 + x^2)(9 - x^2) = (9 + x^2)(3 + x)(3 - x)$

28. Solving the equation:
$$\tfrac{1}{5}x^2 = \tfrac{1}{3}x + \tfrac{2}{15}$$
$$15\left(\tfrac{1}{5}x^2\right) = 15\left(\tfrac{1}{3}x + \tfrac{2}{15}\right)$$
$$3x^2 = 5x + 2$$
$$3x^2 - 5x - 2 = 0$$
$$(3x + 1)(x - 2) = 0$$
$$x = -\tfrac{1}{3}, 2$$

29. Solving the equation:
$$100x^3 = 500x^2$$
$$100x^3 - 500x^2 = 0$$
$$100x^2(x - 5) = 0$$
$$x = 0, 5$$

30. Solving the equation:
$$(x + 1)(x + 2) = 12$$
$$x^2 + 3x + 2 = 12$$
$$x^2 + 3x - 10 = 0$$
$$(x + 5)(x - 2) = 0$$
$$x = -5, 2$$

31. Solving the equation:
$$x^3 + 2x^2 - 16x - 32 = 0$$
$$x^2(x + 2) - 16(x + 2) = 0$$
$$(x + 2)(x^2 - 16) = 0$$
$$(x + 2)(x + 4)(x - 4) = 0$$
$$x = -4, -2, 4$$

32. Let x and $x + 3$ represent the two integers. The equation is:
$$x^2 + (x + 3)^2 = 29$$
$$x^2 + x^2 + 6x + 9 = 29$$
$$2x^2 + 6x - 20 = 0$$
$$x^2 + 3x - 10 = 0$$
$$(x + 5)(x - 2) = 0$$
$$x = -5, 2$$
$$x + 3 = -2, 5$$
The integers are either –5 and –2, or 2 and 5.

33. Let x, $x + 2$, and $x + 4$ represent the three sides. Using the Pythagorean theorem:
$$x^2 + (x + 2)^2 = (x + 4)^2$$
$$x^2 + x^2 + 4x + 4 = x^2 + 8x + 16$$
$$2x^2 + 4x + 4 = x^2 + 8x + 16$$
$$x^2 - 4x - 12 = 0$$
$$(x - 6)(x + 2) = 0$$
$$x = 6 \quad (x = -2 \text{ is impossible})$$
The lengths of the three sides are 6 inches, 8 inches, and 10 inches.

34. The revenue is given by: $R = xp = x(25 - 0.2x) = 25x - 0.2x^2$

35. The profit is given by: $P = R - C = (25x - 0.2x^2) - (2x + 100) = 25x - 0.2x^2 - 2x - 100 = -0.2x^2 + 23x - 100$

36. Substituting $x = 100$ into the revenue equation: $R = 25(100) - 0.2(100)^2 = 2500 - 2000 = \$500$. The revenue is \$500.

37. Substituting $x = 100$ into the cost equation: $C = 2(100) + 100 = 200 + 100 = \$300$. The cost is \$300.

38. Substituting $x = 100$ into the profit equation: $P = -0.2(100)^2 + 23(100) - 100 = -2000 + 2300 - 100 = \$200$
The profit is \$200. Note that this also could have been found by subtracting the answers from problems 36 and 37.

Chapter 6
Rational Expressions and Rational Functions

6.1 Basic Properties and Reducing to Lowest Terms

1. Finding each function value:
$$g(0) = \frac{0+3}{0-1} = \frac{3}{-1} = -3 \qquad g(-3) = \frac{-3+3}{-3-1} = \frac{0}{-4} = 0$$
$$g(3) = \frac{3+3}{3-1} = \frac{6}{2} = 3 \qquad g(-1) = \frac{-1+3}{-1-1} = \frac{2}{-2} = -1$$
$$g(1) = \frac{1+3}{1-1} = \frac{4}{0}, \text{ which is undefined}$$

3. Finding each function value:
$$h(0) = \frac{0-3}{0+1} = \frac{-3}{1} = -3 \qquad h(-3) = \frac{-3-3}{-3+1} = \frac{-6}{-2} = 3$$
$$h(3) = \frac{3-3}{3+1} = \frac{0}{4} = 0 \qquad h(-1) = \frac{-1-3}{-1+1} = \frac{-4}{0}, \text{ which is undefined}$$
$$h(1) = \frac{1-3}{1+1} = \frac{-2}{2} = -1$$

5. The domain is $\{x \mid x \neq 1\}$.

7. Setting the denominator equal to 0:
$$t^2 - 16 = 0$$
$$(t+4)(t-4) = 0$$
$$t = -4, 4$$
The domain is $\{t \mid t \neq -4, t \neq 4\}$.

9. Reducing to lowest terms: $\dfrac{x^2-16}{6x+24} = \dfrac{(x+4)(x-4)}{6(x+4)} = \dfrac{x-4}{6}$

11. Reducing to lowest terms: $\dfrac{12x-9y}{3x^2+3xy} = \dfrac{3(4x-3y)}{3x(x+y)} = \dfrac{4x-3y}{x(x+y)}$

13. Reducing to lowest terms: $\dfrac{a^4-81}{a-3} = \dfrac{\left(a^2+9\right)\left(a^2-9\right)}{a-3} = \dfrac{\left(a^2+9\right)(a+3)(a-3)}{a-3} = \left(a^2+9\right)(a+3)$

15. Reducing to lowest terms: $\dfrac{a^2-4a-12}{a^2+8a+12}=\dfrac{(a-6)(a+2)}{(a+6)(a+2)}=\dfrac{a-6}{a+6}$

17. Reducing to lowest terms: $\dfrac{20y^2-45}{10y^2-5y-15}=\dfrac{5\left(4y^2-9\right)}{5\left(2y^2-y-3\right)}=\dfrac{5(2y+3)(2y-3)}{5(2y-3)(y+1)}=\dfrac{2y+3}{y+1}$

19. Reducing to lowest terms: $\dfrac{a^3+b^3}{a^2-b^2}=\dfrac{(a+b)\left(a^2-ab+b^2\right)}{(a+b)(a-b)}=\dfrac{a^2-ab+b^2}{a-b}$

21. Reducing to lowest terms: $\dfrac{8x^4-8x}{4x^4+4x^3+4x^2}=\dfrac{8x\left(x^3-1\right)}{4x^2\left(x^2+x+1\right)}=\dfrac{8x(x-1)\left(x^2+x+1\right)}{4x^2\left(x^2+x+1\right)}=\dfrac{2(x-1)}{x}$

23. Reducing to lowest terms: $\dfrac{6x^2+7xy-3y^2}{6x^2+xy-y^2}=\dfrac{(2x+3y)(3x-y)}{(2x+y)(3x-y)}=\dfrac{2x+3y}{2x+y}$

25. Reducing to lowest terms: $\dfrac{ax+2x+3a+6}{ay+2y-4a-8}=\dfrac{x(a+2)+3(a+2)}{y(a+2)-4(a+2)}=\dfrac{(a+2)(x+3)}{(a+2)(y-4)}=\dfrac{x+3}{y-4}$

27. Reducing to lowest terms: $\dfrac{x^2+bx-3x-3b}{x^2-2bx-3x+6b}=\dfrac{x(x+b)-3(x+b)}{x(x-2b)-3(x-2b)}=\dfrac{(x+b)(x-3)}{(x-2b)(x-3)}=\dfrac{x+b}{x-2b}$

29. Reducing to lowest terms:

$$\dfrac{x^3+3x^2-4x-12}{x^2+x-6}=\dfrac{x^2(x+3)-4(x+3)}{(x+3)(x-2)}=\dfrac{(x+3)\left(x^2-4\right)}{(x+3)(x-2)}=\dfrac{(x+3)(x+2)(x-2)}{(x+3)(x-2)}=x+2$$

31. Reducing to lowest terms: $\dfrac{4x^4-25}{6x^3-4x^2+15x-10}=\dfrac{\left(2x^2+5\right)\left(2x^2-5\right)}{2x^2(3x-2)+5(3x-2)}=\dfrac{\left(2x^2+5\right)\left(2x^2-5\right)}{(3x-2)\left(2x^2+5\right)}=\dfrac{2x^2-5}{3x-2}$

33. Reducing to lowest terms: $\dfrac{x-4}{4-x}=\dfrac{x-4}{-1(x-4)}=-1$

35. Reducing to lowest terms: $\dfrac{y^2-36}{6-y}=\dfrac{(y+6)(y-6)}{-1(y-6)}=-(y+6)$

37. Reducing to lowest terms: $\dfrac{1-9a^2}{9a^2-6a+1}=\dfrac{-1\left(9a^2-1\right)}{(3a-1)^2}=\dfrac{-1(3a+1)(3a-1)}{(3a-1)^2}=-\dfrac{3a+1}{3a-1}$

39. Evaluating the functions:

$f(0)=\dfrac{0^2-4}{0-2}=\dfrac{-4}{-2}=2$

$g(0)=0+2=2$

41. Evaluating the functions:

$f(2)=\dfrac{2^2-4}{2-2}=\dfrac{0}{0}$, which is undefined

$g(2)=2+2=4$

43. Evaluating the functions:

$f(0)=\dfrac{0^2-1}{0-1}=\dfrac{-1}{-1}=1$

$g(0)=0+1=1$

45. Evaluating the functions:

$f(2)=\dfrac{2^2-1}{2-1}=\dfrac{3}{1}=3$

$g(2)=2+1=3$

47. The graph of $y = x + 2$ contains the point $(2,4)$, while the other graph does not.

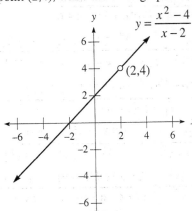

$$y = \frac{x^2 - 4}{x - 2}$$

49. Completing the table:

Weeks x	Weight (lb) $W(x)$
0	200
1	194
4	184
12	173
24	168

51. Dividing: $\dfrac{3.5 \text{ miles}}{29.75 \text{ minutes}} \approx 0.12$ miles per minute

53. Dividing: $\dfrac{175.8 \text{ miles}}{16.3 \text{ gallons}} \approx 10.8$ miles per gallon

55. Dividing: $\dfrac{\pi \cdot 65 \text{ feet}}{30 \text{ seconds}} \approx 6.8$ feet per second

57. **a.** The domain is $\{t \mid 20 \le t \le 50\}$.
 b. Graphing the function:

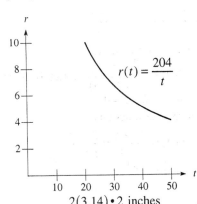

$$r(t) = \frac{204}{t}$$

59. **a.** The domain is $\{d \mid 1 \le d \le 6\}$.
 b. Graphing the function:

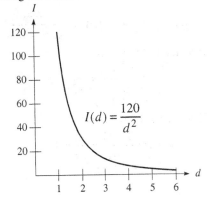

$$I(d) = \frac{120}{d^2}$$

61. Using $r = 2$ inches: $\dfrac{2(3.14) \cdot 2 \text{ inches}}{\frac{1}{300} \text{ minutes}} \approx 3{,}768$ inches per minute

 Using $r = 1.5$ inches: $\dfrac{2(3.14) \cdot 1.5 \text{ inches}}{\frac{1}{300} \text{ minutes}} \approx 2{,}826$ inches per minute

63. Subtracting: $\left(4x^2 - 5x + 5\right) - \left(x^2 + 2x + 1\right) = 4x^2 - 5x + 5 - x^2 - 2x - 1 = 3x^2 - 7x + 4$

65. Subtracting: $(10x - 11) - (10x - 20) = 10x - 11 - 10x + 20 = 9$

67. Subtracting: $\left(4x^3\right) - \left(4x^3 - 8x^2\right) = 4x^3 - 4x^3 + 8x^2 = 8x^2$

69. a. From the graph: $f(2) = 2$ **b.** From the graph: $f(-1) = -4$
   c. From the graph: $f(0)$ is undefined   **d.** From the graph: $g(3) = 2$
   e. From the graph: $g(6) = 1$ **f.** From the graph: $g(-1) = -6$
   g. From the graph: $f(g(6)) = f(1) = 4$ **h.** From the graph: $g(f(-2)) = g(-2) = -3$

6.2   Division of Polynomials

1. Dividing: $\dfrac{4x^3 - 8x^2 + 6x}{2x} = \dfrac{4x^3}{2x} - \dfrac{8x^2}{2x} + \dfrac{6x}{2x} = 2x^2 - 4x + 3$

3. Dividing: $\dfrac{10x^4 + 15x^3 - 20x^2}{-5x^2} = \dfrac{10x^4}{-5x^2} + \dfrac{15x^3}{-5x^2} - \dfrac{20x^2}{-5x^2} = -2x^2 - 3x + 4$

5. Dividing: $\dfrac{8y^5 + 10y^3 - 6y}{4y^3} = \dfrac{8y^5}{4y^3} + \dfrac{10y^3}{4y^3} - \dfrac{6y}{4y^3} = 2y^2 + \dfrac{5}{2} - \dfrac{3}{2y^2}$

7. Dividing: $\dfrac{5x^3 - 8x^2 - 6x}{-2x^2} = \dfrac{5x^3}{-2x^2} - \dfrac{8x^2}{-2x^2} - \dfrac{6x}{-2x^2} = -\dfrac{5}{2}x + 4 + \dfrac{3}{x}$

9. Dividing: $\dfrac{28a^3b^5 + 42a^4b^3}{7a^2b^2} = \dfrac{28a^3b^5}{7a^2b^2} + \dfrac{42a^4b^3}{7a^2b^2} = 4ab^3 + 6a^2b$

11. Dividing: $\dfrac{10x^3y^2 - 20x^2y^3 - 30x^3y^3}{-10x^2y} = \dfrac{10x^3y^2}{-10x^2y} - \dfrac{20x^2y^3}{-10x^2y} - \dfrac{30x^3y^3}{-10x^2y} = -xy + 2y^2 + 3xy^2$

13. Dividing by factoring: $\dfrac{x^2 - x - 6}{x - 3} = \dfrac{(x - 3)(x + 2)}{x - 3} = x + 2$

15. Dividing by factoring: $\dfrac{2a^2 - 3a - 9}{2a + 3} = \dfrac{(2a + 3)(a - 3)}{2a + 3} = a - 3$

17. Dividing by factoring: $\dfrac{5x^2 - 14xy - 24y^2}{x - 4y} = \dfrac{(5x + 6y)(x - 4y)}{x - 4y} = 5x + 6y$

19. Dividing by factoring: $\dfrac{x^3 - y^3}{x - y} = \dfrac{(x - y)(x^2 + xy + y^2)}{x - y} = x^2 + xy + y^2$

21. Dividing by factoring: $\dfrac{y^4 - 16}{y - 2} = \dfrac{(y^2 + 4)(y^2 - 4)}{y - 2} = \dfrac{(y^2 + 4)(y + 2)(y - 2)}{y - 2} = (y^2 + 4)(y + 2)$

23. Dividing by factoring:

$\dfrac{x^3 + 2x^2 - 25x - 50}{x - 5} = \dfrac{x^2(x + 2) - 25(x + 2)}{x - 5} = \dfrac{(x + 2)(x^2 - 25)}{x - 5} = \dfrac{(x + 2)(x + 5)(x - 5)}{x - 5} = (x + 2)(x + 5)$

25. Dividing using long division:

$$\begin{array}{r} x - 7 \\ x + 2 \overline{)\, x^2 - 5x - 7} \\ \underline{x^2 + 2x} \\ -7x - 7 \\ \underline{-7x - 14} \\ 7 \end{array}$$

The quotient is $x - 7 + \dfrac{7}{x + 2}$.

27. Dividing using long division:

$$\begin{array}{r} 2x^2 - 5x + 1 \\ x + 1 \overline{)\, 2x^3 - 3x^2 - 4x + 5} \\ \underline{2x^3 + 2x^2} \\ -5x^2 - 4x \\ \underline{-5x^2 - 5x} \\ x + 5 \\ \underline{x + 1} \\ 4 \end{array}$$

The quotient is $2x^2 - 5x + 1 + \dfrac{4}{x + 1}$.

29. Dividing using long division:

$$
\begin{array}{r}
y^2 - 3y - 13 \\
2y - 3\overline{\smash{\big)}\,2y^3 - 9y^2 - 17y + 39} \\
\underline{2y^3 - 3y^2} \\
-6y^2 - 17y \\
\underline{-6y^2 + 9y} \\
-26y + 39 \\
\underline{-26y + 39} \\
0
\end{array}
$$

The quotient is $y^2 - 3y - 13$.

31. Dividing using long division:

$$
\begin{array}{r}
3y^2 + 6y + 8 \\
2y - 4\overline{\smash{\big)}\,6y^3 + 0y^2 - 8y + 5} \\
\underline{6y^3 - 12y^2} \\
12y^2 - 8y \\
\underline{12y^2 - 24y} \\
16y + 5 \\
\underline{16y - 32} \\
37
\end{array}
$$

The quotient is $3y^2 + 6y + 8 + \dfrac{37}{2y - 4}$.

33. Dividing using long division:

$$
\begin{array}{r}
a^3 + 2a^2 + 4a + 6 \\
a - 2\overline{\smash{\big)}\,a^4 + 0a^3 + 0a^2 - 2a + 5} \\
\underline{a^4 - 2a^3} \\
2a^3 + 0a^2 \\
\underline{2a^3 - 4a^2} \\
4a^2 - 2a \\
\underline{4a^2 - 8a} \\
6a + 5 \\
\underline{6a - 12} \\
17
\end{array}
$$

The quotient is $a^3 + 2a^2 + 4a + 6 + \dfrac{17}{a - 2}$.

35. Dividing using long division:

$$
\begin{array}{r}
y^3 + 2y^2 + 4y + 8 \\
y - 2\overline{\smash{\big)}\,y^4 + 0y^3 + 0y^2 + 0y - 16} \\
\underline{y^4 - 2y^3} \\
2y^3 + 0y^2 \\
\underline{2y^3 - 4y^2} \\
4y^2 + 0y \\
\underline{4y^2 - 8y} \\
8y - 16 \\
\underline{8y - 16} \\
0
\end{array}
$$

The quotient is $y^3 + 2y^2 + 4y + 8$.

37. Evaluating the formula: $\dfrac{f(x) - f(a)}{x - a} = \dfrac{4x - 4a}{x - a} = \dfrac{4(x - a)}{x - a} = 4$

39. Evaluating the formula: $\dfrac{f(x) - f(a)}{x - a} = \dfrac{(5x + 3) - (5a + 3)}{x - a} = \dfrac{5x - 5a}{x - a} = \dfrac{5(x - a)}{x - a} = 5$

41. Evaluating the formula: $\dfrac{f(x) - f(a)}{x - a} = \dfrac{x^2 - a^2}{x - a} = \dfrac{(x + a)(x - a)}{x - a} = x + a$

43. Evaluating the formula: $\dfrac{f(x) - f(a)}{x - a} = \dfrac{\left(x^2 + 1\right) - \left(a^2 + 1\right)}{x - a} = \dfrac{x^2 - a^2}{x - a} = \dfrac{(x + a)(x - a)}{x - a} = x + a$

45. a. Using long division:

$$
\begin{array}{r}
x^2 - x + 3 \\
x - 2\overline{\smash{\big)}\,x^3 - 3x^2 + 5x - 6} \\
\underline{x^3 - 2x^2} \\
-x^2 + 5x \\
\underline{-x^2 + 2x} \\
3x - 6 \\
\underline{3x - 6} \\
0
\end{array}
$$

Since the remainder is 0, $x - 2$ is a factor of $x^3 - 3x^2 + 5x - 6$.

Also note that: $P(2) = (2)^3 - 3(2)^2 + 5(2) - 6 = 8 - 12 + 10 - 6 = 0$

b.   Using long division:

$$x - 5 \overline{\smash{\big)}\ x^4 - 5x^3 - x^2 + 6x - 5}$$

quotient: $x^3 - x + 1$

$$\underline{x^4 - 5x^3}$$
$$-x^2 + 6x$$
$$\underline{-x^2 + 5x}$$
$$x - 5$$
$$\underline{x - 5}$$
$$0$$

Since the remainder is 0, $x - 5$ is a factor of $x^4 - 5x^3 - x^2 + 6x - 5$.

Also note that: $P(5) = (5)^4 - 5(5)^3 - (5)^2 + 6(5) - 5 = 625 - 625 - 25 + 30 - 5 = 0$

47.   a.   Completing the table:

x	1	5	10	15	20
$C(x)$	2.15	2.75	3.50	4.25	5.00

b.   The average cost function is $\overline{C}(x) = \dfrac{2}{x} + 0.15$.

c.   Completing the table:

x	1	5	10	15	20
$\overline{C}(x)$	2.15	0.55	0.35	0.28	0.25

d.   The average cost function decreases.

49.   a.   Substituting the values:

$$T(100) = 4.95 + 0.07(100) = \$11.95$$
$$T(400) = 4.95 + 0.07(400) = \$32.95$$
$$T(500) = 4.95 + 0.07(500) = \$39.95$$

b.   The average cost function is $\overline{T}(m) = \dfrac{4.95}{m} + 0.07$.

c.   Substituting the values:

$$\overline{T}(100) = \dfrac{4.95}{100} + 0.07 = \$0.12 \text{ per minute}$$
$$\overline{T}(400) = \dfrac{4.95}{400} + 0.07 = \$0.08 \text{ per minute}$$
$$\overline{T}(500) = \dfrac{4.95}{500} + 0.07 = \$0.08 \text{ per minute}$$

51.   Dividing: $\frac{3}{5} \div \frac{2}{7} = \frac{3}{5} \cdot \frac{7}{2} = \frac{21}{10}$

53.   Dividing: $\frac{3}{4} \div \frac{6}{11} = \frac{3}{4} \cdot \frac{11}{6} = \frac{11}{8}$

55.   Dividing: $\frac{4}{9} \div 8 = \frac{4}{9} \cdot \frac{1}{8} = \frac{1}{18}$

57.   Dividing: $8 \div \frac{1}{4} = 8 \cdot 4 = 32$

59.   Simplifying: $\left(\frac{1}{3}\right)^{-2} + \left(\frac{1}{2}\right)^{-3} = 3^2 + 2^3 = 9 + 8 = 17$

61.   Simplifying: $\left(9x^{-4}y^9\right)^{-2}\left(3x^2y^{-1}\right)^4 = 9^{-2}x^8y^{-18} \cdot 3^4 x^8 y^{-4} = \frac{1}{81} \cdot 81x^{16}y^{-22} = \dfrac{x^{16}}{y^{22}}$

63. Using long division:

$$
x^2-5 \overline{\smash{\big)}\, 4x^5 - x^4 - 20x^3 + 8x^2 - 15} \quad \overset{\displaystyle 4x^3 - x^2 + 3}{}
$$

$$
\begin{array}{r}
4x^5 \qquad\quad -20x^3 \\ \hline
-x^4 \qquad\qquad +8x^2 \\
-x^4 \qquad\qquad +5x^2 \\ \hline
3x^2 - 15 \\
3x^2 - 15 \\ \hline
0
\end{array}
$$

The quotient is $4x^3 - x^2 + 3$.

67. Using long division:

$$
2x+4 \overline{\smash{\big)}\, 3x^2 + x - 9} \quad \overset{\displaystyle \frac{3}{2}x - \frac{5}{2}}{}
$$

$$
\begin{array}{r}
3x^2 + 6x \\ \hline
-5x - 9 \\
-5x - 10 \\ \hline
1
\end{array}
$$

The quotient is $\frac{3}{2}x - \frac{5}{2} + \dfrac{1}{2x+4}$.

65. Using long division:

$$
x+0.2 \overline{\smash{\big)}\, 0.5x^3 - 0.3x^2 + 0.22x + 0.06} \quad \overset{\displaystyle 0.5x^2 - 0.4x + 0.3}{}
$$

$$
\begin{array}{r}
0.5x^3 + 0.1x^2 \\ \hline
-0.4x^2 + 0.22x \\
-0.4x^2 - 0.08x \\ \hline
0.3x + 0.06 \\
0.3x + 0.06 \\ \hline
0
\end{array}
$$

The quotient is $0.5x^2 - 0.4x + 0.3$.

69. Using long division:

$$
3x-1 \overline{\smash{\big)}\, 2x^2 + \tfrac{1}{3}x + \tfrac{5}{3}} \quad \overset{\displaystyle \frac{2}{3}x + \frac{1}{3}}{}
$$

$$
\begin{array}{r}
2x^2 - \tfrac{2}{3}x \\ \hline
x + \tfrac{5}{3} \\
x - \tfrac{1}{3} \\ \hline
2
\end{array}
$$

The quotient is $\frac{2}{3}x + \frac{1}{3} + \dfrac{2}{3x-1}$.

6.3 Multiplication and Division of Rational Expressions

1. Performing the operations: $\dfrac{2}{9} \cdot \dfrac{3}{4} = \dfrac{2}{3 \cdot 3} \cdot \dfrac{3}{2 \cdot 2} = \dfrac{1}{2 \cdot 3} = \dfrac{1}{6}$

3. Performing the operations: $\dfrac{3}{4} \div \dfrac{1}{3} = \dfrac{3}{4} \cdot \dfrac{3}{1} = \dfrac{9}{4}$

5. Performing the operations: $\dfrac{3}{7} \cdot \dfrac{14}{24} \div \dfrac{1}{2} = \dfrac{1}{4} \div \dfrac{1}{2} = \dfrac{1}{4} \cdot \dfrac{2}{1} = \dfrac{2}{4} = \dfrac{1}{2}$

7. Performing the operations: $\dfrac{10x^2}{5y^2} \cdot \dfrac{15y^3}{2x^4} = \dfrac{150x^2y^3}{10x^4y^2} = \dfrac{15y}{x^2}$

9. Performing the operations: $\dfrac{11a^2b}{5ab^2} \div \dfrac{22a^3b^2}{10ab^4} = \dfrac{11a^2b}{5ab^2} \cdot \dfrac{10ab^4}{22a^3b^2} = \dfrac{110a^3b^5}{110a^4b^4} = \dfrac{b}{a}$

11. Performing the operations: $\dfrac{6x^2}{5y^3} \cdot \dfrac{11z^2}{2x^2} \div \dfrac{33z^5}{10y^8} = \dfrac{33z^2}{5y^3} \cdot \dfrac{10y^8}{33z^5} = \dfrac{2y^8z^2}{y^3z^5} = \dfrac{2y^5}{z^3}$

13. Performing the operations: $\dfrac{x^2-9}{x^2-4} \cdot \dfrac{x-2}{x-3} = \dfrac{(x+3)(x-3)}{(x+2)(x-2)} \cdot \dfrac{x-2}{x-3} = \dfrac{x+3}{x+2}$

15. Performing the operations: $\dfrac{y^2-1}{y+2} \cdot \dfrac{y^2+5y+6}{y^2+2y-3} = \dfrac{(y+1)(y-1)}{y+2} \cdot \dfrac{(y+2)(y+3)}{(y+3)(y-1)} = y+1$

17. Performing the operations: $\dfrac{3x-12}{x^2-4} \cdot \dfrac{x^2+6x+8}{x-4} = \dfrac{3(x-4)}{(x+2)(x-2)} \cdot \dfrac{(x+4)(x+2)}{x-4} = \dfrac{3(x+4)}{x-2}$

19. Performing the operations: $\dfrac{5x+2y}{25x^2-5xy-6y^2} \cdot \dfrac{20x^2-7xy-3y^2}{4x+y} = \dfrac{5x+2y}{(5x+2y)(5x-3y)} \cdot \dfrac{(5x-3y)(4x+y)}{4x+y} = 1$

21.   Performing the operations:
$$\frac{a^2-5a+6}{a^2-2a-3} \div \frac{a-5}{a^2+3a+2} = \frac{a^2-5a+6}{a^2-2a-3} \cdot \frac{a^2+3a+2}{a-5} = \frac{(a-3)(a-2)}{(a-3)(a+1)} \cdot \frac{(a+2)(a+1)}{a-5} = \frac{(a-2)(a+2)}{a-5}$$

23.   Performing the operations:
$$\frac{4t^2-1}{6t^2+t-2} \div \frac{8t^3+1}{27t^3+8} = \frac{4t^2-1}{6t^2+t-2} \cdot \frac{27t^3+8}{8t^3+1} = \frac{(2t+1)(2t-1)}{(3t+2)(2t-1)} \cdot \frac{(3t+2)\left(9t^2-6t+4\right)}{(2t+1)\left(4t^2-2t+1\right)} = \frac{9t^2-6t+4}{4t^2-2t+1}$$

25.   Performing the operations:
$$\frac{2x^2-5x-12}{4x^2+8x+3} \div \frac{x^2-16}{2x^2+7x+3} = \frac{2x^2-5x-12}{4x^2+8x+3} \cdot \frac{2x^2+7x+3}{x^2-16} = \frac{(2x+3)(x-4)}{(2x+1)(2x+3)} \cdot \frac{(2x+1)(x+3)}{(x+4)(x-4)} = \frac{x+3}{x+4}$$

27.   Performing the operations:
$$\frac{6a^2b+2ab^2-20b^3}{4a^2b-16b^3} \cdot \frac{10a^2-22ab+4b^2}{27a^3-125b^3} = \frac{2b\left(3a^2+ab-10b^2\right)}{4b\left(a^2-4b^2\right)} \cdot \frac{2\left(5a^2-11ab+2b^2\right)}{(3a-5b)\left(9a^2+15ab+25b^2\right)}$$
$$= \frac{2b(3a-5b)(a+2b)}{4b(a+2b)(a-2b)} \cdot \frac{2(5a-b)(a-2b)}{(3a-5b)\left(9a^2+15ab+25b^2\right)}$$
$$= \frac{5a-b}{9a^2+15ab+25b^2}$$

29.   Performing the operations:
$$\frac{360x^3-490x}{36x^2+84x+49} \cdot \frac{30x^2+83x+56}{150x^3+65x^2-280x} = \frac{10x\left(36x^2-49\right)}{(6x+7)^2} \cdot \frac{(6x+7)(5x+8)}{5x\left(30x^2+13x-56\right)}$$
$$= \frac{10x(6x+7)(6x-7)}{(6x+7)^2} \cdot \frac{(6x+7)(5x+8)}{5x(6x-7)(5x+8)}$$
$$= 2$$

31.   Performing the operations:
$$\frac{x^5-x^2}{5x^5-5x} \cdot \frac{10x^4-10x^2}{2x^4+2x^3+2x^2} = \frac{x^2\left(x^3-1\right)}{5x\left(x^4-1\right)} \cdot \frac{10x^2\left(x^2-1\right)}{2x^2\left(x^2+x+1\right)}$$
$$= \frac{x^2(x-1)\left(x^2+x+1\right)}{5x\left(x^2+1\right)(x-1)(x-1)} \cdot \frac{10x^2(x+1)(x-1)}{2x^2\left(x^2+x+1\right)}$$
$$= \frac{x(x-1)}{x^2+1}$$

33.   Performing the operations:
$$\frac{a^2-16b^2}{a^2-8ab+16b^2} \cdot \frac{a^2-9ab+20b^2}{a^2-7ab+12b^2} \div \frac{a^2-25b^2}{a^2-6ab+9b^2}$$
$$= \frac{a^2-16b^2}{a^2-8ab+16b^2} \cdot \frac{a^2-9ab+20b^2}{a^2-7ab+12b^2} \cdot \frac{a^2-6ab+9b^2}{a^2-25b^2}$$
$$= \frac{(a+4b)(a-4b)}{(a-4b)^2} \cdot \frac{(a-5b)(a-4b)}{(a-3b)(a-4b)} \cdot \frac{(a-3b)^2}{(a+5b)(a-5b)}$$
$$= \frac{(a+4b)(a-3b)}{(a-4b)(a+5b)}$$

35. Performing the operations:

$$\frac{2y^2-7y-15}{42y^2-29y-5}\cdot\frac{12y^2-16y+5}{7y^2-36y+5}\div\frac{4y^2-9}{49y^2-1}=\frac{2y^2-7y-15}{42y^2-29y-5}\cdot\frac{12y^2-16y+5}{7y^2-36y+5}\cdot\frac{49y^2-1}{4y^2-9}$$

$$=\frac{(2y+3)(y-5)}{(6y-5)(7y+1)}\cdot\frac{(6y-5)(2y-1)}{(7y-1)(y-5)}\cdot\frac{(7y+1)(7y-1)}{(2y+3)(2y-3)}$$

$$=\frac{2y-1}{2y-3}$$

37. Performing the operations:

$$\frac{xy-2x+3y-6}{xy+2x-4y-8}\cdot\frac{xy+x-4y-4}{xy-x+3y-3}=\frac{x(y-2)+3(y-2)}{x(y+2)-4(y+2)}\cdot\frac{x(y+1)-4(y+1)}{x(y-1)+3(y-1)}$$

$$=\frac{(y-2)(x+3)}{(y+2)(x-4)}\cdot\frac{(y+1)(x-4)}{(y-1)(x+3)}$$

$$=\frac{(y-2)(y+1)}{(y+2)(y-1)}$$

39. Performing the operations:

$$\frac{xy^2-y^2+4xy-4y}{xy-3y+4x-12}\div\frac{xy^3+2xy^2+y^3+2y^2}{xy^2-3y^2+2xy-6y}=\frac{xy^2-y^2+4xy-4y}{xy-3y+4x-12}\cdot\frac{xy^2-3y^2+2xy-6y}{xy^3+2xy^2+y^3+2y^2}$$

$$=\frac{y^2(x-1)+4y(x-1)}{y(x-3)+4(x-3)}\cdot\frac{y^2(x-3)+2y(x-3)}{xy^2(y+2)+y^2(y+2)}$$

$$=\frac{y(x-1)(y+4)}{(x-3)(y+4)}\cdot\frac{y(x-3)(y+2)}{y^2(y+2)(x+1)}$$

$$=\frac{x-1}{x+1}$$

41. Performing the operations:

$$\frac{2x^3+10x^2-8x-40}{x^3+4x^2-9x-36}\cdot\frac{x^2+x-12}{2x^2+14x+20}=\frac{2x^2(x+5)-8(x+5)}{x^2(x+4)-9(x+4)}\cdot\frac{(x+4)(x-3)}{2(x^2+7x+10)}$$

$$=\frac{2(x+5)(x^2-4)}{(x+4)(x^2-9)}\cdot\frac{(x+4)(x-3)}{2(x+5)(x+2)}$$

$$=\frac{2(x+5)(x+2)(x-2)}{(x+4)(x+3)(x-3)}\cdot\frac{(x+4)(x-3)}{2(x+5)(x+2)}$$

$$=\frac{x-2}{x+3}$$

43. Finding the product: $(3x-6)\cdot\dfrac{x}{x-2}=\dfrac{3(x-2)}{1}\cdot\dfrac{x}{x-2}=3x$

45. Finding the product: $\left(x^2-25\right)\cdot\dfrac{2}{x-5}=\dfrac{(x+5)(x-5)}{1}\cdot\dfrac{2}{x-5}=2(x+5)$

47. Finding the product: $\left(x^2-3x+2\right)\cdot\dfrac{3}{3x-3}=\dfrac{(x-2)(x-1)}{1}\cdot\dfrac{3}{3(x-1)}=x-2$

49. Finding the product: $(y-3)(y-4)(y+3)\cdot\dfrac{-1}{y^2-9}=\dfrac{(y-3)(y-4)(y+3)}{1}\cdot\dfrac{-1}{(y+3)(y-3)}=-(y-4)$

51. Finding the product: $a(a+5)(a-5)\cdot\dfrac{a+1}{a^2+5a}=\dfrac{a(a+5)(a-5)}{1}\cdot\dfrac{a+1}{a(a+5)}=(a-5)(a+1)$

53. Completing the table:

Number of Copies x	Price per Copy (\$) $p(x)$
1	20.33
10	9.33
20	6.40
50	4.00
100	3.05

55. Finding the revenue: $R = 100 \cdot \dfrac{2(100+60)}{100+5} = 100 \cdot \dfrac{320}{105} \approx \$305.00$

57. First use long division to find the length of the square base:

$$
\begin{array}{r}
x - 3 \\
x^2 + x + 1 \overline{\smash{)}\; x^3 - 2x^2 - 2x - 3} \\
\underline{x^3 + x^2 + x } \\
-3x^2 - 3x - 3 \\
\underline{-3x^2 - 3x - 3} \\
0
\end{array}
$$

The area of the base is therefore: $A = (x-3)^2 = x^2 - 6x + 9$

59. Multiplying the polynomials: $2x^2 \left(5x^3 + 4x - 3\right) = 2x^2 \cdot 5x^3 + 2x^2 \cdot 4x - 2x^2 \cdot 3 = 10x^5 + 8x^3 - 6x^2$

61. Multiplying the polynomials: $(3a-1)(4a+5) = 12a^2 + 15a - 4a - 5 = 12a^2 + 11a - 5$

63. Multiplying the polynomials: $(3x+7)(4y-2) = 12xy + 28y - 6x - 14$

65. Multiplying the polynomials: $\left(3 - t^2\right)^2 = (3)^2 - 2(3)\left(t^2\right) + \left(t^2\right)^2 = 9 - 6t^2 + t^4$

67. Multiplying the polynomials:

$3(x+1)(x+2)(x+3) = 3(x+1)\left(x^2 + 5x + 6\right) = 3\left(x^3 + 6x^2 + 11x + 6\right) = 3x^3 + 18x^2 + 33x + 18$

69. Adding the fractions: $\frac{3}{14} + \frac{7}{30} = \frac{3}{14} \cdot \frac{15}{15} + \frac{7}{30} \cdot \frac{7}{7} = \frac{45}{210} + \frac{49}{210} = \frac{94}{210} = \frac{47}{105}$

71. Subtracting the fractions: $\frac{4}{15} - \frac{5}{21} = \frac{4}{15} \cdot \frac{7}{7} - \frac{5}{21} \cdot \frac{5}{5} = \frac{28}{105} - \frac{25}{105} = \frac{3}{105} = \frac{1}{35}$

6.4 Addition and Subtraction of Rational Expressions

1. Combining the fractions: $\frac{3}{4} + \frac{1}{2} = \frac{3}{4} + \frac{1}{2} \cdot \frac{2}{2} = \frac{3}{4} + \frac{2}{4} = \frac{5}{4}$

3. Combining the fractions: $\frac{2}{5} - \frac{1}{15} = \frac{2}{5} \cdot \frac{3}{3} - \frac{1}{15} = \frac{6}{15} - \frac{1}{15} = \frac{5}{15} = \frac{1}{3}$

5. Combining the fractions: $\frac{5}{6} + \frac{7}{8} = \frac{5}{6} \cdot \frac{4}{4} + \frac{7}{8} \cdot \frac{3}{3} = \frac{20}{24} + \frac{21}{24} = \frac{41}{24}$

7. Combining the fractions: $\frac{9}{48} - \frac{3}{54} = \frac{9}{48} \cdot \frac{9}{9} - \frac{3}{54} \cdot \frac{8}{8} = \frac{81}{432} - \frac{24}{432} = \frac{57}{432} = \frac{19}{144}$

9. Combining the fractions: $\frac{3}{4} - \frac{1}{8} + \frac{2}{3} = \frac{3}{4} \cdot \frac{6}{6} - \frac{1}{8} \cdot \frac{3}{3} + \frac{2}{3} \cdot \frac{8}{8} = \frac{18}{24} - \frac{3}{24} + \frac{16}{24} = \frac{31}{24}$

11. Combining the rational expressions: $\dfrac{x}{x+3} + \dfrac{3}{x+3} = \dfrac{x+3}{x+3} = 1$

13. Combining the rational expressions: $\dfrac{4}{y-4} - \dfrac{y}{y-4} = \dfrac{4-y}{y-4} = \dfrac{-1(y-4)}{y-4} = -1$

15. Combining the rational expressions: $\dfrac{x}{x^2-y^2} - \dfrac{y}{x^2-y^2} = \dfrac{x-y}{x^2-y^2} = \dfrac{x-y}{(x+y)(x-y)} = \dfrac{1}{x+y}$

17. Combining the rational expressions: $\dfrac{2x-3}{x-2} - \dfrac{x-1}{x-2} = \dfrac{2x-3-x+1}{x-2} = \dfrac{x-2}{x-2} = 1$

19. Combining the rational expressions: $\dfrac{1}{a}+\dfrac{2}{a^2}-\dfrac{3}{a^3}=\dfrac{1}{a}\cdot\dfrac{a^2}{a^2}+\dfrac{2}{a^2}\cdot\dfrac{a}{a}-\dfrac{3}{a^3}=\dfrac{a^2+2a-3}{a^3}$

21. Combining the rational expressions: $\dfrac{7x-2}{2x+1}-\dfrac{5x-3}{2x+1}=\dfrac{7x-2-5x+3}{2x+1}=\dfrac{2x+1}{2x+1}=1$

23. Combining the rational expressions: $\dfrac{2}{t^2}-\dfrac{3}{2t}=\dfrac{2}{t^2}\cdot\dfrac{2}{2}-\dfrac{3}{2t}\cdot\dfrac{t}{t}=\dfrac{4}{2t^2}-\dfrac{3t}{2t^2}=\dfrac{4-3t}{2t^2}$

25. Combining the rational expressions:

$$\frac{3x+1}{2x-6}-\frac{x+2}{x-3}=\frac{3x+1}{2(x-3)}-\frac{x+2}{x-3}\cdot\frac{2}{2}=\frac{3x+1}{2(x-3)}-\frac{2x+4}{2(x-3)}=\frac{3x+1-2x-4}{2(x-3)}=\frac{x-3}{2(x-3)}=\frac{1}{2}$$

27. Combining the rational expressions:

$$\frac{6x+5}{5x-25}-\frac{x+2}{x-5}=\frac{6x+5}{5(x-5)}-\frac{x+2}{x-5}\cdot\frac{5}{5}=\frac{6x+5}{5(x-5)}-\frac{5x+10}{5(x-5)}=\frac{6x+5-5x-10}{5(x-5)}=\frac{x-5}{5(x-5)}=\frac{1}{5}$$

29. Combining the rational expressions:

$$\frac{x+1}{2x-2}-\frac{2}{x^2-1}=\frac{x+1}{2(x-1)}\cdot\frac{x+1}{x+1}-\frac{2}{(x+1)(x-1)}\cdot\frac{2}{2}$$

$$=\frac{x^2+2x+1}{2(x+1)(x-1)}-\frac{4}{2(x+1)(x-1)}$$

$$=\frac{x^2+2x-3}{2(x+1)(x-1)}$$

$$=\frac{(x+3)(x-1)}{2(x+1)(x-1)}$$

$$=\frac{x+3}{2(x+1)}$$

31. Combining the rational expressions:

$$\frac{1}{a-b}-\frac{3ab}{a^3-b^3}=\frac{1}{a-b}\cdot\frac{a^2+ab+b^2}{a^2+ab+b^2}-\frac{3ab}{a^3-b^3}$$

$$=\frac{a^2+ab+b^2}{a^3-b^3}-\frac{3ab}{a^3-b^3}$$

$$=\frac{a^2-2ab+b^2}{a^3-b^3}$$

$$=\frac{(a-b)^2}{(a-b)\left(a^2+ab+b^2\right)}$$

$$=\frac{a-b}{a^2+ab+b^2}$$

33. Combining the rational expressions:

$$\frac{1}{2y-3}-\frac{18y}{8y^3-27}=\frac{1}{2y-3}\cdot\frac{4y^2+6y+9}{4y^2+6y+9}-\frac{18y}{8y^3-27}$$

$$=\frac{4y^2+6y+9}{8y^3-27}-\frac{18y}{8y^3-27}$$

$$=\frac{4y^2-12y+9}{8y^3-27}$$

$$=\frac{(2y-3)^2}{(2y-3)\left(4y^2+6y+9\right)}$$

$$=\frac{2y-3}{4y^2+6y+9}$$

35.   Combining the rational expressions:

$$\frac{x}{x^2-5x+6}-\frac{3}{3-x}=\frac{x}{(x-2)(x-3)}+\frac{3}{x-3}\cdot\frac{x-2}{x-2}$$

$$=\frac{x}{(x-2)(x-3)}+\frac{3x-6}{(x-2)(x-3)}$$

$$=\frac{4x-6}{(x-2)(x-3)}$$

$$=\frac{2(2x-3)}{(x-3)(x-2)}$$

37.   Combining the rational expressions:

$$\frac{2}{4t-5}+\frac{9}{8t^2-38t+35}=\frac{2}{4t-5}\cdot\frac{2t-7}{2t-7}+\frac{9}{(4t-5)(2t-7)}$$

$$=\frac{4t-14}{(4t-5)(2t-7)}+\frac{9}{(4t-5)(2t-7)}$$

$$=\frac{4t-5}{(4t-5)(2t-7)}$$

$$=\frac{1}{2t-7}$$

39.   Combining the rational expressions:

$$\frac{1}{a^2-5a+6}+\frac{3}{a^2-a-2}=\frac{1}{(a-2)(a-3)}\cdot\frac{a+1}{a+1}+\frac{3}{(a-2)(a+1)}\cdot\frac{a-3}{a-3}$$

$$=\frac{a+1}{(a-2)(a-3)(a+1)}+\frac{3a-9}{(a-2)(a-3)(a+1)}$$

$$=\frac{4a-8}{(a-2)(a-3)(a+1)}$$

$$=\frac{4(a-2)}{(a-2)(a-3)(a+1)}$$

$$=\frac{4}{(a-3)(a+1)}$$

41.   Combining the rational expressions:

$$\frac{1}{8x^3-1}-\frac{1}{4x^2-1}=\frac{1}{(2x-1)\left(4x^2+2x+1\right)}\cdot\frac{2x+1}{2x+1}-\frac{1}{(2x+1)(2x-1)}\cdot\frac{4x^2+2x+1}{4x^2+2x+1}$$

$$=\frac{2x+1}{(2x+1)(2x-1)\left(4x^2+2x+1\right)}-\frac{4x^2+2x+1}{(2x+1)(2x-1)\left(4x^2+2x+1\right)}$$

$$=\frac{2x+1-4x^2-2x-1}{(2x+1)(2x-1)\left(4x^2+2x+1\right)}$$

$$=\frac{-4x^2}{(2x+1)(2x-1)\left(4x^2+2x+1\right)}$$

43. Combining the rational expressions:

$$\frac{4}{4x^2-9}-\frac{6}{8x^2-6x-9}=\frac{4}{(2x+3)(2x-3)}\cdot\frac{4x+3}{4x+3}-\frac{6}{(2x-3)(4x+3)}\cdot\frac{2x+3}{2x+3}$$

$$=\frac{16x+12}{(2x+3)(2x-3)(4x+3)}-\frac{12x+18}{(2x+3)(2x-3)(4x+3)}$$

$$=\frac{16x+12-12x-18}{(2x+3)(2x-3)(4x+3)}$$

$$=\frac{4x-6}{(2x+3)(2x-3)(4x+3)}$$

$$=\frac{2(2x-3)}{(2x+3)(2x-3)(4x+3)}$$

$$=\frac{2}{(2x+3)(4x+3)}$$

45. Combining the rational expressions:

$$\frac{4a}{a^2+6a+5}-\frac{3a}{a^2+5a+4}=\frac{4a}{(a+5)(a+1)}\cdot\frac{a+4}{a+4}-\frac{3a}{(a+4)(a+1)}\cdot\frac{a+5}{a+5}$$

$$=\frac{4a^2+16a}{(a+4)(a+5)(a+1)}-\frac{3a^2+15a}{(a+4)(a+5)(a+1)}$$

$$=\frac{4a^2+16a-3a^2-15a}{(a+4)(a+5)(a+1)}$$

$$=\frac{a^2+a}{(a+4)(a+5)(a+1)}$$

$$=\frac{a(a+1)}{(a+4)(a+5)(a+1)}$$

$$=\frac{a}{(a+4)(a+5)}$$

47. Combining the rational expressions:

$$\frac{2x-1}{x^2+x-6}-\frac{x+2}{x^2+5x+6}=\frac{2x-1}{(x+3)(x-2)}\cdot\frac{x+2}{x+2}-\frac{x+2}{(x+3)(x+2)}\cdot\frac{x-2}{x-2}$$

$$=\frac{2x^2+3x-2}{(x+3)(x+2)(x-2)}-\frac{x^2-4}{(x+3)(x+2)(x-2)}$$

$$=\frac{2x^2+3x-2-x^2+4}{(x+3)(x+2)(x-2)}$$

$$=\frac{x^2+3x+2}{(x+3)(x+2)(x-2)}$$

$$=\frac{(x+2)(x+1)}{(x+3)(x+2)(x-2)}$$

$$=\frac{x+1}{(x-2)(x+3)}$$

49. Combining the rational expressions:

$$\frac{2x-8}{3x^2+8x+4}+\frac{x+3}{3x^2+5x+2}=\frac{2x-8}{(3x+2)(x+2)}+\frac{x+3}{(3x+2)(x+1)}$$

$$=\frac{2x-8}{(3x+2)(x+2)}\cdot\frac{x+1}{x+1}+\frac{x+3}{(3x+2)(x+1)}\cdot\frac{x+2}{x+2}$$

$$=\frac{2x^2-6x-8}{(3x+2)(x+2)(x+1)}+\frac{x^2+5x+6}{(3x+2)(x+2)(x+1)}$$

$$=\frac{3x^2-x-2}{(3x+2)(x+2)(x+1)}$$

$$=\frac{(3x+2)(x-1)}{(3x+2)(x+2)(x+1)}$$

$$=\frac{x-1}{(x+1)(x+2)}$$

51. Combining the rational expressions:

$$\frac{2}{x^2+5x+6}-\frac{4}{x^2+4x+3}+\frac{3}{x^2+3x+2}=\frac{2}{(x+3)(x+2)}-\frac{4}{(x+3)(x+1)}+\frac{3}{(x+2)(x+1)}$$

$$=\frac{2}{(x+3)(x+2)}\cdot\frac{x+1}{x+1}-\frac{4}{(x+3)(x+1)}\cdot\frac{x+2}{x+2}+\frac{3}{(x+2)(x+1)}\cdot\frac{x+3}{x+3}$$

$$=\frac{2x+2}{(x+3)(x+2)(x+1)}-\frac{4x+8}{(x+3)(x+2)(x+1)}+\frac{3x+9}{(x+3)(x+2)(x+1)}$$

$$=\frac{2x+2-4x-8+3x+9}{(x+3)(x+2)(x+1)}$$

$$=\frac{x+3}{(x+3)(x+2)(x+1)}$$

$$=\frac{1}{(x+2)(x+1)}$$

53. Combining the rational expressions:

$$\frac{2x+8}{x^2+5x+6}-\frac{x+5}{x^2+4x+3}-\frac{x-1}{x^2+3x+2}=\frac{2x+8}{(x+3)(x+2)}-\frac{x+5}{(x+3)(x+1)}-\frac{x-1}{(x+2)(x+1)}$$

$$=\frac{2x+8}{(x+3)(x+2)}\cdot\frac{x+1}{x+1}-\frac{x+5}{(x+3)(x+1)}\cdot\frac{x+2}{x+2}-\frac{x-1}{(x+2)(x+1)}\cdot\frac{x+3}{x+3}$$

$$=\frac{2x^2+10x+8}{(x+3)(x+2)(x+1)}-\frac{x^2+7x+10}{(x+3)(x+2)(x+1)}-\frac{x^2+2x-3}{(x+3)(x+2)(x+1)}$$

$$=\frac{2x^2+10x+8-x^2-7x-10-x^2-2x+3}{(x+3)(x+2)(x+1)}$$

$$=\frac{x+1}{(x+3)(x+2)(x+1)}$$

$$=\frac{1}{(x+2)(x+3)}$$

55. Combining the rational expressions: $2+\dfrac{3}{2x+1}=\dfrac{2}{1}\cdot\dfrac{2x+1}{2x+1}+\dfrac{3}{2x+1}=\dfrac{4x+2}{2x+1}+\dfrac{3}{2x+1}=\dfrac{4x+5}{2x+1}$

57. Combining the rational expressions: $5+\dfrac{2}{4-t}=\dfrac{5}{1}\cdot\dfrac{4-t}{4-t}+\dfrac{2}{4-t}=\dfrac{20-5t}{4-t}+\dfrac{2}{4-t}=\dfrac{22-5t}{4-t}$

59. Combining the rational expressions: $x-\dfrac{4}{2x+3}=\dfrac{x}{1}\cdot\dfrac{2x+3}{2x+3}-\dfrac{4}{2x+3}=\dfrac{2x^2+3x}{2x+3}-\dfrac{4}{2x+3}=\dfrac{2x^2+3x-4}{2x+3}$

61. Combining the rational expressions:

$$\frac{x}{x+2}+\frac{1}{2x+4}-\frac{3}{x^2+2x} = \frac{x}{x+2}\cdot\frac{2x}{2x}+\frac{1}{2(x+2)}\cdot\frac{x}{x}-\frac{3}{x(x+2)}\cdot\frac{2}{2}$$

$$=\frac{2x^2}{2x(x+2)}+\frac{x}{2x(x+2)}-\frac{6}{2x(x+2)}$$

$$=\frac{2x^2+x-6}{2x(x+2)}$$

$$=\frac{(2x-3)(x+2)}{2x(x+2)}$$

$$=\frac{2x-3}{2x}$$

63. Combining the rational expressions:

$$\frac{1}{x}+\frac{x}{2x+4}-\frac{2}{x^2+2x} = \frac{1}{x}\cdot\frac{2(x+2)}{2(x+2)}+\frac{x}{2(x+2)}\cdot\frac{x}{x}-\frac{2}{x(x+2)}\cdot\frac{2}{2}$$

$$=\frac{2x+4}{2x(x+2)}+\frac{x^2}{2x(x+2)}-\frac{4}{2x(x+2)}$$

$$=\frac{x^2+2x}{2x(x+2)}$$

$$=\frac{x(x+2)}{2x(x+2)}$$

$$=\frac{1}{2}$$

65. Writing the expression and simplifying: $x+\dfrac{4}{x}=\dfrac{x^2+4}{x}$

67. Substituting the values: $P=\dfrac{1}{10}+\dfrac{1}{0.2}=0.1+5=5.1$

69. a. Substituting the values:

$$\frac{1}{T}=\frac{1}{24}-\frac{1}{30}=\frac{5}{120}-\frac{4}{120}=\frac{1}{120}$$
$$T=120$$

The two objects will meet in 120 months.

   b. If $t_A=t_B$, then $\dfrac{1}{T}=\dfrac{1}{t_A}-\dfrac{1}{t_A}=0$. Since this is impossible, the two objects will never meet.

71. Writing in scientific notation: $54,000=5.4\times10^4$ **73.** Writing in scientific notation: $0.00034=3.4\times10^{-4}$

75. Writing in expanded form: $6.44\times10^3=6,440$ **77.** Writing in expanded form: $6.44\times10^{-3}=0.00644$

79. Simplifying: $\left(3\times10^8\right)\left(4\times10^{-5}\right)=12\times10^3=1.2\times10^4$

81. Simplifying the expression:

$$\left(1-\frac{1}{x}\right)\left(1-\frac{1}{x+1}\right)\left(1-\frac{1}{x+2}\right)\left(1-\frac{1}{x+3}\right)=\left(\frac{x}{x}-\frac{1}{x}\right)\left(\frac{x+1}{x+1}-\frac{1}{x+1}\right)\left(\frac{x+2}{x+2}-\frac{1}{x+2}\right)\left(\frac{x+3}{x+3}-\frac{1}{x+3}\right)$$

$$=\left(\frac{x-1}{x}\right)\left(\frac{x}{x+1}\right)\left(\frac{x+1}{x+2}\right)\left(\frac{x+2}{x+3}\right)$$

$$=\frac{x-1}{x+3}$$

6.5   Complex Fractions

1. Simplifying the complex fraction: $\dfrac{\frac{3}{4}}{\frac{2}{3}} = \dfrac{\frac{3}{4} \cdot 12}{\frac{2}{3} \cdot 12} = \dfrac{9}{8}$

3. Simplifying the complex fraction: $\dfrac{\frac{1}{3} - \frac{1}{4}}{\frac{1}{2} + \frac{1}{8}} = \dfrac{\left(\frac{1}{3} - \frac{1}{4}\right) \cdot 24}{\left(\frac{1}{2} + \frac{1}{8}\right) \cdot 24} = \dfrac{8 - 6}{12 + 3} = \dfrac{2}{15}$

5. Simplifying the complex fraction: $\dfrac{3 + \frac{2}{5}}{1 - \frac{3}{7}} = \dfrac{\left(3 + \frac{2}{5}\right) \cdot 35}{\left(1 - \frac{3}{7}\right) \cdot 35} = \dfrac{105 + 14}{35 - 15} = \dfrac{119}{20}$

7. Simplifying the complex fraction: $\dfrac{\frac{1}{x}}{1 + \frac{1}{x}} = \dfrac{\left(\frac{1}{x}\right) \cdot x}{\left(1 + \frac{1}{x}\right) \cdot x} = \dfrac{1}{x + 1}$

9. Simplifying the complex fraction: $\dfrac{1 + \frac{1}{a}}{1 - \frac{1}{a}} = \dfrac{\left(1 + \frac{1}{a}\right) \cdot a}{\left(1 - \frac{1}{a}\right) \cdot a} = \dfrac{a + 1}{a - 1}$

11. Simplifying the complex fraction: $\dfrac{\frac{1}{x} - \frac{1}{y}}{\frac{1}{x} + \frac{1}{y}} = \dfrac{\left(\frac{1}{x} - \frac{1}{y}\right) \cdot xy}{\left(\frac{1}{x} + \frac{1}{y}\right) \cdot xy} = \dfrac{y - x}{y + x}$

13. Simplifying the complex fraction: $\dfrac{\frac{x-5}{x^2-4}}{\frac{x^2-25}{x+2}} = \dfrac{\frac{x-5}{(x+2)(x-2)} \cdot (x+2)(x-2)}{\frac{(x+5)(x-5)}{x+2} \cdot (x+2)(x-2)} = \dfrac{x-5}{(x+5)(x-5)(x-2)} = \dfrac{1}{(x+5)(x-2)}$

15. Simplifying the complex fraction:

$$\dfrac{\frac{4a}{2a^3+2}}{\frac{8a}{4a+4}} = \dfrac{\frac{4a}{2(a+1)\left(a^2-a+1\right)} \cdot 2(a+1)\left(a^2-a+1\right)}{\frac{8a}{4(a+1)} \cdot 2(a+1)\left(a^2-a+1\right)} = \dfrac{4a}{4a\left(a^2-a+1\right)} = \dfrac{1}{a^2-a+1}$$

17. Simplifying the complex fraction: $\dfrac{1 - \frac{9}{x^2}}{1 - \frac{1}{x} - \frac{6}{x^2}} = \dfrac{\left(1 - \frac{9}{x^2}\right) \cdot x^2}{\left(1 - \frac{1}{x} - \frac{6}{x^2}\right) \cdot x^2} = \dfrac{x^2 - 9}{x^2 - x - 6} = \dfrac{(x+3)(x-3)}{(x+2)(x-3)} = \dfrac{x+3}{x+2}$

19. Simplifying the complex fraction: $\dfrac{2 + \frac{5}{a} - \frac{3}{a^2}}{2 - \frac{5}{a} + \frac{2}{a^2}} = \dfrac{\left(2 + \frac{5}{a} - \frac{3}{a^2}\right) \cdot a^2}{\left(2 - \frac{5}{a} + \frac{2}{a^2}\right) \cdot a^2} = \dfrac{2a^2 + 5a - 3}{2a^2 - 5a + 2} = \dfrac{(2a-1)(a+3)}{(2a-1)(a-2)} = \dfrac{a+3}{a-2}$

21. Simplifying the complex fraction: $\dfrac{1 + \frac{1}{x+3}}{1 - \frac{1}{x+3}} = \dfrac{\left(1 + \frac{1}{x+3}\right) \cdot (x+3)}{\left(1 - \frac{1}{x+3}\right) \cdot (x+3)} = \dfrac{x+3+1}{x+3-1} = \dfrac{x+4}{x+2}$

23. Simplifying the complex fraction:

$$\frac{1-\dfrac{1}{a+1}}{1+\dfrac{1}{a-1}} = \frac{\left(1-\dfrac{1}{a+1}\right)\bullet(a+1)(a-1)}{\left(1+\dfrac{1}{a-1}\right)\bullet(a+1)(a-1)} = \frac{(a+1)(a-1)-(a-1)}{(a+1)(a-1)+(a+1)} = \frac{(a-1)(a+1-1)}{(a+1)(a-1+1)} = \frac{a(a-1)}{a(a+1)} = \frac{a-1}{a+1}$$

25. Simplifying the complex fraction: $\dfrac{\dfrac{1}{x+3}+\dfrac{1}{x-3}}{\dfrac{1}{x+3}-\dfrac{1}{x-3}} = \dfrac{\left(\dfrac{1}{x+3}+\dfrac{1}{x-3}\right)\bullet(x+3)(x-3)}{\left(\dfrac{1}{x+3}-\dfrac{1}{x-3}\right)\bullet(x+3)(x-3)} = \dfrac{(x-3)+(x+3)}{(x-3)-(x+3)} = \dfrac{2x}{-6} = -\dfrac{x}{3}$

27. Simplifying the complex fraction:

$$\frac{\dfrac{y+1}{y-1}+\dfrac{y-1}{y+1}}{\dfrac{y+1}{y-1}-\dfrac{y-1}{y+1}} = \frac{\left(\dfrac{y+1}{y-1}+\dfrac{y-1}{y+1}\right)\bullet(y+1)(y-1)}{\left(\dfrac{y+1}{y-1}-\dfrac{y-1}{y+1}\right)\bullet(y+1)(y-1)}$$

$$= \frac{(y+1)^2+(y-1)^2}{(y+1)^2-(y-1)^2}$$

$$= \frac{y^2+2y+1+y^2-2y+1}{y^2+2y+1-y^2+2y-1}$$

$$= \frac{2y^2+2}{4y}$$

$$= \frac{2(y^2+1)}{4y}$$

$$= \frac{y^2+1}{2y}$$

29. Simplifying the complex fraction: $1-\dfrac{x}{1-\dfrac{1}{x}} = 1-\dfrac{x\bullet x}{\left(1-\dfrac{1}{x}\right)\bullet x} = 1-\dfrac{x^2}{x-1} = \dfrac{x-1-x^2}{x-1} = \dfrac{-x^2+x-1}{x-1}$

31. Simplifying the complex fraction: $1+\dfrac{1}{1+\dfrac{1}{1+1}} = 1+\dfrac{1}{1+\frac{1}{2}} = 1+\dfrac{1}{\frac{3}{2}} = 1+\frac{2}{3} = \frac{5}{3}$

33. **a.** Simplifying the difference quotient: $\dfrac{f(x)-f(a)}{x-a} = \dfrac{\frac{4}{x}-\frac{4}{a}}{x-a} = \dfrac{\left(\frac{4}{x}-\frac{4}{a}\right)ax}{(x-a)ax} = \dfrac{4a-4x}{ax(x-a)} = \dfrac{-4(x-a)}{ax(x-a)} = -\dfrac{4}{ax}$

b. Simplifying the difference quotient:

$$\frac{f(x)-f(a)}{x-a} = \frac{\frac{1}{x+1}-\frac{1}{a+1}}{x-a}$$

$$= \frac{\left(\frac{1}{x+1}-\frac{1}{a+1}\right)(x+1)(a+1)}{(x-a)(x+1)(a+1)x}$$

$$= \frac{a+1-x-1}{x(x-a)(x+1)(a+1)}$$

$$= \frac{a-x}{x(x-a)(x+1)(a+1)}$$

$$= -\frac{1}{x(x+1)(a+1)}$$

c. Simplifying the difference quotient:

$$\frac{f(x)-f(a)}{x-a} = \frac{\frac{1}{x^2}-\frac{1}{a^2}}{x-a} = \frac{\left(\frac{1}{x^2}-\frac{1}{a^2}\right)a^2x^2}{a^2x^2(x-a)} = \frac{a^2-x^2}{a^2x^2(x-a)} = \frac{(a+x)(a-x)}{a^2x^2(x-a)} = -\frac{a+x}{a^2x^2}$$

35. Rewriting without negative exponents: $f = \left(a^{-1}+b^{-1}\right)^{-1} = \dfrac{1}{a^{-1}+b^{-1}} = \dfrac{1}{\frac{1}{a}+\frac{1}{b}} = \dfrac{1 \cdot ab}{\left(\frac{1}{a}+\frac{1}{b}\right) \cdot ab} = \dfrac{ab}{a+b}$

37. **a.** As v approaches 0, the denominator approaches 1.

   b. Solving v:

$$h = \frac{f}{1+\frac{v}{s}}$$

$$h = \frac{f \cdot s}{\left(1+\frac{v}{s}\right)s}$$

$$h = \frac{fs}{s+v}$$

$$h(s+v) = fs$$

$$s+v = \frac{fs}{h}$$

$$v = \frac{fs}{h} - s$$

39. Solving the equation:
$$3x + 60 = 15$$
$$3x = -45$$
$$x = -15$$

41. Solving the equation:
$$3(y-3) = 2(y-2)$$
$$3y - 9 = 2y - 4$$
$$y = 5$$

43. Solving the equation:
$$10 - 2(x+3) = x+1$$
$$10 - 2x - 6 = x+1$$
$$-2x + 4 = x+1$$
$$-3x = -3$$
$$x = 1$$

45. Solving the equation:

$$x^2 - x - 12 = 0$$
$$(x-4)(x+3) = 0$$
$$x = -3, 4$$

47. Solving the equation:
$$(x+1)(x-6) = -12$$
$$x^2 - 5x - 6 = -12$$
$$x^2 - 5x + 6 = 0$$
$$(x-2)(x-3) = 0$$
$$x = 2, 3$$

6.6 Equations Involving Rational Expressions

1. Solving the equation:
$$\frac{x}{5} + 4 = \frac{5}{3}$$
$$15\left(\frac{x}{5} + 4\right) = 15\left(\frac{5}{3}\right)$$
$$3x + 60 = 25$$
$$3x = -35$$
$$x = -\frac{35}{3}$$

3. Solving the equation:
$$\frac{a}{3} + 2 = \frac{4}{5}$$
$$15\left(\frac{a}{3} + 2\right) = 15\left(\frac{4}{5}\right)$$
$$5a + 30 = 12$$
$$5a = -18$$
$$a = -\frac{18}{5}$$

5. Solving the equation:
$$\frac{y}{2} + \frac{y}{4} + \frac{y}{6} = 3$$
$$12\left(\frac{y}{2} + \frac{y}{4} + \frac{y}{6}\right) = 12(3)$$
$$6y + 3y + 2y = 36$$
$$11y = 36$$
$$y = \frac{36}{11}$$

7. Solving the equation:
$$\frac{5}{2x} = \frac{1}{x} + \frac{3}{4}$$
$$4x\left(\frac{5}{2x}\right) = 4x\left(\frac{1}{x} + \frac{3}{4}\right)$$
$$10 = 4 + 3x$$
$$3x = 6$$
$$x = 2$$

9. Solving the equation:
$$\frac{1}{x} = \frac{1}{3} - \frac{2}{3x}$$
$$3x\left(\frac{1}{x}\right) = 3x\left(\frac{1}{3} - \frac{2}{3x}\right)$$
$$3 = x - 2$$
$$x = 5$$

11. Solving the equation:
$$\frac{2x}{x-3} + 2 = \frac{2}{x-3}$$
$$(x-3)\left(\frac{2x}{x-3} + 2\right) = (x-3)\left(\frac{2}{x-3}\right)$$
$$2x + 2(x-3) = 2$$
$$2x + 2x - 6 = 2$$
$$4x = 8$$
$$x = 2$$

13. Solving the equation:
$$1 - \frac{1}{x} = \frac{12}{x^2}$$
$$x^2\left(1 - \frac{1}{x}\right) = x^2\left(\frac{12}{x^2}\right)$$
$$x^2 - x = 12$$
$$x^2 - x - 12 = 0$$
$$(x+3)(x-4) = 0$$
$$x = -3, 4$$

15. Solving the equation:
$$y - \frac{4}{3y} = -\frac{1}{3}$$
$$3y\left(y - \frac{4}{3y}\right) = 3y\left(-\frac{1}{3}\right)$$
$$3y^2 - 4 = -y$$
$$3y^2 + y - 4 = 0$$
$$(3y+4)(y-1) = 0$$
$$x = -\frac{4}{3}, 1$$

17. Solving the equation:

$$\frac{x+2}{x+1} = \frac{1}{x+1} + 2$$

$$(x+1)\left(\frac{x+2}{x+1}\right) = (x+1)\left(\frac{1}{x+1} + 2\right)$$

$$x+2 = 1 + 2(x+1)$$

$$x+2 = 1 + 2x + 2$$

$$x+2 = 2x + 3$$

$$x = -1 \quad \text{(does not check)}$$

There is no solution (−1 does not check).

21. Solving the equation:

$$6 - \frac{5}{x^2} = \frac{7}{x}$$

$$x^2\left(6 - \frac{5}{x^2}\right) = x^2\left(\frac{7}{x}\right)$$

$$6x^2 - 5 = 7x$$

$$6x^2 - 7x - 5 = 0$$

$$(2x+1)(3x-5) = 0$$

$$x = -\frac{1}{2}, \frac{5}{3}$$

23. Solving the equation:

$$\frac{1}{x-1} - \frac{1}{x+1} = \frac{3x}{x^2-1}$$

$$(x+1)(x-1)\left(\frac{1}{x-1} - \frac{1}{x+1}\right) = (x+1)(x-1)\left(\frac{3x}{(x+1)(x-1)}\right)$$

$$(x+1) - (x-1) = 3x$$

$$x+1-x+1 = 3x$$

$$3x = 2$$

$$x = \frac{2}{3}$$

25. Solving the equation:

$$\frac{2}{x-3} + \frac{x}{x^2-9} = \frac{4}{x+3}$$

$$(x+3)(x-3)\left(\frac{2}{x-3} + \frac{x}{(x+3)(x-3)}\right) = (x+3)(x-3)\left(\frac{4}{x+3}\right)$$

$$2(x+3) + x = 4(x-3)$$

$$2x+6+x = 4x-12$$

$$3x+6 = 4x-12$$

$$-x = -18$$

$$x = 18$$

19. Solving the equation:

$$\frac{3}{a-2} = \frac{2}{a-3}$$

$$(a-2)(a-3)\left(\frac{3}{a-2}\right) = (a-2)(a-3)\left(\frac{2}{a-3}\right)$$

$$3(a-3) = 2(a-2)$$

$$3a-9 = 2a-4$$

$$a = 5$$

27. Solving the equation:
$$\frac{3}{2} - \frac{1}{x-4} = \frac{-2}{2x-8}$$
$$2(x-4)\left(\frac{3}{2} - \frac{1}{x-4}\right) = 2(x-4)\left(\frac{-2}{2(x-4)}\right)$$
$$3(x-4) - 2 = -2$$
$$3x - 12 - 2 = -2$$
$$3x - 14 = -2$$
$$3x = 12$$
$$x = 4 \quad (\text{does not check})$$
There is no solution (4 does not check).

29. Solving the equation:
$$\frac{t-4}{t^2-3t} = \frac{-2}{t^2-9}$$
$$t(t+3)(t-3) \cdot \frac{t-4}{t(t-3)} = t(t+3)(t-3) \cdot \frac{-2}{(t+3)(t-3)}$$
$$(t+3)(t-4) = -2t$$
$$t^2 - t - 12 = -2t$$
$$t^2 + t - 12 = 0$$
$$(t+4)(t-3) = 0$$
$$t = -4 \quad (t = 3 \text{ does not check})$$
The solution is -4 (3 does not check).

31. Solving the equation:
$$\frac{3}{y-4} - \frac{2}{y+1} = \frac{5}{y^2-3y-4}$$
$$(y-4)(y+1)\left(\frac{3}{y-4} - \frac{2}{y+1}\right) = (y-4)(y+1)\left(\frac{5}{(y-4)(y+1)}\right)$$
$$3(y+1) - 2(y-4) = 5$$
$$3y + 3 - 2y + 8 = 5$$
$$y + 11 = 5$$
$$y = -6$$

33. Solving the equation:
$$\frac{2}{1+a} = \frac{3}{1-a} + \frac{5}{a}$$
$$a(1+a)(1-a)\left(\frac{2}{1+a}\right) = a(1+a)(1-a)\left(\frac{3}{1-a} + \frac{5}{a}\right)$$
$$2a(1-a) = 3a(1+a) + 5(1+a)(1-a)$$
$$2a - 2a^2 = 3a + 3a^2 + 5 - 5a^2$$
$$-2a^2 + 2a = -2a^2 + 3a + 5$$
$$2a = 3a + 5$$
$$-a = 5$$
$$a = -5$$

35.   Solving the equation:

$$\frac{3}{2x-6}-\frac{x+1}{4x-12}=4$$

$$4(x-3)\left(\frac{3}{2(x-3)}-\frac{x+1}{4(x-3)}\right)=4(x-3)(4)$$

$$6-(x+1)=16x-48$$

$$5-x=16x-48$$

$$-17x=-53$$

$$x=\frac{53}{17}$$

37.   Solving the equation:

$$\frac{y+2}{y^2-y}-\frac{6}{y^2-1}=0$$

$$y(y+1)(y-1)\left(\frac{y+2}{y(y-1)}-\frac{6}{(y+1)(y-1)}\right)=y(y+1)(y-1)(0)$$

$$(y+1)(y+2)-6y=0$$

$$y^2+3y+2-6y=0$$

$$y^2-3y+2=0$$

$$(y-1)(y-2)=0$$

$$y=2\quad(y=1\text{ does not check})$$

The solution is 2 (1 does not check).

39.   Solving the equation:

$$\frac{4}{2x-6}-\frac{12}{4x+12}=\frac{12}{x^2-9}$$

$$4(x+3)(x-3)\left(\frac{4}{2(x-3)}-\frac{12}{4(x+3)}\right)=4(x+3)(x-3)\left(\frac{12}{(x+3)(x-3)}\right)$$

$$8(x+3)-12(x-3)=48$$

$$8x+24-12x+36=48$$

$$-4x+60=48$$

$$-4x=-12$$

$$x=3\quad(x=3\text{ does not check})$$

There is no solution (3 does not check).

41.   Solving the equation:

$$\frac{2}{y^2-7y+12}-\frac{1}{y^2-9}=\frac{4}{y^2-y-12}$$

$$(y+3)(y-3)(y-4)\left(\frac{2}{(y-3)(y-4)}-\frac{1}{(y+3)(y-3)}\right)=(y+3)(y-3)(y-4)\left(\frac{4}{(y-4)(y+3)}\right)$$

$$2(y+3)-(y-4)=4(y-3)$$

$$2y+6-y+4=4y-12$$

$$y+10=4y-12$$

$$-3y=-22$$

$$y=\frac{22}{3}$$

43. Solving the equation:

$$6x^{-1} + 4 = 7$$
$$x\left(6x^{-1} + 4\right) = x(7)$$
$$6 + 4x = 7x$$
$$6 = 3x$$
$$x = 2$$

45. Solving the equation:

$$1 + 5x^{-2} = 6x^{-1}$$
$$x^2\left(1 + 5x^{-2}\right) = x^2\left(6x^{-1}\right)$$
$$x^2 + 5 = 6x$$
$$x^2 - 6x + 5 = 0$$
$$(x-1)(x-5) = 0$$
$$x = 1, 5$$

47. Solving for x:

$$\frac{1}{x} = \frac{1}{b} - \frac{1}{a}$$
$$abx\left(\frac{1}{x}\right) = abx\left(\frac{1}{b} - \frac{1}{a}\right)$$
$$ab = ax - bx$$
$$ab = x(a - b)$$
$$x = \frac{ab}{a - b}$$

49. Solving for x:

$$\frac{1}{R} = \frac{1}{R_1} + \frac{1}{R_2}$$
$$RR_1R_2\left(\frac{1}{R}\right) = RR_1R_2\left(\frac{1}{R_1} + \frac{1}{R_2}\right)$$
$$R_1R_2 = RR_2 + RR_1$$
$$R_1R_2 = R\left(R_1 + R_2\right)$$
$$R = \frac{R_1R_2}{R_1 + R_2}$$

51. Solving for y:

$$x = \frac{y - 3}{y - 1}$$
$$x(y - 1) = y - 3$$
$$xy - x = y - 3$$
$$xy - y = x - 3$$
$$y(x - 1) = x - 3$$
$$y = \frac{x - 3}{x - 1}$$

53. Solving for y:

$$x = \frac{2y + 1}{3y + 1}$$
$$x(3y + 1) = 2y + 1$$
$$3xy + x = 2y + 1$$
$$3xy - 2y = -x + 1$$
$$y(3x - 2) = -x + 1$$
$$y = \frac{1 - x}{3x - 2}$$

55. Simplifying the left-hand side: $\dfrac{2}{x - y} - \dfrac{1}{y - x} = \dfrac{2}{x - y} - \dfrac{-1}{x - y} = \dfrac{2}{x - y} + \dfrac{1}{x - y} = \dfrac{3}{x - y}$

57. Completing the table:

Time t (sec)	Speed of Kayak Relative to the Water v (m / sec)	Current of the River c (m / sec)
240	4	1
300	4	2
514	4	3
338	3	1
540	3	2
impossible	3	3

59. Let x represent the number. The equation is:

$$2(x + 3) = 16$$
$$2x + 6 = 16$$
$$2x = 10$$
$$x = 5$$

The number is 5.

61. Let w represent the width and $2w - 3$ represent the length. The equation is:
$$2w + 2(2w - 3) = 42$$
$$2w + 4w - 6 = 42$$
$$6w = 48$$
$$w = 8$$
$$2w - 3 = 13$$
The width is 8 meters and the length is 13 meters.

63. Let x and $x + 1$ represent the two integers. The equation is:
$$x^2 + (x + 1)^2 = 61$$
$$x^2 + x^2 + 2x + 1 = 61$$
$$2x^2 + 2x - 60 = 0$$
$$x^2 + x - 30 = 0$$
$$(x + 6)(x - 5) = 0$$
$$x = -6, 5$$
$$x + 1 = -5, 6$$
The integers are either –6 and –5, or 5 and 6.

65. Let x, $x + 1$, and $x + 2$ represent the three sides. The equation is:
$$x^2 + (x + 1)^2 = (x + 2)^2$$
$$x^2 + x^2 + 2x + 1 = x^2 + 4x + 4$$
$$x^2 - 2x - 3 = 0$$
$$(x + 1)(x - 3) = 0$$
$$x = -1, 3$$
$$x + 1 = 0, 4$$
$$x + 2 = 1, 5$$
Since the sides of the triangle must be positive, the sides are 3, 4, and 5.

67. Solving the equation:
$$\frac{12}{x} + \frac{8}{x^2} - \frac{75}{x^3} - \frac{50}{x^4} = 0$$
$$x^4\left(\frac{12}{x} + \frac{8}{x^2} - \frac{75}{x^3} - \frac{50}{x^4}\right) = x^4(0)$$
$$12x^3 + 8x^2 - 75x - 50 = 0$$
$$4x^2(3x + 2) - 25(3x + 2) = 0$$
$$(3x + 2)(4x^2 - 25) = 0$$
$$(3x + 2)(2x + 5)(2x - 5) = 0$$
$$x = -\frac{5}{2}, -\frac{2}{3}, \frac{5}{2}$$

69. Solving the equation:
$$\frac{1}{x^3} - \frac{1}{3x^2} - \frac{1}{4x} + \frac{1}{12} = 0$$
$$12x^3\left(\frac{1}{x^3} - \frac{1}{3x^2} - \frac{1}{4x} + \frac{1}{12}\right) = 12x^3(0)$$
$$12 - 4x - 3x^2 + x^3 = 0$$
$$x^2(x - 3) - 4(x - 3) = 0$$
$$(x - 3)(x^2 - 4) = 0$$
$$(x - 3)(x + 2)(x - 2) = 0$$
$$x = -2, 2, 3$$

6.7   Applications

1.   Let x and $3x$ represent the two numbers. The equation is:

$$\frac{1}{x} + \frac{1}{3x} = \frac{20}{3}$$

$$3x\left(\frac{1}{x} + \frac{1}{3x}\right) = 3x\left(\frac{20}{3}\right)$$

$$3 + 1 = 20x$$

$$20x = 4$$

$$x = \frac{1}{5}$$

The numbers are $\frac{1}{5}$ and $\frac{3}{5}$.

3.   Let x represent the number. The equation is:

$$x + \frac{1}{x} = \frac{10}{3}$$

$$3x\left(x + \frac{1}{x}\right) = 3x\left(\frac{10}{3}\right)$$

$$3x^2 + 3 = 10x$$

$$3x^2 - 10x + 3 = 0$$

$$(3x - 1)(x - 3) = 0$$

$$x = \frac{1}{3}, 3$$

The number is either 3 or $\frac{1}{3}$.

5.   Let x and $x + 1$ represent the two integers. The equation is:

$$\frac{1}{x} + \frac{1}{x+1} = \frac{7}{12}$$

$$12x(x+1)\left(\frac{1}{x} + \frac{1}{x+1}\right) = 12x(x+1)\left(\frac{7}{12}\right)$$

$$12(x+1) + 12x = 7x(x+1)$$

$$12x + 12 + 12x = 7x^2 + 7x$$

$$0 = 7x^2 - 17x - 12$$

$$0 = (7x + 4)(x - 3)$$

$$x = 3 \quad \left(x = -\frac{4}{7} \text{ is not an integer}\right)$$

The two integers are 3 and 4.

7.   Let x represent the number. The equation is:

$$\frac{7 + x}{9 + x} = \frac{5}{6}$$

$$6(9 + x)\left(\frac{7 + x}{9 + x}\right) = 6(9 + x)\left(\frac{5}{6}\right)$$

$$6(7 + x) = 5(9 + x)$$

$$42 + 6x = 45 + 5x$$

$$x = 3$$

The number is 3.

9.   Let x represent the speed of the current. Setting the times equal:

$$\frac{3}{5+x} = \frac{1.5}{5-x}$$
$$3(5-x) = 1.5(5+x)$$
$$15 - 3x = 7.5 + 1.5x$$
$$7.5 = 4.5x$$
$$x = \frac{75}{45} = \frac{5}{3}$$

The speed of the current is $\frac{5}{3}$ mph.

11.   Let x represent the speed of the boat. Since the total time is 3 hours:

$$\frac{8}{x-2} + \frac{8}{x+2} = 3$$
$$(x+2)(x-2)\left(\frac{8}{x-2} + \frac{8}{x+2}\right) = 3(x+2)(x-2)$$
$$8(x+2) + 8(x-2) = 3x^2 - 12$$
$$16x = 3x^2 - 12$$
$$0 = 3x^2 - 16x - 12$$
$$0 = (3x+2)(x-6)$$
$$x = 6 \quad \left(x = -\frac{2}{3} \text{ is impossible}\right)$$

The speed of the boat is 6 mph.

13.   Let r represent the speed of train B and $r + 15$ represent the speed of train A. Since the times are equal:

$$\frac{150}{r+15} = \frac{120}{r}$$
$$150r = 120(r+15)$$
$$150r = 120r + 1800$$
$$30r = 1800$$
$$r = 60$$

The speed of train A is 75 mph and the speed of train B is 60 mph.

15.   The smaller plane makes the trip in 3 hours, so the 747 must take $1\frac{1}{2}$ hours to complete the trip. Thus the average

speed is given by: $\dfrac{810 \text{ miles}}{1\frac{1}{2} \text{ hours}} = 540$ miles per hour

17.   Let r represent the bus's usual speed. The difference of the two times is $\frac{1}{2}$ hour, therefore:

$$\frac{270}{r} - \frac{270}{r+6} = \frac{1}{2}$$
$$2r(r+6)\left(\frac{270}{r} - \frac{270}{r+6}\right) = 2r(r+6)\left(\frac{1}{2}\right)$$
$$540(r+6) - 540(r) = r(r+6)$$
$$540r + 3240 - 540r = r^2 + 6r$$
$$0 = r^2 + 6r - 3240$$
$$0 = (r-54)(r+60)$$
$$r = 54 \quad (r = -60 \text{ is impossible})$$

The usual speed is 54 mph.

19. Let x represent the time to fill the tank if both pipes are open. The rate equation is:

$$\frac{1}{8} - \frac{1}{16} = \frac{1}{x}$$

$$16x\left(\frac{1}{8} - \frac{1}{16}\right) = 16x\left(\frac{1}{x}\right)$$

$$2x - x = 16$$

$$x = 16$$

It will take 16 hours to fill the tank if both pipes are open.

21. Let x represent the time to fill the pool with both pipes open. The rate equation is:

$$\frac{1}{10} - \frac{1}{15} = \frac{1}{2} \cdot \frac{1}{x}$$

$$30x\left(\frac{1}{10} - \frac{1}{15}\right) = 30x\left(\frac{1}{2x}\right)$$

$$3x - 2x = 15$$

$$x = 15$$

It will take 15 hours to fill the pool with both pipes open.

23. Let x represent the time to fill the sink with the hot water faucet. The rate equation is:

$$\frac{1}{3.5} + \frac{1}{x} = \frac{1}{2.1}$$

$$7.35x\left(\frac{1}{3.5} + \frac{1}{x}\right) = 7.35x\left(\frac{1}{2.1}\right)$$

$$2.1x + 7.35 = 3.5x$$

$$7.35 = 1.4x$$

$$x = 5.25$$

It will take $5\frac{1}{4}$ minutes to fill the sink with the hot water faucet.

25. Converting to acres: $\dfrac{2,224,750 \text{ sq. ft.}}{43,560 \text{ sq. ft. / acre}} \approx 51.1$ acres

27. Converting the speed: $\dfrac{5750 \text{ feet}}{11 \text{ minutes}} \cdot \dfrac{1 \text{ mile}}{5280 \text{ feet}} \cdot \dfrac{60 \text{ minutes}}{1 \text{ hour}} \approx 5.9$ mph

29. Converting the speed: $\dfrac{100 \text{ meters}}{10.8 \text{ seconds}} \cdot \dfrac{3.28 \text{ feet}}{1 \text{ meter}} \cdot \dfrac{1 \text{ mile}}{5280 \text{ feet}} \cdot \dfrac{60 \text{ seconds}}{1 \text{ minute}} \cdot \dfrac{60 \text{ minutes}}{1 \text{ hour}} \approx 20.7$ mph

31. Converting the speed: $\dfrac{\pi \cdot 65 \text{ feet}}{30 \text{ seconds}} \cdot \dfrac{1 \text{ mile}}{5280 \text{ feet}} \cdot \dfrac{60 \text{ seconds}}{1 \text{ minute}} \cdot \dfrac{60 \text{ minutes}}{1 \text{ hour}} \approx 4.6$ mph

33. Converting the speed: $\dfrac{2\pi \cdot 2 \text{ inches}}{\frac{1}{300} \text{ minutes}} \cdot \dfrac{1 \text{ foot}}{12 \text{ inches}} \cdot \dfrac{1 \text{ mile}}{5280 \text{ feet}} \cdot \dfrac{60 \text{ minutes}}{1 \text{ hour}} \approx 3.6$ mph

35. Solving the equation:

$$\frac{1}{3}\left[\left(x + \frac{2}{3}x\right) + \frac{1}{3}\left(x + \frac{2}{3}x\right)\right] = 10$$

$$\left(x + \frac{2}{3}x\right) + \frac{1}{3}\left(x + \frac{2}{3}x\right) = 30$$

$$x + \frac{2}{3}x + \frac{1}{3}x + \frac{2}{9}x = 30$$

$$\frac{20}{9}x = 30$$

$$20x = 270$$

$$x = \frac{27}{2}$$

37. Performing the operations: $\dfrac{2a+10}{a^3} \cdot \dfrac{a^2}{3a+15} = \dfrac{2(a+5)}{a^3} \cdot \dfrac{a^2}{3(a+5)} = \dfrac{2}{3a}$

39. Performing the operations: $\left(x^2 - 9\right)\left(\dfrac{x+2}{x+3}\right) = (x+3)(x-3)\left(\dfrac{x+2}{x+3}\right) = (x-3)(x+2)$

41. Performing the operations: $\dfrac{2x-7}{x-2} - \dfrac{x-5}{x-2} = \dfrac{2x-7-x+5}{x-2} = \dfrac{x-2}{x-2} = 1$

43. Simplifying the expression: $\dfrac{\frac{1}{x} - \frac{1}{3}}{\frac{1}{x} + \frac{1}{3}} = \dfrac{\left(\frac{1}{x} - \frac{1}{3}\right) \cdot 3x}{\left(\frac{1}{x} + \frac{1}{3}\right) \cdot 3x} = \dfrac{3 - x}{3 + x}$

45. Solving the equation:

$$\frac{x}{x-3} + \frac{3}{2} = \frac{3}{x-3}$$

$$2(x-3)\left(\frac{x}{x-3} + \frac{3}{2}\right) = 2(x-3)\left(\frac{3}{x-3}\right)$$

$$2x + 3(x-3) = 6$$

$$2x + 3x - 9 = 6$$

$$5x = 15$$

$$x = 3 \quad (\text{does not check})$$

There is no solution (3 does not check).

Chapter 6 Review

1. Reducing the fraction: $\dfrac{125x^4 yz^3}{35x^2 y^4 z^3} = \dfrac{25x^2}{7y^3}$

3. Reducing the fraction: $\dfrac{x^2 - 25}{x^2 + 10x + 25} = \dfrac{(x+5)(x-5)}{(x+5)^2} = \dfrac{x-5}{x+5}$

5. Dividing: $\dfrac{12x^3 + 8x^2 + 16x}{4x^2} = \dfrac{12x^3}{4x^2} + \dfrac{8x^2}{4x^2} + \dfrac{16x}{4x^2} = 3x + 2 + \dfrac{4}{x}$

7. Dividing: $\dfrac{x^{6n} - x^{5n}}{x^{3n}} = \dfrac{x^{6n}}{x^{3n}} - \dfrac{x^{5n}}{x^{3n}} = x^{3n} - x^{2n}$

9. Dividing by factoring: $\dfrac{5x^2 - 14xy - 24y^2}{x - 4y} = \dfrac{(5x + 6y)(x - 4y)}{x - 4y} = 5x + 6y$

11. Dividing using long division:

$$
\begin{array}{r}
4x + 1 \\
2x - 7 \overline{)8x^2 - 26x - 9} \\
\underline{8x^2 - 28x} \\
2x - 9 \\
\underline{2x - 7} \\
-2
\end{array}
$$

The quotient is $4x + 1 - \dfrac{2}{2x - 7}$.

13. Performing the operations: $\frac{3}{4} \cdot \frac{12}{15} \div \frac{1}{3} = \frac{3}{4} \cdot \frac{12}{15} \cdot \frac{3}{1} = \frac{9}{5}$

15. Performing the operations: $\dfrac{x^3 - 1}{x^4 - 1} \cdot \dfrac{x^2 - 1}{x^2 + x + 1} = \dfrac{(x-1)(x^2 + x + 1)}{(x^2 + 1)(x^2 - 1)} \cdot \dfrac{x^2 - 1}{x^2 + x + 1} = \dfrac{x - 1}{x^2 + 1}$

17. Performing the operations:
$$\frac{ax+bx+2a+2b}{ax-3a+bx-3b} \div \frac{ax-bx-2a+2b}{ax-bx-3a+3b} = \frac{ax+bx+2a+2b}{ax-3a+bx-3b} \cdot \frac{ax-bx-3a+3b}{ax-bx-2a+2b}$$
$$= \frac{x(a+b)+2(a+b)}{a(x-3)+b(x-3)} \cdot \frac{x(a-b)-3(a-b)}{x(a-b)-2(a-b)}$$
$$= \frac{(a+b)(x+2)}{(x-3)(a+b)} \cdot \frac{(a-b)(x-3)}{(a-b)(x-2)}$$
$$= \frac{x+2}{x-2}$$

19. Performing the operations: $\frac{3}{5} - \frac{1}{10} + \frac{8}{15} = \frac{3}{5} \cdot \frac{6}{6} - \frac{1}{10} \cdot \frac{3}{3} + \frac{8}{15} \cdot \frac{2}{2} = \frac{18}{30} - \frac{3}{30} + \frac{16}{30} = \frac{31}{30}$

21. Performing the operations: $\frac{1}{x} + \frac{1}{x^2} + \frac{1}{x^3} = \frac{1}{x} \cdot \frac{x^2}{x^2} + \frac{1}{x^2} \cdot \frac{x}{x} + \frac{1}{x^3} = \frac{x^2+x+1}{x^3}$

23. Performing the operations:
$$\frac{x-2}{x^2+5x+4} - \frac{x-4}{2x^2+12x+16} = \frac{x-2}{(x+4)(x+1)} \cdot \frac{2(x+2)}{2(x+2)} - \frac{x-4}{2(x+4)(x+2)} \cdot \frac{x+1}{x+1}$$
$$= \frac{2x^2-8}{2(x+4)(x+1)(x+2)} - \frac{x^2-3x-4}{2(x+4)(x+1)(x+2)}$$
$$= \frac{2x^2-8-x^2+3x+4}{2(x+4)(x+1)(x+2)}$$
$$= \frac{x^2+3x-4}{2(x+4)(x+1)(x+2)}$$
$$= \frac{(x+4)(x-1)}{2(x+4)(x+1)(x+2)}$$
$$= \frac{x-1}{2(x+1)(x+2)}$$

25. Simplifying the complex fraction: $\dfrac{1+\frac{2}{3}}{1-\frac{2}{3}} = \dfrac{\left(1+\frac{2}{3}\right) \cdot 3}{\left(1-\frac{2}{3}\right) \cdot 3} = \dfrac{3+2}{3-2} = 5$

27. Simplifying the complex fraction: $1 + \dfrac{1}{x+\frac{1}{x}} = 1 + \dfrac{1 \cdot x}{\left(x+\frac{1}{x}\right) \cdot x} = 1 + \dfrac{x}{x^2+1} = \dfrac{x^2+x+1}{x^2+1}$

29. Solving the equation:

$$\frac{3}{x-1} = \frac{3}{5}$$
$$5(x-1)\left(\frac{3}{x-1}\right) = 5(x-1)\left(\frac{3}{5}\right)$$
$$15 = 3x-3$$
$$18 = 3x$$
$$x = 6$$

31. Solving the equation:

$$\frac{5}{y+1} = \frac{4}{y+2}$$
$$5(y+2) = 4(y+1)$$
$$5y+10 = 4y+4$$
$$y = -6$$

33.   Solving the equation:

$$\frac{4}{x^2-x-12}+\frac{1}{x^2-9}=\frac{2}{x^2-7x+12}$$

$$(x-4)(x+3)(x-3)\left(\frac{4}{(x-4)(x+3)}+\frac{1}{(x+3)(x-3)}\right)=(x-4)(x+3)(x-3)\left(\frac{2}{(x-4)(x-3)}\right)$$

$$4(x-3)+x-4=2(x+3)$$
$$4x-12+x-4=2x+6$$
$$5x-16=2x+6$$
$$3x=22$$
$$x=\frac{22}{3}$$

35.   Let x represent the rate of the truck and $x+10$ represent the rate of the car. The equation is:

$$\frac{120}{x}-\frac{120}{x+10}=2$$

$$x(x+10)\left(\frac{120}{x}-\frac{120}{x+10}\right)=2x(x+10)$$

$$120(x+10)-120x=2x^2+20x$$

$$1200=2x^2+20x$$

$$0=2\left(x^2+10x-600\right)$$

$$0=2(x+30)(x-20)$$

$$x=20\quad(x=-30\text{ is impossible})$$

The car's rate is 30 mph and the truck's rate is 20 mph.

37.   Converting the speed: $\dfrac{1088\text{ feet}}{1\text{ second}}\cdot\dfrac{1\text{ mile}}{5280\text{ feet}}\cdot\dfrac{60\text{ seconds}}{1\text{ minute}}\cdot\dfrac{60\text{ minutes}}{1\text{ hour}}\approx742\text{ mph}$

Chapters 1-6 Cumulative Review

1.   Simplifying: $11-(-9)-7-(-5)=11+9-7+5=18$ **3.**   Simplifying: $\dfrac{x^{-5}}{x^{-8}}=x^{-5+8}=x^3$

5.   Simplifying: $-3(5x+4)+12x=-15x-12+12x=-3x-12$

7.   Simplifying: $(x+3)^2-(x-3)^2=x^2+6x+9-x^2+6x-9=12x$

9.   Computing the value: $-3\cdot\frac{5}{12}-\frac{3}{4}=-\frac{5}{4}-\frac{3}{4}=-\frac{8}{4}=-2$ **11.** $\{1,7\}$

13.   These are the commutative and associative properties of addition.

15.   Subtracting: $\dfrac{y}{x^2-y^2}-\dfrac{x}{x^2-y^2}=\dfrac{y-x}{x^2-y^2}=\dfrac{-1(x-y)}{(x+y)(x-y)}=-\dfrac{1}{x+y}$

17.   Multiplying: $\dfrac{x^4-16}{x^3-8}\cdot\dfrac{x^2+2x+4}{x^2+4}=\dfrac{\left(x^2+4\right)(x+2)(x-2)}{(x-2)\left(x^2+2x+4\right)}\cdot\dfrac{x^2+2x+4}{x^2+4}=x+2$

19. Dividing using long division:

$$a+2 \overline{) a^4 + a^3 + 0a^2 + 0a - 1}$$

quotient: $a^3 - a^2 + 2a - 4$

$$\underline{a^4 + 2a^3}$$
$$-a^3 + 0a^2$$
$$\underline{-a^3 - 2a^2}$$
$$2a^2 + 0a$$
$$\underline{2a^2 + 4a}$$
$$-4a - 1$$
$$\underline{-4a - 8}$$
$$7$$

The quotient is $a^3 - a^2 + 2a - 4 + \dfrac{7}{a+2}$.

21. Solving the equation:

$$7y - 6 = 2y + 9$$
$$5y = 15$$
$$y = 3$$

23. Solving the equation:

$$|a| - 5 = 7$$
$$|a| = 12$$
$$a = -12, 12$$

25. Solving the equation:

$$\frac{3}{y-2} = \frac{2}{y-3}$$
$$3(y-3) = 2(y-2)$$
$$3y - 9 = 2y - 4$$
$$y = 5$$

27. Substitute into the first equation:

$$5x - 2(3x + 2) = -1$$
$$5x - 6x - 4 = -1$$
$$-x = 3$$
$$x = -3$$

The solution is $(-3, -7)$.

29. Multiply the second equation by 3:

$$-5x + 3y = 1$$
$$5x - 3y = 6$$

Adding yields $0 = 7$, which is false. There is no solution (lines are parallel).

31. Evaluating the determinants:

$$D = \begin{vmatrix} 7 & -9 \\ -3 & 11 \end{vmatrix} = 7(11) - (-9)(-3) = 77 + 27 = 50$$

$$D_x = \begin{vmatrix} 2 & -9 \\ 1 & 11 \end{vmatrix} = 2(11) - (-9)(1) = 22 + 9 = 31$$

$$D_y = \begin{vmatrix} 7 & 2 \\ -3 & 1 \end{vmatrix} = 7(1) - 2(-3) = 7 + 6 = 13$$

Using Cramer's rule: $x = \dfrac{D_x}{D} = \dfrac{31}{50}, y = \dfrac{D_y}{D} = \dfrac{13}{50}$. The solution is $\left(\dfrac{31}{50}, \dfrac{13}{50} \right)$.

33.   Solving the inequality:
$$-3(3x-1) \le -2(3x-3)$$
$$-9x+3 \le -6x+6$$
$$-3x \le 3$$
$$x \ge -1$$
Graphing the solution set:

35.   Solving for y:
$$x+y=-3$$
$$y=-x-3$$
The slope is -1.

37.   Graphing the line:

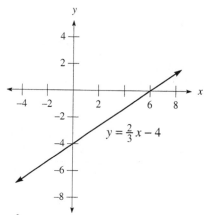

39.   Factoring into primes: $168 = 8 \cdot 21 = 2^3 \cdot 3 \cdot 7$

41.   Factoring: $x^2 + 10x + 25 - y^2 = (x+5)^2 - y^2 = (x+5-y)(x+5+y)$

43.   Writing in scientific notation: $9,270,000.00 = 9.27 \times 10^6$

45.   Evaluating the function: $f(-2) - g(3) = \left[(-2)^2 - 2(-2)\right] - [3+5] = 4 + 4 - 8 = 0$

47.   Since the slope is undefined, the line is vertical and its equation is $x = -2$.

49.   Let b represent the base and $2b - 5$ represent the height. Using the area formula:
$$\tfrac{1}{2}b(2b-5) = 75$$
$$2b^2 - 5b = 150$$
$$2b^2 - 5b - 150 = 0$$
$$(2b+15)(b-10) = 0$$
$$b = 10 \quad \left(b = -\tfrac{15}{2} \text{ is impossible}\right)$$
The base is 10 feet and the height is 15 feet.

Chapter 6 Test

1. Reducing the fraction: $\dfrac{x^2-y^2}{x-y}=\dfrac{(x+y)(x-y)}{x-y}=x+y$

2. Reducing the fraction: $\dfrac{2x^2-5x+3}{2x^2-x-3}=\dfrac{(2x-3)(x-1)}{(2x-3)(x+1)}=\dfrac{x-1}{x+1}$

3. Dividing: $\dfrac{24x^3y+12x^2y^2-16xy^3}{4xy}=\dfrac{24x^3y}{4xy}+\dfrac{12x^2y^2}{4xy}-\dfrac{16xy^3}{4xy}=6x^2+3xy-4y^2$

4. Dividing using long division:

$$
\begin{array}{r}
x^2-4x-2 \\
2x-1\overline{\smash{\big)}\,2x^3-9x^2+0x+10} \\
\underline{2x^3-\ \ x^2} \\
-8x^2+0x \\
\underline{-8x^2+4x} \\
-4x+10 \\
\underline{-4x+2} \\
8
\end{array}
$$

The quotient is $x^2-4x-2+\dfrac{8}{2x-1}$.

5. Performing the operations: $\dfrac{a^2-16}{5a-15}\cdot\dfrac{10(a-3)^2}{a^2-7a+12}=\dfrac{(a+4)(a-4)}{5(a-3)}\cdot\dfrac{10(a-3)^2}{(a-4)(a-3)}=2(a+4)$

6. Performing the operations:

$$\dfrac{a^4-81}{a^2+9}\div\dfrac{a^2-8a+15}{4a-20}=\dfrac{a^4-81}{a^2+9}\cdot\dfrac{4a-20}{a^2-8a+15}=\dfrac{\left(a^2+9\right)(a+3)(a-3)}{a^2+9}\cdot\dfrac{4(a-5)}{(a-5)(a-3)}=4(a+3)$$

7. Performing the operations:

$$\dfrac{x^3-8}{2x^2-9x+10}\div\dfrac{x^2+2x+4}{2x^2+x-15}=\dfrac{x^3-8}{2x^2-9x+10}\cdot\dfrac{2x^2+x-15}{x^2+2x+4}=\dfrac{(x-2)\left(x^2+2x+4\right)}{(2x-5)(x-2)}\cdot\dfrac{(2x-5)(x+3)}{x^2+2x+4}=x+3$$

8. Performing the operations: $\dfrac{4}{21}+\dfrac{6}{35}=\dfrac{4}{21}\cdot\dfrac{5}{5}+\dfrac{6}{35}\cdot\dfrac{3}{3}=\dfrac{20}{105}+\dfrac{18}{105}=\dfrac{38}{105}$

9. Performing the operations: $\dfrac{3}{4}-\dfrac{1}{2}+\dfrac{5}{8}=\dfrac{3}{4}\cdot\dfrac{2}{2}-\dfrac{1}{2}\cdot\dfrac{4}{4}+\dfrac{5}{8}=\dfrac{6}{8}-\dfrac{4}{8}+\dfrac{5}{8}=\dfrac{7}{8}$

10. Performing the operations: $\dfrac{a}{a^2-9}+\dfrac{3}{a^2-9}=\dfrac{a+3}{a^2-9}=\dfrac{a+3}{(a+3)(a-3)}=\dfrac{1}{a-3}$

11. Performing the operations: $\dfrac{1}{x}+\dfrac{2}{x-3}=\dfrac{1}{x}\cdot\dfrac{x-3}{x-3}+\dfrac{2}{x-3}\cdot\dfrac{x}{x}=\dfrac{x-3}{x(x-3)}+\dfrac{2x}{x(x-3)}=\dfrac{3x-3}{x(x-3)}=\dfrac{3(x-1)}{x(x-3)}$

12. Performing the operations:

$$\frac{4x}{x^2+6x+5}-\frac{3x}{x^2+5x+4}=\frac{4x}{(x+5)(x+1)}-\frac{3x}{(x+4)(x+1)}$$

$$=\frac{4x}{(x+5)(x+1)}\cdot\frac{x+4}{x+4}-\frac{3x}{(x+4)(x+1)}\cdot\frac{x+5}{x+5}$$

$$=\frac{4x^2+16x}{(x+5)(x+1)(x+4)}-\frac{3x^2+15x}{(x+5)(x+1)(x+4)}$$

$$=\frac{4x^2+16x-3x^2-15x}{(x+5)(x+1)(x+4)}$$

$$=\frac{x^2+x}{(x+5)(x+1)(x+4)}$$

$$=\frac{x(x+1)}{(x+5)(x+1)(x+4)}$$

$$=\frac{x}{(x+4)(x+5)}$$

13. Performing the operations:

$$\frac{2x+8}{x^2+4x+3}-\frac{x+4}{x^2+5x+6}=\frac{2x+8}{(x+3)(x+1)}-\frac{x+4}{(x+3)(x+2)}$$

$$=\frac{2x+8}{(x+3)(x+1)}\cdot\frac{x+2}{x+2}-\frac{x+4}{(x+3)(x+2)}\cdot\frac{x+1}{x+1}$$

$$=\frac{2x^2+12x+16}{(x+1)(x+2)(x+3)}-\frac{x^2+5x+4}{(x+1)(x+2)(x+3)}$$

$$=\frac{2x^2+12x+16-x^2-5x-4}{(x+1)(x+2)(x+3)}$$

$$=\frac{x^2+7x+12}{(x+1)(x+2)(x+3)}$$

$$=\frac{(x+3)(x+4)}{(x+1)(x+2)(x+3)}$$

$$=\frac{x+4}{(x+1)(x+2)}$$

14. Simplifying the complex fraction: $\dfrac{3-\dfrac{1}{a+3}}{3+\dfrac{1}{a+3}}=\dfrac{\left(3-\dfrac{1}{a+3}\right)(a+3)}{\left(3+\dfrac{1}{a+3}\right)(a+3)}=\dfrac{3(a+3)-1}{3(a+3)+1}=\dfrac{3a+9-1}{3a+9+1}=\dfrac{3a+8}{3a+10}$

15. Simplifying the complex fraction: $\dfrac{1-\dfrac{9}{x^2}}{1+\dfrac{1}{x}-\dfrac{6}{x^2}}=\dfrac{\left(1-\dfrac{9}{x^2}\right)\cdot x^2}{\left(1+\dfrac{1}{x}-\dfrac{6}{x^2}\right)\cdot x^2}=\dfrac{x^2-9}{x^2+x-6}=\dfrac{(x+3)(x-3)}{(x+3)(x-2)}=\dfrac{x-3}{x-2}$

16. Solving the equation:
$$\frac{1}{x} + 3 = \frac{4}{3}$$
$$3x\left(\frac{1}{x} + 3\right) = 3x\left(\frac{4}{3}\right)$$
$$3 + 9x = 4x$$
$$5x = -3$$
$$x = -\frac{3}{5}$$

17. Solving the equation:
$$\frac{x}{x-3} + 3 = \frac{3}{x-3}$$
$$(x-3)\left(\frac{x}{x-3} + 3\right) = (x-3)\left(\frac{3}{x-3}\right)$$
$$x + 3(x-3) = 3$$
$$x + 3x - 9 = 3$$
$$4x = 12$$
$$x = 3 \quad (x = 3 \text{ does not check})$$
There is no solution (3 does not check).

18. Solving the equation:
$$\frac{y+3}{2y} + \frac{5}{y-1} = \frac{1}{2}$$
$$2y(y-1)\left(\frac{y+3}{2y} + \frac{5}{y-1}\right) = 2y(y-1)\left(\frac{1}{2}\right)$$
$$(y-1)(y+3) + 10y = y(y-1)$$
$$y^2 + 2y - 3 + 10y = y^2 - y$$
$$12y - 3 = -y$$
$$13y = 3$$
$$y = \frac{3}{13}$$

19. Solving the equation:
$$1 - \frac{1}{x} = \frac{6}{x^2}$$
$$x^2\left(1 - \frac{1}{x}\right) = x^2\left(\frac{6}{x^2}\right)$$
$$x^2 - x = 6$$
$$x^2 - x - 6 = 0$$
$$(x+2)(x-3) = 0$$
$$x = -2, 3$$

20. Let x represent the number. The equation is:
$$\frac{10}{23-x} = \frac{1}{3}$$
$$30 = 23 - x$$
$$x = -7$$
The number is -7.

21. Let x represent the speed of the boat. Since the total time is 3 hours:
$$\frac{8}{x-2} + \frac{8}{x+2} = 3$$
$$3(x-2)(x+2)\left(\frac{8}{x-2} + \frac{8}{x+2}\right) = 3(x-2)(x+2) \cdot 3$$
$$24(x+2) + 24(x-2) = 9\left(x^2 - 4\right)$$
$$48x = 9x^2 - 36$$
$$0 = 9x^2 - 48x - 36$$
$$0 = 3\left(3x^2 - 16x - 12\right)$$
$$0 = 3(3x+2)(x-6)$$
$$x = 6 \quad \left(x = -\frac{2}{3} \text{ is impossible}\right)$$

The speed of the boat is 6 mph.

22. Let x represent the time to fill the pool with both the pipe and drain open. The rate equation is:

$$\frac{1}{10} - \frac{1}{15} = \frac{1}{2} \cdot \frac{1}{x}$$

$$30x\left(\frac{1}{10} - \frac{1}{15}\right) = 30x\left(\frac{1}{2x}\right)$$

$$3x - 2x = 15$$

$$x = 15$$

The pool can be filled in 15 hours with both the pipe and drain open.

23. Converting the height: $14,494 \text{ feet} \cdot \dfrac{1 \text{ mile}}{5280 \text{ feet}} \approx 2.7 \text{ miles}$

24. Converting the speed: $\dfrac{4,750 \text{ feet}}{3.2 \text{ seconds}} \cdot \dfrac{1 \text{ mile}}{5280 \text{ feet}} \cdot \dfrac{60 \text{ seconds}}{1 \text{ minute}} \cdot \dfrac{60 \text{ minutes}}{1 \text{ hour}} \approx 1012 \text{ miles per hour}$

Chapter 7
Rational Exponents and Roots

7.1 Rational Exponents

1. Finding the root: $\sqrt{144} = 12$

3. Finding the root: $\sqrt{-144}$ is not a real number

5. Finding the root: $-\sqrt{49} = -7$

7. Finding the root: $\sqrt[3]{-27} = -3$

9. Finding the root: $\sqrt[4]{16} = 2$

11. Finding the root: $\sqrt[4]{-16}$ is not a real number

13. Finding the root: $\sqrt{0.04} = 0.2$

15. Finding the root: $\sqrt[3]{0.008} = 0.2$

17. Simplifying: $\sqrt{36a^8} = 6a^4$

19. Simplifying: $\sqrt[3]{27a^{12}} = 3a^4$

21. Simplifying: $\sqrt[5]{32x^{10}y^5} = 2x^2 y$

23. Simplifying: $\sqrt[4]{16a^{12}b^{20}} = 2a^3 b^5$

25. Writing as a root and simplifying: $36^{1/2} = \sqrt{36} = 6$

27. Writing as a root and simplifying: $-9^{1/2} = -\sqrt{9} = -3$

29. Writing as a root and simplifying: $8^{1/3} = \sqrt[3]{8} = 2$

31. Writing as a root and simplifying: $(-8)^{1/3} = \sqrt[3]{-8} = -2$

33. Writing as a root and simplifying: $32^{1/5} = \sqrt[5]{32} = 2$

35. Writing as a root and simplifying: $\left(\frac{81}{25}\right)^{1/2} = \sqrt{\frac{81}{25}} = \frac{9}{5}$

37. Simplifying: $27^{2/3} = \left(27^{1/3}\right)^2 = 3^2 = 9$

39. Simplifying: $25^{3/2} = \left(25^{1/2}\right)^3 = 5^3 = 125$

41. Simplifying: $27^{-1/3} = \left(27^{1/3}\right)^{-1} = 3^{-1} = \frac{1}{3}$

43. Simplifying: $81^{-3/4} = \left(81^{1/4}\right)^{-3} = 3^{-3} = \frac{1}{3^3} = \frac{1}{27}$

45. Simplifying: $\left(\frac{25}{36}\right)^{-1/2} = \left(\frac{36}{25}\right)^{1/2} = \frac{6}{5}$

47. Simplifying: $\left(\frac{81}{16}\right)^{-3/4} = \left(\frac{16}{81}\right)^{3/4} = \left[\left(\frac{16}{81}\right)^{1/4}\right]^3 = \left(\frac{2}{3}\right)^3 = \frac{8}{27}$

49. Simplifying: $16^{1/2} + 27^{1/3} = 4 + 3 = 7$

51. Simplifying: $8^{-2/3} + 4^{-1/2} = \left(8^{1/3}\right)^{-2} + \left(4^{1/2}\right)^{-1} = 2^{-2} + 2^{-1} = \frac{1}{4} + \frac{1}{2} = \frac{3}{4}$

53. Using properties of exponents: $x^{3/5} \cdot x^{1/5} = x^{3/5+1/5} = x^{4/5}$

55. Using properties of exponents: $\left(a^{3/4}\right)^{4/3} = a^{3/4 \cdot 4/3} = a$

57. Using properties of exponents: $\dfrac{x^{1/5}}{x^{3/5}} = x^{1/5-3/5} = x^{-2/5} = \dfrac{1}{x^{2/5}}$

59. Using properties of exponents: $\dfrac{x^{5/6}}{x^{2/3}} = x^{5/6-2/3} = x^{5/6-4/6} = x^{1/6}$

61. Using properties of exponents: $\left(x^{3/5}y^{5/6}z^{1/3}\right)^{3/5} = x^{3/5\cdot3/5}y^{5/6\cdot3/5}z^{1/3\cdot3/5} = x^{9/25}y^{1/2}z^{1/5}$

63. Using properties of exponents: $\dfrac{a^{3/4}b^2}{a^{7/8}b^{1/4}} = a^{3/4-7/8}b^{2-1/4} = a^{6/8-7/8}b^{8/4-1/4} = a^{-1/8}b^{7/4} = \dfrac{b^{7/4}}{a^{1/8}}$

65. Using properties of exponents: $\dfrac{\left(y^{2/3}\right)^{3/4}}{\left(y^{1/3}\right)^{3/5}} = \dfrac{y^{1/2}}{y^{1/5}} = y^{1/2-1/5} = y^{5/10-2/10} = y^{3/10}$

67. Using properties of exponents: $\left(\dfrac{a^{-1/4}}{b^{1/2}}\right)^8 = \dfrac{a^{-1/4\cdot8}}{b^{1/2\cdot8}} = \dfrac{a^{-2}}{b^4} = \dfrac{1}{a^2b^4}$

69. Using properties of exponents: $\dfrac{\left(r^{-2}s^{1/3}\right)^6}{r^8s^{3/2}} = \dfrac{r^{-12}s^2}{r^8s^{3/2}} = r^{-12-8}s^{2-3/2} = r^{-20}s^{1/2} = \dfrac{s^{1/2}}{r^{20}}$

71. Using properties of exponents: $\dfrac{\left(25a^6b^4\right)^{1/2}}{\left(8a^{-9}b^3\right)^{-1/3}} = \dfrac{25^{1/2}a^3b^2}{8^{-1/3}a^3b^{-1}} = \dfrac{5}{1/2}a^{3-3}b^{2+1} = 10b^3$

73. Substituting $r = 250$: $v = \left(\dfrac{5\cdot250}{2}\right)^{1/2} = 625^{1/2} = 25$. The maximum speed is 25 mph.

75. **a.**   The length of the side is $126 + 86 + 86 + 126 = 424$ pm
     b.   Let d represent the diagonal. Using the Pythagorean theorem:
$$d^2 = 424^2 + 424^2 = 359552$$
$$d = \sqrt{359552} \approx 600 \text{ pm}$$
     c.   Converting to meters: $600 \text{ pm} \cdot \dfrac{1 \text{ m}}{10^{12} \text{ pm}} = 6 \times 10^{-10}$ m

77. **a.**   This graph is B.   **b.**   This graph is A.
     c.   This graph is C.   **d.**   The points of intersection are (0,0) and (1,1).

79. Multiplying: $x^2\left(x^4 - x\right) = x^2 \cdot x^4 - x^2 \cdot x = x^6 - x^3$

81. Multiplying: $(x - 3)(x + 5) = x^2 + 5x - 3x - 15 = x^2 + 2x - 15$

83. Multiplying: $\left(x^2 - 5\right)^2 = \left(x^2\right)^2 - 2\left(x^2\right)(5) + 5^2 = x^4 - 10x^2 + 25$

85. Multiplying: $(x - 3)\left(x^2 + 3x + 9\right) = x^3 + 3x^2 + 9x - 3x^2 - 9x - 27 = x^3 - 27$

87. Simplifying each expression:
$$\left(9^{1/2} + 4^{1/2}\right)^2 = (3 + 2)^2 = 5^2 = 25$$
$$9 + 4 = 13$$
     Note that the values are not equal.

89. Rewriting with exponents: $\sqrt{\sqrt{a}} = \sqrt{a^{1/2}} = \left(a^{1/2}\right)^{1/2} = a^{1/4} = \sqrt[4]{a}$

7.2 More Expressions Involving Rational Exponents

1. Multiplying: $x^{2/3}\left(x^{1/3} + x^{4/3}\right) = x^{2/3} \cdot x^{1/3} + x^{2/3} \cdot x^{4/3} = x + x^2$

3. Multiplying: $a^{1/2}\left(a^{3/2} - a^{1/2}\right) = a^{1/2} \cdot a^{3/2} - a^{1/2} \cdot a^{1/2} = a^2 - a$

5. Multiplying: $2x^{1/3}\left(3x^{8/3} - 4x^{5/3} + 5x^{2/3}\right) = 2x^{1/3} \cdot 3x^{8/3} - 2x^{1/3} \cdot 4x^{5/3} + 2x^{1/3} \cdot 5x^{2/3} = 6x^3 - 8x^2 + 10x$

7. Multiplying:

$$4x^{1/2}y^{3/5}\left(3x^{3/2}y^{-3/5} - 9x^{-1/2}y^{7/5}\right) = 4x^{1/2}y^{3/5} \cdot 3x^{3/2}y^{-3/5} - 4x^{1/2}y^{3/5} \cdot 9x^{-1/2}y^{7/5} = 12x^2 - 36y^2$$

9. Multiplying: $\left(x^{2/3} - 4\right)\left(x^{2/3} + 2\right) = x^{2/3} \cdot x^{2/3} + 2x^{2/3} - 4x^{2/3} - 8 = x^{4/3} - 2x^{2/3} - 8$

11. Multiplying: $\left(a^{1/2} - 3\right)\left(a^{1/2} - 7\right) = a^{1/2} \cdot a^{1/2} - 7a^{1/2} - 3a^{1/2} + 21 = a - 10a^{1/2} + 21$

13. Multiplying: $\left(4y^{1/3} - 3\right)\left(5y^{1/3} + 2\right) = 20y^{2/3} + 8y^{1/3} - 15y^{1/3} - 6 = 20y^{2/3} - 7y^{1/3} - 6$

15. Multiplying: $\left(5x^{2/3} + 3y^{1/2}\right)\left(2x^{2/3} + 3y^{1/2}\right) = 10x^{4/3} + 15x^{2/3}y^{1/2} + 6x^{2/3}y^{1/2} + 9y = 10x^{4/3} + 21x^{2/3}y^{1/2} + 9y$

17. Multiplying: $\left(t^{1/2} + 5\right)^2 = \left(t^{1/2} + 5\right)\left(t^{1/2} + 5\right) = t + 5t^{1/2} + 5t^{1/2} + 25 = t + 10t^{1/2} + 25$

19. Multiplying: $\left(x^{3/2} + 4\right)^2 = \left(x^{3/2} + 4\right)\left(x^{3/2} + 4\right) = x^3 + 4x^{3/2} + 4x^{3/2} + 16 = x^3 + 8x^{3/2} + 16$

21. Multiplying: $\left(a^{1/2} - b^{1/2}\right)^2 = \left(a^{1/2} - b^{1/2}\right)\left(a^{1/2} - b^{1/2}\right) = a - a^{1/2}b^{1/2} - a^{1/2}b^{1/2} + b = a - 2a^{1/2}b^{1/2} + b$

23. Multiplying:

$$\left(2x^{1/2} - 3y^{1/2}\right)^2 = \left(2x^{1/2} - 3y^{1/2}\right)\left(2x^{1/2} - 3y^{1/2}\right)$$
$$= 4x - 6x^{1/2}y^{1/2} - 6x^{1/2}y^{1/2} + 9y$$
$$= 4x - 12x^{1/2}y^{1/2} + 9y$$

25. Multiplying: $\left(a^{1/2} - 3^{1/2}\right)\left(a^{1/2} + 3^{1/2}\right) = \left(a^{1/2}\right)^2 - \left(3^{1/2}\right)^2 = a - 3$

27. Multiplying: $\left(x^{3/2} + y^{3/2}\right)\left(x^{3/2} - y^{3/2}\right) = \left(x^{3/2}\right)^2 - \left(y^{3/2}\right)^2 = x^3 - y^3$

29. Multiplying: $\left(t^{1/2} - 2^{3/2}\right)\left(t^{1/2} + 2^{3/2}\right) = \left(t^{1/2}\right)^2 - \left(2^{3/2}\right)^2 = t - 2^3 = t - 8$

31. Multiplying: $\left(2x^{3/2} + 3^{1/2}\right)\left(2x^{3/2} - 3^{1/2}\right) = \left(2x^{3/2}\right)^2 - \left(3^{1/2}\right)^2 = 4x^3 - 3$

33. Multiplying: $\left(x^{1/3} + y^{1/3}\right)\left(x^{2/3} - x^{1/3}y^{1/3} + y^{2/3}\right) = \left(x^{1/3}\right)^3 + \left(y^{1/3}\right)^3 = x + y$

35. Multiplying: $\left(a^{1/3} - 2\right)\left(a^{2/3} + 2a^{1/3} + 4\right) = \left(a^{1/3}\right)^3 - (2)^3 = a - 8$

37. Multiplying: $\left(2x^{1/3} + 1\right)\left(4x^{2/3} - 2x^{1/3} + 1\right) = \left(2x^{1/3}\right)^3 + (1)^3 = 8x + 1$

39. Multiplying: $\left(t^{1/4} - 1\right)\left(t^{1/4} + 1\right)\left(t^{1/2} + 1\right) = \left(t^{1/2} - 1\right)\left(t^{1/2} + 1\right) = t - 1$

41. Dividing: $\dfrac{18x^{3/4} + 27x^{1/4}}{9x^{1/4}} = \dfrac{18x^{3/4}}{9x^{1/4}} + \dfrac{27x^{1/4}}{9x^{1/4}} = 2x^{1/2} + 3$

43. Dividing: $\dfrac{12x^{2/3}y^{1/3} - 16x^{1/3}y^{2/3}}{4x^{1/3}y^{1/3}} = \dfrac{12x^{2/3}y^{1/3}}{4x^{1/3}y^{1/3}} - \dfrac{16x^{1/3}y^{2/3}}{4x^{1/3}y^{1/3}} = 3x^{1/3} - 4y^{1/3}$

45. Dividing: $\dfrac{21a^{7/5}b^{3/5} - 14a^{2/5}b^{8/5}}{7a^{2/5}b^{3/5}} = \dfrac{21a^{7/5}b^{3/5}}{7a^{2/5}b^{3/5}} - \dfrac{14a^{2/5}b^{8/5}}{7a^{2/5}b^{3/5}} = 3a - 2b$

47. Factoring: $12(x-2)^{3/2} - 9(x-2)^{1/2} = 3(x-2)^{1/2}\left[4(x-2) - 3\right] = 3(x-2)^{1/2}(4x - 8 - 3) = 3(x-2)^{1/2}(4x - 11)$

49. Factoring: $5(x-3)^{12/5} - 15(x-3)^{7/5} = 5(x-3)^{7/5}\left[(x-3) - 3\right] = 5(x-3)^{7/5}(x - 6)$

51. Factoring: $9x(x+1)^{3/2} + 6(x+1)^{1/2} = 3(x+1)^{1/2}\left[3x(x+1) + 2\right] = 3(x+1)^{1/2}\left(3x^2 + 3x + 2\right)$

53. Factoring: $x^{2/3} - 5x^{1/3} + 6 = \left(x^{1/3} - 2\right)\left(x^{1/3} - 3\right)$ **55.** Factoring: $a^{2/5} - 2a^{1/5} - 8 = \left(a^{1/5} - 4\right)\left(a^{1/5} + 2\right)$

57. Factoring: $2y^{2/3} - 5y^{1/3} - 3 = \left(2y^{1/3} + 1\right)\left(y^{1/3} - 3\right)$ **59.** Factoring: $9t^{2/5} - 25 = \left(3t^{1/5} + 5\right)\left(3t^{1/5} - 5\right)$

61. Factoring: $4x^{2/7} + 20x^{1/7} + 25 = \left(2x^{1/7} + 5\right)^2$

63. Writing as a single fraction: $\dfrac{3}{x^{1/2}} + x^{1/2} = \dfrac{3}{x^{1/2}} + x^{1/2} \cdot \dfrac{x^{1/2}}{x^{1/2}} = \dfrac{3+x}{x^{1/2}}$

65. Writing as a single fraction: $x^{2/3} + \dfrac{5}{x^{1/3}} = x^{2/3} \cdot \dfrac{x^{1/3}}{x^{1/3}} + \dfrac{5}{x^{1/3}} = \dfrac{x+5}{x^{1/3}}$

67. Writing as a single fraction:

$$\frac{3x^2}{\left(x^3 + 1\right)^{1/2}} + \left(x^3 + 1\right)^{1/2} = \frac{3x^2}{\left(x^3 + 1\right)^{1/2}} + \left(x^3 + 1\right)^{1/2} \cdot \frac{\left(x^3 + 1\right)^{1/2}}{\left(x^3 + 1\right)^{1/2}} = \frac{3x^2 + x^3 + 1}{\left(x^3 + 1\right)^{1/2}} = \frac{x^3 + 3x^2 + 1}{\left(x^3 + 1\right)^{1/2}}$$

69. Writing as a single fraction:

$$\frac{x^2}{\left(x^2 + 4\right)^{1/2}} - \left(x^2 + 4\right)^{1/2} = \frac{x^2}{\left(x^2 + 4\right)^{1/2}} - \left(x^2 + 4\right)^{1/2} \cdot \frac{\left(x^2 + 4\right)^{1/2}}{\left(x^2 + 4\right)^{1/2}}$$

$$= \frac{x^2 - \left(x^2 + 4\right)}{\left(x^2 + 4\right)^{1/2}}$$

$$= \frac{x^2 - x^2 - 4}{\left(x^2 + 4\right)^{1/2}}$$

$$= \frac{-4}{\left(x^2 + 4\right)^{1/2}}$$

71. Using the formula $r = \left(\dfrac{A}{P}\right)^{1/t} - 1$ to find the annual rate of return: $r = \left(\dfrac{900}{500}\right)^{1/4} - 1 \approx 0.158$

The annual rate of return is approximately 15.8%.

73. Substituting $k = 0.0408$ and $d = 240$: $v = \left(\dfrac{240}{0.0408}\right)^{1/2} \approx 76.7$ mph

75. Reducing to lowest terms: $\dfrac{x^2 - 9}{x^4 - 81} = \dfrac{x^2 - 9}{\left(x^2 + 9\right)\left(x^2 - 9\right)} = \dfrac{1}{x^2 + 9}$

77. Dividing: $\dfrac{15x^2 y - 20x^4 y^2}{5xy} = \dfrac{15x^2 y}{5xy} - \dfrac{20x^4 y^2}{5xy} = 3x - 4x^3 y$

79. Dividing using long division:

$$\begin{array}{r}
5x - 4 \\
2x + 3 \overline{\smash{)}10x^2 + 7x - 12} \\
\underline{10x^2 + 15x} \\
-8x - 12 \\
\underline{-8x - 12} \\
0
\end{array}$$

81. Dividing using long division:

$$\begin{array}{r}
x^2 + 5x + 25 \\
x - 5 \overline{\smash{)}x^3 + 0x^2 + 0x - 125} \\
\underline{x^3 - 5x^2} \\
5x^2 + 0x \\
\underline{5x^2 - 25x} \\
25x - 125 \\
\underline{25x - 125} \\
0
\end{array}$$

7.3   Simplified Form for Radicals

1. Simplifying the radical: $\sqrt{8} = \sqrt{4 \cdot 2} = 2\sqrt{2}$

3. Simplifying the radical: $\sqrt{98} = \sqrt{49 \cdot 2} = 7\sqrt{2}$

5. Simplifying the radical: $\sqrt{288} = \sqrt{144 \cdot 2} = 12\sqrt{2}$

7. Simplifying the radical: $\sqrt{80} = \sqrt{16 \cdot 5} = 4\sqrt{5}$

9. Simplifying the radical: $\sqrt{48} = \sqrt{16 \cdot 3} = 4\sqrt{3}$

11. Simplifying the radical: $\sqrt{675} = \sqrt{225 \cdot 3} = 15\sqrt{3}$

13. Simplifying the radical: $\sqrt[3]{54} = \sqrt[3]{27 \cdot 2} = 3\sqrt[3]{2}$

15. Simplifying the radical: $\sqrt[3]{128} = \sqrt[3]{64 \cdot 2} = 4\sqrt[3]{2}$

17. Simplifying the radical: $\sqrt[3]{432} = \sqrt[3]{216 \cdot 2} = 6\sqrt[3]{2}$

19. Simplifying the radical: $\sqrt[5]{64} = \sqrt[5]{32 \cdot 2} = 2\sqrt[5]{2}$

21. Simplifying the radical: $\sqrt{18x^3} = \sqrt{9x^2 \cdot 2x} = 3x\sqrt{2x}$

23. Simplifying the radical: $\sqrt[4]{32y^7} = \sqrt[4]{16y^4 \cdot 2y^3} = 2y\sqrt[4]{2y^3}$

25. Simplifying the radical: $\sqrt[3]{40x^4y^7} = \sqrt[3]{8x^3y^6 \cdot 5xy} = 2xy^2\sqrt[3]{5xy}$

27. Simplifying the radical: $\sqrt{48a^2b^3c^4} = \sqrt{16a^2b^2c^4 \cdot 3b} = 4abc^2\sqrt{3b}$

29. Simplifying the radical: $\sqrt[3]{48a^2b^3c^4} = \sqrt[3]{8b^3c^3 \cdot 6a^2c} = 2bc\sqrt[3]{6a^2c}$

31. Simplifying the radical: $\sqrt[5]{64x^8y^{12}} = \sqrt[5]{32x^5y^{10} \cdot 2x^3y^2} = 2xy^2\sqrt[5]{2x^3y^2}$

33. Simplifying the radical: $\sqrt[5]{243x^7y^{10}z^5} = \sqrt[5]{243x^5y^{10}z^5 \cdot x^2} = 3xy^2z\sqrt[5]{x^2}$

35. Substituting into the expression: $\sqrt{b^2 - 4ac} = \sqrt{(-6)^2 - 4(2)(3)} = \sqrt{36 - 24} = \sqrt{12} = 2\sqrt{3}$

37. Substituting into the expression: $\sqrt{b^2 - 4ac} = \sqrt{(2)^2 - 4(1)(6)} = \sqrt{4 - 24} = \sqrt{-20}$, which is not a real number

39. Substituting into the expression: $\sqrt{b^2 - 4ac} = \sqrt{\left(-\frac{1}{2}\right)^2 - 4\left(\frac{1}{2}\right)\left(-\frac{5}{4}\right)} = \sqrt{\frac{1}{4} + \frac{5}{2}} = \sqrt{\frac{11}{4}} = \frac{\sqrt{11}}{2}$

41. Rationalizing the denominator: $\dfrac{2}{\sqrt{3}} = \dfrac{2}{\sqrt{3}} \cdot \dfrac{\sqrt{3}}{\sqrt{3}} = \dfrac{2\sqrt{3}}{3}$

43. Rationalizing the denominator: $\dfrac{5}{\sqrt{6}} = \dfrac{5}{\sqrt{6}} \cdot \dfrac{\sqrt{6}}{\sqrt{6}} = \dfrac{5\sqrt{6}}{6}$

45. Rationalizing the denominator: $\sqrt{\dfrac{1}{2}} = \dfrac{1}{\sqrt{2}} \cdot \dfrac{\sqrt{2}}{\sqrt{2}} = \dfrac{\sqrt{2}}{2}$

47. Rationalizing the denominator: $\sqrt{\dfrac{1}{5}} = \dfrac{1}{\sqrt{5}} \cdot \dfrac{\sqrt{5}}{\sqrt{5}} = \dfrac{\sqrt{5}}{5}$

49. Rationalizing the denominator: $\dfrac{4}{\sqrt[3]{2}} = \dfrac{4}{\sqrt[3]{2}} \cdot \dfrac{\sqrt[3]{4}}{\sqrt[3]{4}} = \dfrac{4\sqrt[3]{4}}{2} = 2\sqrt[3]{4}$

51. Rationalizing the denominator: $\dfrac{2}{\sqrt[3]{9}} = \dfrac{2}{\sqrt[3]{9}} \cdot \dfrac{\sqrt[3]{3}}{\sqrt[3]{3}} = \dfrac{2\sqrt[3]{3}}{3}$

53. Rationalizing the denominator: $\sqrt[4]{\dfrac{3}{2x^2}} = \dfrac{\sqrt[4]{3}}{\sqrt[4]{2x^2}} \cdot \dfrac{\sqrt[4]{8x^2}}{\sqrt[4]{8x^2}} = \dfrac{\sqrt[4]{24x^2}}{2x}$

55. Rationalizing the denominator: $\sqrt[4]{\dfrac{8}{y}} = \dfrac{\sqrt[4]{8}}{\sqrt[4]{y}} \cdot \dfrac{\sqrt[4]{y^3}}{\sqrt[4]{y^3}} = \dfrac{\sqrt[4]{8y^3}}{y}$

57. Rationalizing the denominator: $\sqrt[3]{\dfrac{4x}{3y}} = \dfrac{\sqrt[3]{4x}}{\sqrt[3]{3y}} \cdot \dfrac{\sqrt[3]{9y^2}}{\sqrt[3]{9y^2}} = \dfrac{\sqrt[3]{36xy^2}}{3y}$

59. Rationalizing the denominator: $\sqrt[3]{\dfrac{2x}{9y}} = \dfrac{\sqrt[3]{2x}}{\sqrt[3]{9y}} \cdot \dfrac{\sqrt[3]{3y^2}}{\sqrt[3]{3y^2}} = \dfrac{\sqrt[3]{6xy^2}}{3y}$

61. Rationalizing the denominator: $\sqrt[4]{\dfrac{1}{8x^3}} = \dfrac{1}{\sqrt[4]{8x^3}} \cdot \dfrac{\sqrt[4]{2x}}{\sqrt[4]{2x}} = \dfrac{\sqrt[4]{2x}}{2x}$

63. Simplifying: $\sqrt{\dfrac{27x^3}{5y}} = \dfrac{\sqrt{27x^3}}{\sqrt{5y}} \cdot \dfrac{\sqrt{5y}}{\sqrt{5y}} = \dfrac{\sqrt{135x^3y}}{5y} = \dfrac{3x\sqrt{15xy}}{5y}$

65. Simplifying: $\sqrt{\dfrac{75x^3y^2}{2z}} = \dfrac{\sqrt{75x^3y^2}}{\sqrt{2z}} \cdot \dfrac{\sqrt{2z}}{\sqrt{2z}} = \dfrac{\sqrt{150x^3y^2z}}{2z} = \dfrac{5xy\sqrt{6xz}}{2z}$

67. Simplifying: $\sqrt[3]{\dfrac{16a^4b^3}{9c}} = \dfrac{\sqrt[3]{16a^4b^3}}{\sqrt[3]{9c}} \cdot \dfrac{\sqrt[3]{3c^2}}{\sqrt[3]{3c^2}} = \dfrac{\sqrt[3]{48a^4b^3c^2}}{3c} = \dfrac{2ab\sqrt[3]{6ac^2}}{3c}$

69. Simplifying: $\sqrt[3]{\dfrac{8x^3y^6}{9z}} = \dfrac{\sqrt[3]{8x^3y^6}}{\sqrt[3]{9z}} \cdot \dfrac{\sqrt[3]{3z^2}}{\sqrt[3]{3z^2}} = \dfrac{\sqrt[3]{24x^3y^6z^2}}{3z} = \dfrac{2xy^2\sqrt[3]{3z^2}}{3z}$

71. Simplifying: $\sqrt{25x^2} = 5|x|$ **73.** Simplifying: $\sqrt{27x^3y^2} = \sqrt{9x^2y^2 \cdot 3x} = 3|xy|\sqrt{3x}$

75. Simplifying: $\sqrt{x^2 - 10x + 25} = \sqrt{(x-5)^2} = |x-5|$ **77.** Simplifying: $\sqrt{4x^2 + 12x + 9} = \sqrt{(2x+3)^2} = |2x+3|$

79. Simplifying: $\sqrt{4a^4 + 16a^3 + 16a^2} = \sqrt{4a^2(a^2 + 4a + 4)} = \sqrt{4a^2(a+2)^2} = 2|a(a+2)|$

81. Simplifying: $\sqrt{4x^3 - 8x^2} = \sqrt{4x^2(x-2)} = 2|x|\sqrt{x-2}$

83. Substituting $a = 9$ and $b = 16$:
$$\sqrt{a+b} = \sqrt{9+16} = \sqrt{25} = 5$$
$$\sqrt{a} + \sqrt{b} = \sqrt{9} + \sqrt{16} = 3 + 4 = 7$$
Thus $\sqrt{a+b} \neq \sqrt{a} + \sqrt{b}$.

85. Substituting $w = 10$ and $l = 15$: $d = \sqrt{l^2 + w^2} = \sqrt{15^2 + 10^2} = \sqrt{225 + 100} = \sqrt{325} = \sqrt{25 \cdot 13} = 5\sqrt{13}$ feet

87. **a.** Substituting $w = 3$, $l = 4$, and $h = 12$:
$$d = \sqrt{l^2 + w^2 + h^2} = \sqrt{4^2 + 3^2 + 12^2} = \sqrt{16 + 9 + 144} = \sqrt{169} = 13 \text{ feet}$$

 b. Substituting $w = 2$, $l = 4$, and $h = 6$:
$$d = \sqrt{l^2 + w^2 + h^2} = \sqrt{4^2 + 2^2 + 6^2} = \sqrt{16 + 4 + 36} = \sqrt{56} = 2\sqrt{14} \approx 7.5 \text{ feet}$$

89. Answers will vary.

91. Finding the first six terms:
$$f(1) = \sqrt{1^2 + 1} = \sqrt{2}$$
$$f(f(1)) = \sqrt{(\sqrt{2})^2 + 1} = \sqrt{2+1} = \sqrt{3}$$
$$f(f(f(1))) = \sqrt{(\sqrt{3})^2 + 1} = \sqrt{3+1} = 2$$
$$f(f(f(f(1)))) = \sqrt{(2)^2 + 1} = \sqrt{4+1} = \sqrt{5}$$
$$f(f(f(f(f(1))))) = \sqrt{(\sqrt{5})^2 + 1} = \sqrt{5+1} = \sqrt{6}$$
$$f\Big(f\big(f(f(f(f(1))))\big)\Big) = \sqrt{(\sqrt{6})^2 + 1} = \sqrt{6+1} = \sqrt{7}$$
Following this pattern, the 10th term will be $\sqrt{11}$ and the 100th term will be $\sqrt{101}$.

93. Performing the operations: $\dfrac{8xy^3}{9x^2y} \div \dfrac{16x^2y^2}{18xy^3} = \dfrac{8xy^3}{9x^2y} \cdot \dfrac{18xy^3}{16x^2y^2} = \dfrac{144x^2y^6}{144x^4y^3} = \dfrac{y^3}{x^2}$

95. Performing the operations: $\dfrac{12a^2 - 4a - 5}{2a+1} \cdot \dfrac{7a+3}{42a^2 - 17a - 15} = \dfrac{(6a-5)(2a+1)}{2a+1} \cdot \dfrac{7a+3}{(7a+3)(6a-5)} = 1$

97. Performing the operations:

$$\frac{8x^3+27}{27x^3+1} \div \frac{6x^2+7x-3}{9x^2-1} = \frac{8x^3+27}{27x^3+1} \cdot \frac{9x^2-1}{6x^2+7x-3}$$

$$= \frac{(2x+3)(4x^2-6x+9)}{(3x+1)(9x^2-3x+1)} \cdot \frac{(3x+1)(3x-1)}{(2x+3)(3x-1)}$$

$$= \frac{4x^2-6x+9}{9x^2-3x+1}$$

99. The prime factorization of 8,640 is: $8640 = 2^6 \cdot 3^3 \cdot 5$. Therefore: $\sqrt[3]{8640} = \sqrt[3]{2^6 \cdot 3^3 \cdot 5} = 2^2 \cdot 3\sqrt[3]{5} = 12\sqrt[3]{5}$

101. The prime factorization of 10,584 is: $10584 = 2^3 \cdot 3^3 \cdot 7^2$. Therefore: $\sqrt[3]{10584} = \sqrt[3]{2^3 \cdot 3^3 \cdot 7^2} = 2 \cdot 3\sqrt[3]{7^2} = 6\sqrt[3]{49}$

103. Rationalizing the denominator: $\dfrac{1}{\sqrt[10]{a^3}} = \dfrac{1}{\sqrt[10]{a^3}} \cdot \dfrac{\sqrt[10]{a^7}}{\sqrt[10]{a^7}} = \dfrac{\sqrt[10]{a^7}}{a}$

105. Rationalizing the denominator: $\dfrac{1}{\sqrt[20]{a^{11}}} = \dfrac{1}{\sqrt[20]{a^{11}}} \cdot \dfrac{\sqrt[20]{a^9}}{\sqrt[20]{a^9}} = \dfrac{\sqrt[20]{a^9}}{a}$

107. Graphing each function:

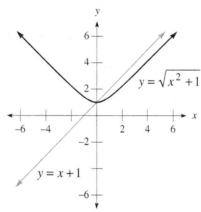

Note that the two graphs are not the same.

109. When $x = 2$, the distance apart is: $(2+1) - \sqrt{2^2+1} = 3 - \sqrt{5} \approx 0.76$

111. The two expressions are equal when $x = 0$.

7.4 Addition and Subtraction of Radical Expressions

1. Combining radicals: $3\sqrt{5} + 4\sqrt{5} = 7\sqrt{5}$

3. Combining radicals: $3x\sqrt{7} - 4x\sqrt{7} = -x\sqrt{7}$

5. Combining radicals: $5\sqrt[3]{10} - 4\sqrt[3]{10} = \sqrt[3]{10}$

7. Combining radicals: $8\sqrt[5]{6} - 2\sqrt[5]{6} + 3\sqrt[5]{6} = 9\sqrt[5]{6}$

9. Combining radicals: $3x\sqrt{2} - 4x\sqrt{2} + x\sqrt{2} = 0$

11. Combining radicals: $\sqrt{20} - \sqrt{80} + \sqrt{45} = 2\sqrt{5} - 4\sqrt{5} + 3\sqrt{5} = \sqrt{5}$

13. Combining radicals: $4\sqrt{8} - 2\sqrt{50} - 5\sqrt{72} = 8\sqrt{2} - 10\sqrt{2} - 30\sqrt{2} = -32\sqrt{2}$

15. Combining radicals: $5x\sqrt{8} + 3\sqrt{32x^2} - 5\sqrt{50x^2} = 10x\sqrt{2} + 12x\sqrt{2} - 25x\sqrt{2} = -3x\sqrt{2}$

17. Combining radicals: $5\sqrt[3]{16} - 4\sqrt[3]{54} = 10\sqrt[3]{2} - 12\sqrt[3]{2} = -2\sqrt[3]{2}$

19. Combining radicals: $\sqrt[3]{x^4y^2} + 7x\sqrt[3]{xy^2} = x\sqrt[3]{xy^2} + 7x\sqrt[3]{xy^2} = 8x\sqrt[3]{xy^2}$

21. Combining radicals: $5a^2\sqrt{27ab^3} - 6b\sqrt{12a^5b} = 15a^2b\sqrt{3ab} - 12a^2b\sqrt{3ab} = 3a^2b\sqrt{3ab}$

23. Combining radicals: $b\sqrt[3]{24a^5b} + 3a\sqrt[3]{81a^2b^4} = 2ab\sqrt[3]{3a^2b} + 9ab\sqrt[3]{3a^2b} = 11ab\sqrt[3]{3a^2b}$

25. Combining radicals: $5x\sqrt[4]{3y^5} + y\sqrt[4]{243x^4y} + \sqrt[4]{48x^4y^5} = 5xy\sqrt[4]{3y} + 3xy\sqrt[4]{3y} + 2xy\sqrt[4]{3y} = 10xy\sqrt[4]{3y}$

27. Combining radicals: $\dfrac{\sqrt{2}}{2}+\dfrac{1}{\sqrt{2}}=\dfrac{\sqrt{2}}{2}+\dfrac{1}{\sqrt{2}}\cdot\dfrac{\sqrt{2}}{\sqrt{2}}=\dfrac{\sqrt{2}}{2}+\dfrac{\sqrt{2}}{2}=\sqrt{2}$

29. Combining radicals: $\dfrac{\sqrt{5}}{3}+\dfrac{1}{\sqrt{5}}=\dfrac{\sqrt{5}}{3}+\dfrac{1}{\sqrt{5}}\cdot\dfrac{\sqrt{5}}{\sqrt{5}}=\dfrac{\sqrt{5}}{3}+\dfrac{\sqrt{5}}{5}=\dfrac{5\sqrt{5}}{15}+\dfrac{3\sqrt{5}}{15}=\dfrac{8\sqrt{5}}{15}$

31. Combining radicals: $\sqrt{x}-\dfrac{1}{\sqrt{x}}=\sqrt{x}-\dfrac{1}{\sqrt{x}}\cdot\dfrac{\sqrt{x}}{\sqrt{x}}=\sqrt{x}-\dfrac{\sqrt{x}}{x}=\dfrac{x\sqrt{x}}{x}-\dfrac{\sqrt{x}}{x}=\dfrac{(x-1)\sqrt{x}}{x}$

33. Combining radicals: $\dfrac{\sqrt{18}}{6}+\sqrt{\dfrac{1}{2}}+\dfrac{\sqrt{2}}{2}=\dfrac{3\sqrt{2}}{6}+\dfrac{1}{\sqrt{2}}\cdot\dfrac{\sqrt{2}}{\sqrt{2}}+\dfrac{\sqrt{2}}{2}=\dfrac{\sqrt{2}}{2}+\dfrac{\sqrt{2}}{2}+\dfrac{\sqrt{2}}{2}=\dfrac{3\sqrt{2}}{2}$

35. Combining radicals: $\sqrt{6}-\sqrt{\dfrac{2}{3}}+\sqrt{\dfrac{1}{6}}=\sqrt{6}-\dfrac{\sqrt{2}}{\sqrt{3}}\cdot\dfrac{\sqrt{3}}{\sqrt{3}}+\dfrac{1}{\sqrt{6}}\cdot\dfrac{\sqrt{6}}{\sqrt{6}}=\sqrt{6}-\dfrac{\sqrt{6}}{3}+\dfrac{\sqrt{6}}{6}=\dfrac{6\sqrt{6}}{6}-\dfrac{2\sqrt{6}}{6}+\dfrac{\sqrt{6}}{6}=\dfrac{5\sqrt{6}}{6}$

37. Combining radicals: $\sqrt[3]{25}+\dfrac{3}{\sqrt[3]{5}}=\sqrt[3]{25}+\dfrac{3}{\sqrt[3]{5}}\cdot\dfrac{\sqrt[3]{25}}{\sqrt[3]{25}}=\sqrt[3]{25}+\dfrac{3\sqrt[3]{25}}{5}=\dfrac{5\sqrt[3]{25}}{5}+\dfrac{3\sqrt[3]{25}}{5}=\dfrac{8\sqrt[3]{25}}{5}$

39. Using a calculator:
$$\sqrt{12}\approx 3.464 \qquad 2\sqrt{3}\approx 3.464$$

41. It is equal to the decimal approximation for $\sqrt{50}$:
$$\sqrt{8}+\sqrt{18}\approx 7.071\approx\sqrt{50} \qquad \sqrt{26}\approx 5.099$$

43. The correct statement is: $3\sqrt{2x}+5\sqrt{2x}=8\sqrt{2x}$

45. The correct statement is: $\sqrt{9+16}=\sqrt{25}=5$

47. Answers will vary.

49. Answers will vary.

51. First use the Pythagorean theorem to find the height h:
$$x^2+h^2=(2x)^2$$
$$x^2+h^2=4x^2$$
$$h^2=3x^2$$
$$h=x\sqrt{3}$$

Therefore the ratio is: $\dfrac{x\sqrt{3}}{2x}=\dfrac{\sqrt{3}}{2}$

53. Combining the fractions: $\dfrac{2a-4}{a+2}-\dfrac{a-6}{a+2}=\dfrac{2a-4-a+6}{a+2}=\dfrac{a+2}{a+2}=1$

55. Combining the fractions: $3+\dfrac{4}{3-t}=3\cdot\dfrac{3-t}{3-t}+\dfrac{4}{3-t}=\dfrac{9-3t+4}{3-t}=\dfrac{13-3t}{3-t}$

57. Combining the fractions:
$$\dfrac{3}{2x-5}-\dfrac{39}{8x^2-14x-15}=\dfrac{3}{2x-5}\cdot\dfrac{4x+3}{4x+3}-\dfrac{39}{(2x-5)(4x+3)}$$
$$=\dfrac{12x+9-39}{(2x-5)(4x+3)}$$
$$=\dfrac{12x-30}{(2x-5)(4x+3)}$$
$$=\dfrac{6(2x-5)}{(2x-5)(4x+3)}$$
$$=\dfrac{6}{4x+3}$$

59.   Combining the fractions:

$$\frac{1}{x-y} - \frac{3xy}{x^3-y^3} = \frac{1}{x-y} \cdot \frac{x^2+xy+y^2}{x^2+xy+y^2} - \frac{3xy}{(x-y)(x^2+xy+y^2)}$$

$$= \frac{x^2+xy+y^2-3xy}{(x-y)(x^2+xy+y^2)}$$

$$= \frac{x^2-2xy+y^2}{(x-y)(x^2+xy+y^2)}$$

$$= \frac{(x-y)^2}{(x-y)(x^2+xy+y^2)}$$

$$= \frac{x-y}{x^2+xy+y^2}$$

7.5   Multiplication and Division of Radical Expressions

1.   Multiplying: $\sqrt{6}\sqrt{3} = \sqrt{18} = 3\sqrt{2}$ **3.**   Multiplying: $\left(2\sqrt{3}\right)\left(5\sqrt{7}\right) = 10\sqrt{21}$

5.   Multiplying: $\left(4\sqrt{6}\right)\left(2\sqrt{15}\right)\left(3\sqrt{10}\right) = 24\sqrt{900} = 24 \cdot 30 = 720$

7.   Multiplying: $\left(3\sqrt[3]{3}\right)\left(6\sqrt[3]{9}\right) = 18\sqrt[3]{27} = 18 \cdot 3 = 54$ **9.**   Multiplying: $\sqrt{3}\left(\sqrt{2} - 3\sqrt{3}\right) = \sqrt{6} - 3\sqrt{9} = \sqrt{6} - 9$

11.   Multiplying: $6\sqrt[3]{4}\left(2\sqrt[3]{2} + 1\right) = 12\sqrt[3]{8} + 6\sqrt[3]{4} = 24 + 6\sqrt[3]{4}$

13.   Multiplying: $\left(\sqrt{3} + \sqrt{2}\right)\left(3\sqrt{3} - \sqrt{2}\right) = 3\sqrt{9} - \sqrt{6} + 3\sqrt{6} - \sqrt{4} = 9 + 2\sqrt{6} - 2 = 7 + 2\sqrt{6}$

15.   Multiplying: $\left(\sqrt{x} + 5\right)\left(\sqrt{x} - 3\right) = x - 3\sqrt{x} + 5\sqrt{x} - 15 = x + 2\sqrt{x} - 15$

17.   Multiplying: $\left(3\sqrt{6} + 4\sqrt{2}\right)\left(\sqrt{6} + 2\sqrt{2}\right) = 3\sqrt{36} + 4\sqrt{12} + 6\sqrt{12} + 8\sqrt{4} = 18 + 8\sqrt{3} + 12\sqrt{3} + 16 = 34 + 20\sqrt{3}$

19.   Multiplying: $\left(\sqrt{3} + 4\right)^2 = \left(\sqrt{3} + 4\right)\left(\sqrt{3} + 4\right) = \sqrt{9} + 4\sqrt{3} + 4\sqrt{3} + 16 = 19 + 8\sqrt{3}$

21.   Multiplying: $\left(\sqrt{x} - 3\right)^2 = \left(\sqrt{x} - 3\right)\left(\sqrt{x} - 3\right) = x - 3\sqrt{x} - 3\sqrt{x} + 9 = x - 6\sqrt{x} + 9$

23.   Multiplying: $\left(2\sqrt{a} - 3\sqrt{b}\right)^2 = \left(2\sqrt{a} - 3\sqrt{b}\right)\left(2\sqrt{a} - 3\sqrt{b}\right) = 4a - 6\sqrt{ab} - 6\sqrt{ab} + 9b = 4a - 12\sqrt{ab} + 9b$

25.   Multiplying: $\left(\sqrt{x-4} + 2\right)^2 = \left(\sqrt{x-4} + 2\right)\left(\sqrt{x-4} + 2\right) = x - 4 + 2\sqrt{x-4} + 2\sqrt{x-4} + 4 = x + 4\sqrt{x-4}$

27.   Multiplying: $\left(\sqrt{x-5} - 3\right)^2 = \left(\sqrt{x-5} - 3\right)\left(\sqrt{x-5} - 3\right) = x - 5 - 3\sqrt{x-5} - 3\sqrt{x-5} + 9 = x + 4 - 6\sqrt{x-5}$

29.   Multiplying: $\left(\sqrt{3} - \sqrt{2}\right)\left(\sqrt{3} + \sqrt{2}\right) = \left(\sqrt{3}\right)^2 - \left(\sqrt{2}\right)^2 = 3 - 2 = 1$

31.   Multiplying: $\left(\sqrt{a} + 7\right)\left(\sqrt{a} - 7\right) = \left(\sqrt{a}\right)^2 - (7)^2 = a - 49$

33.   Multiplying: $\left(5 - \sqrt{x}\right)\left(5 + \sqrt{x}\right) = (5)^2 - \left(\sqrt{x}\right)^2 = 25 - x$

35.   Multiplying: $\left(\sqrt{x-4} + 2\right)\left(\sqrt{x-4} - 2\right) = \left(\sqrt{x-4}\right)^2 - (2)^2 = x - 4 - 4 = x - 8$

37.   Multiplying: $\left(\sqrt{3} + 1\right)^3 = \left(\sqrt{3} + 1\right)\left(3 + 2\sqrt{3} + 1\right) = \left(\sqrt{3} + 1\right)\left(4 + 2\sqrt{3}\right) = 4\sqrt{3} + 4 + 6 + 2\sqrt{3} = 10 + 6\sqrt{3}$

39.   Rationalizing the denominator: $\dfrac{\sqrt{2}}{\sqrt{6} - \sqrt{2}} = \dfrac{\sqrt{2}}{\sqrt{6} - \sqrt{2}} \cdot \dfrac{\sqrt{6} + \sqrt{2}}{\sqrt{6} + \sqrt{2}} = \dfrac{\sqrt{12} + 2}{6 - 2} = \dfrac{2\sqrt{3} + 2}{4} = \dfrac{1 + \sqrt{3}}{2}$

41.   Rationalizing the denominator: $\dfrac{\sqrt{5}}{\sqrt{5} + 1} = \dfrac{\sqrt{5}}{\sqrt{5} + 1} \cdot \dfrac{\sqrt{5} - 1}{\sqrt{5} - 1} = \dfrac{5 - \sqrt{5}}{5 - 1} = \dfrac{5 - \sqrt{5}}{4}$

43. Rationalizing the denominator: $\dfrac{\sqrt{x}}{\sqrt{x}-3} = \dfrac{\sqrt{x}}{\sqrt{x}-3} \cdot \dfrac{\sqrt{x}+3}{\sqrt{x}+3} = \dfrac{x+3\sqrt{x}}{x-9}$

45. Rationalizing the denominator: $\dfrac{\sqrt{5}}{2\sqrt{5}-3} = \dfrac{\sqrt{5}}{2\sqrt{5}-3} \cdot \dfrac{2\sqrt{5}+3}{2\sqrt{5}+3} = \dfrac{2\sqrt{25}+3\sqrt{5}}{20-9} = \dfrac{10+3\sqrt{5}}{11}$

47. Rationalizing the denominator: $\dfrac{3}{\sqrt{x}-\sqrt{y}} = \dfrac{3}{\sqrt{x}-\sqrt{y}} \cdot \dfrac{\sqrt{x}+\sqrt{y}}{\sqrt{x}+\sqrt{y}} = \dfrac{3\sqrt{x}+3\sqrt{y}}{x-y}$

49. Rationalizing the denominator: $\dfrac{\sqrt{6}+\sqrt{2}}{\sqrt{6}-\sqrt{2}} = \dfrac{\sqrt{6}+\sqrt{2}}{\sqrt{6}-\sqrt{2}} \cdot \dfrac{\sqrt{6}+\sqrt{2}}{\sqrt{6}+\sqrt{2}} = \dfrac{6+2\sqrt{12}+2}{6-2} = \dfrac{8+4\sqrt{3}}{4} = 2+\sqrt{3}$

51. Rationalizing the denominator: $\dfrac{\sqrt{7}-2}{\sqrt{7}+2} = \dfrac{\sqrt{7}-2}{\sqrt{7}+2} \cdot \dfrac{\sqrt{7}-2}{\sqrt{7}-2} = \dfrac{7-4\sqrt{7}+4}{7-4} = \dfrac{11-4\sqrt{7}}{3}$

53. Rationalizing the denominator: $\dfrac{\sqrt{a}+\sqrt{b}}{\sqrt{a}-\sqrt{b}} = \dfrac{\sqrt{a}+\sqrt{b}}{\sqrt{a}-\sqrt{b}} \cdot \dfrac{\sqrt{a}+\sqrt{b}}{\sqrt{a}+\sqrt{b}} = \dfrac{a+2\sqrt{ab}+b}{a-b}$

55. Rationalizing the denominator: $\dfrac{\sqrt{x}+2}{\sqrt{x}-2} = \dfrac{\sqrt{x}+2}{\sqrt{x}-2} \cdot \dfrac{\sqrt{x}+2}{\sqrt{x}+2} = \dfrac{x+4\sqrt{x}+4}{x-4}$

57. Rationalizing the denominator: $\dfrac{2\sqrt{3}-\sqrt{7}}{3\sqrt{3}+\sqrt{7}} = \dfrac{2\sqrt{3}-\sqrt{7}}{3\sqrt{3}+\sqrt{7}} \cdot \dfrac{3\sqrt{3}-\sqrt{7}}{3\sqrt{3}-\sqrt{7}} = \dfrac{18-3\sqrt{21}-2\sqrt{21}+7}{27-7} = \dfrac{25-5\sqrt{21}}{20} = \dfrac{5-\sqrt{21}}{4}$

59. Rationalizing the denominator: $\dfrac{3\sqrt{x}+2}{1+\sqrt{x}} = \dfrac{3\sqrt{x}+2}{1+\sqrt{x}} \cdot \dfrac{1-\sqrt{x}}{1-\sqrt{x}} = \dfrac{3\sqrt{x}+2-3x-2\sqrt{x}}{1-x} = \dfrac{\sqrt{x}-3x+2}{1-x}$

61. Simplifying the product: $\left(\sqrt[3]{2}+\sqrt[3]{3}\right)\left(\sqrt[3]{4}-\sqrt[3]{6}+\sqrt[3]{9}\right) = \sqrt[3]{8}-\sqrt[3]{12}+\sqrt[3]{18}+\sqrt[3]{12}-\sqrt[3]{18}+\sqrt[3]{27} = 2+3 = 5$

63. The correct statement is: $5\left(2\sqrt{3}\right) = 10\sqrt{3}$

65. The correct statement is: $\left(\sqrt{x}+3\right)^2 = \left(\sqrt{x}+3\right)\left(\sqrt{x}+3\right) = x+6\sqrt{x}+9$

67. The correct statement is: $\left(5\sqrt{3}\right)^2 = \left(5\sqrt{3}\right)\left(5\sqrt{3}\right) = 25 \cdot 3 = 75$

69. Substituting $h = 50$: $t = \dfrac{\sqrt{100-50}}{4} = \dfrac{\sqrt{50}}{4} = \dfrac{5\sqrt{2}}{4}$ second

Substituting $h = 0$: $t = \dfrac{\sqrt{100-0}}{4} = \dfrac{\sqrt{100}}{4} = \dfrac{10}{4} = \dfrac{5}{2}$ second

71. Substituting $t = 1.25$:

$$1.25 = \dfrac{\sqrt{100-h}}{4}$$
$$5 = \sqrt{100-h}$$
$$100-h = 25$$
$$h = 75 \text{ feet}$$

73. Since the large rectangle is a golden rectangle and $AC = 2x$, then $CE = 2x\left(\dfrac{1+\sqrt{5}}{2}\right) = x\left(1+\sqrt{5}\right)$. Since $CD = 2x$, then

$DE = x\left(1+\sqrt{5}\right) - 2x = x\left(-1+\sqrt{5}\right)$. Now computing the ratio:

$$\dfrac{EF}{DE} = \dfrac{2x}{x\left(-1+\sqrt{5}\right)} = \dfrac{2}{-1+\sqrt{5}} \cdot \dfrac{-1-\sqrt{5}}{-1-\sqrt{5}} = \dfrac{-2\left(\sqrt{5}+1\right)}{1-5} = \dfrac{-2\left(\sqrt{5}+1\right)}{-4} = \dfrac{1+\sqrt{5}}{2}$$

Therefore the smaller rectangle $BDEF$ is also a golden rectangle.

75. Simplifying the complex fraction: $\dfrac{\frac{1}{4}-\frac{1}{3}}{\frac{1}{2}+\frac{1}{6}} = \dfrac{\left(\frac{1}{4}-\frac{1}{3}\right)\cdot 12}{\left(\frac{1}{2}+\frac{1}{6}\right)\cdot 12} = \dfrac{3-4}{6+2} = -\dfrac{1}{8}$

77. Simplifying the complex fraction: $\dfrac{1-\frac{2}{y}}{1+\frac{2}{y}} = \dfrac{\left(1-\frac{2}{y}\right)\cdot y}{\left(1+\frac{2}{y}\right)\cdot y} = \dfrac{y-2}{y+2}$

79. Simplifying the complex fraction: $\dfrac{4+\frac{4}{x}+\frac{1}{x^2}}{4-\frac{1}{x^2}} = \dfrac{\left(4+\frac{4}{x}+\frac{1}{x^2}\right)\cdot x^2}{\left(4-\frac{1}{x^2}\right)\cdot x^2} = \dfrac{4x^2+4x+1}{4x^2-1} = \dfrac{(2x+1)^2}{(2x+1)(2x-1)} = \dfrac{2x+1}{2x-1}$

7.6   Equations With Radicals

1. Solving the equation:
$$\sqrt{2x+1}=3$$
$$\left(\sqrt{2x+1}\right)^2=3^2$$
$$2x+1=9$$
$$2x=8$$
$$x=4$$

3. Solving the equation:
$$\sqrt{4x+1}=-5$$
$$\left(\sqrt{4x+1}\right)^2=(-5)^2$$
$$4x+1=25$$
$$4x=24$$
$$x=6$$
Since this value does not check, the solution set is $\varnothing$.

5. Solving the equation:
$$\sqrt{2y-1}=3$$
$$\left(\sqrt{2y-1}\right)^2=3^2$$
$$2y-1=9$$
$$2y=10$$
$$y=5$$

7. Solving the equation:
$$\sqrt{5x-7}=-1$$
$$\left(\sqrt{5x-7}\right)^2=(-1)^2$$
$$5x-7=1$$
$$5x=8$$
$$x=\frac{8}{5}$$
Since this value does not check, the solution set is $\varnothing$.

9. Solving the equation:
$$\sqrt{2x-3}-2=4$$
$$\sqrt{2x-3}=6$$
$$\left(\sqrt{2x-3}\right)^2=6^2$$
$$2x-3=36$$
$$2x=39$$
$$x=\frac{39}{2}$$

11. Solving the equation:
$$\sqrt{4a+1}+3=2$$
$$\sqrt{4a+1}=-1$$
$$\left(\sqrt{4a+1}\right)^2=(-1)^2$$
$$4a+1=1$$
$$4a=0$$
$$a=0$$
Since this value does not check, the solution set is $\varnothing$.

13. Solving the equation:
$$\sqrt[4]{3x+1}=2$$
$$\left(\sqrt[4]{3x+1}\right)^4=2^4$$
$$3x+1=16$$
$$3x=15$$
$$x=5$$

15. Solving the equation:
$$\sqrt[3]{2x-5}=1$$
$$\left(\sqrt[3]{2x-5}\right)^3=1^3$$
$$2x-5=1$$
$$2x=6$$
$$x=3$$

17. Solving the equation:
$$\sqrt[3]{3a+5} = -3$$
$$\left(\sqrt[3]{3a+5}\right)^3 = (-3)^3$$
$$3a+5 = -27$$
$$3a = -32$$
$$a = -\frac{32}{3}$$

19. Solving the equation:
$$\sqrt{y-3} = y-3$$
$$\left(\sqrt{y-3}\right)^2 = (y-3)^2$$
$$y-3 = y^2 - 6y + 9$$
$$0 = y^2 - 7y + 12$$
$$0 = (y-3)(y-4)$$
$$y = 3, 4$$

21. Solving the equation:
$$\sqrt{a+2} = a+2$$
$$\left(\sqrt{a+2}\right)^2 = (a+2)^2$$
$$a+2 = a^2 + 4a + 4$$
$$0 = a^2 + 3a + 2$$
$$0 = (a+2)(a+1)$$
$$a = -2, -1$$

23. Solving the equation:
$$\sqrt{2x+4} = \sqrt{1-x}$$
$$\left(\sqrt{2x+4}\right)^2 = \left(\sqrt{1-x}\right)^2$$
$$2x+4 = 1-x$$
$$3x = -3$$
$$x = -1$$

25. Solving the equation:
$$\sqrt{4a+7} = -\sqrt{a+2}$$
$$\left(\sqrt{4a+7}\right)^2 = \left(-\sqrt{a+2}\right)^2$$
$$4a+7 = a+2$$
$$3a = -5$$
$$a = -\frac{5}{3}$$
Since this value does not check, the solution set is $\varnothing$.

27. Solving the equation:
$$\sqrt[4]{5x-8} = \sqrt[4]{4x-1}$$
$$\left(\sqrt[4]{5x-8}\right)^4 = \left(\sqrt[4]{4x-1}\right)^4$$
$$5x-8 = 4x-1$$
$$x = 7$$

29. Solving the equation:
$$x+1 = \sqrt{5x+1}$$
$$(x+1)^2 = \left(\sqrt{5x+1}\right)^2$$
$$x^2 + 2x + 1 = 5x + 1$$
$$x^2 - 3x = 0$$
$$x(x-3) = 0$$
$$x = 0, 3$$

31. Solving the equation:
$$t+5 = \sqrt{2t+9}$$
$$(t+5)^2 = \left(\sqrt{2t+9}\right)^2$$
$$t^2 + 10t + 25 = 2t + 9$$
$$t^2 + 8t + 16 = 0$$
$$(t+4)^2 = 0$$
$$t = -4$$

33. Solving the equation:
$$\sqrt{y-8} = \sqrt{8-y}$$
$$\left(\sqrt{y-8}\right)^2 = \left(\sqrt{8-y}\right)^2$$
$$y-8 = 8-y$$
$$2y = 16$$
$$y = 8$$

35. Solving the equation:
$$\sqrt[3]{3x+5} = \sqrt[3]{5-2x}$$
$$\left(\sqrt[3]{3x+5}\right)^3 = \left(\sqrt[3]{5-2x}\right)^3$$
$$3x+5 = 5-2x$$
$$5x = 0$$
$$x = 0$$

37. Solving the equation:
$$\sqrt{x-8} = \sqrt{x} - 2$$
$$\left(\sqrt{x-8}\right)^2 = \left(\sqrt{x}-2\right)^2$$
$$x-8 = x - 4\sqrt{x} + 4$$
$$-12 = -4\sqrt{x}$$
$$\sqrt{x} = 3$$
$$x = 9$$

39. Solving the equation:
$$\sqrt{x+1} = \sqrt{x} + 1$$
$$\left(\sqrt{x+1}\right)^2 = \left(\sqrt{x}+1\right)^2$$
$$x+1 = x + 2\sqrt{x} + 1$$
$$0 = 2\sqrt{x}$$
$$\sqrt{x} = 0$$
$$x = 0$$

41. Solving the equation:
$$\sqrt{x+8} = \sqrt{x-4} + 2$$
$$\left(\sqrt{x+8}\right)^2 = \left(\sqrt{x-4}+2\right)^2$$
$$x+8 = x-4+4\sqrt{x-4}+4$$
$$8 = 4\sqrt{x-4}$$
$$\sqrt{x-4} = 2$$
$$x-4 = 4$$
$$x = 8$$

43. Solving the equation:
$$\sqrt{x-5} - 3 = \sqrt{x-8}$$
$$\left(\sqrt{x-5}-3\right)^2 = \left(\sqrt{x-8}\right)^2$$
$$x-5-6\sqrt{x-5}+9 = x-8$$
$$-6\sqrt{x-5} = -12$$
$$\sqrt{x-5} = 2$$
$$x-5 = 4$$
$$x = 9$$
Since this value does not check, the solution set is $\varnothing$.

45. Solving the equation:
$$\sqrt{x+4} = 2 - \sqrt{2x}$$
$$\left(\sqrt{x+4}\right)^2 = \left(2-\sqrt{2x}\right)^2$$
$$x+4 = 4-4\sqrt{2x}+2x$$
$$-x = -4\sqrt{2x}$$
$$\left(-x\right)^2 = \left(-4\sqrt{2x}\right)^2$$
$$x^2 = 32x$$
$$x^2 - 32x = 0$$
$$x(x-32) = 0$$
$$x = 0, 32$$
The solution is 0 (32 does not check).

47. Solving the equation:
$$\sqrt{2x+4} = \sqrt{x+3} + 1$$
$$\left(\sqrt{2x+4}\right)^2 = \left(\sqrt{x+3}+1\right)^2$$
$$2x+4 = x+3+2\sqrt{x+3}+1$$
$$x = 2\sqrt{x+3}$$
$$x^2 = \left(2\sqrt{x+3}\right)^2$$
$$x^2 = 4x+12$$
$$x^2 - 4x - 12 = 0$$
$$(x-6)(x+2) = 0$$
$$x = -2, 6$$
The solution is 6 (–2 does not check).

49. Graphing the equation:

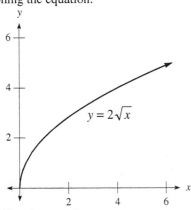

51. Graphing the equation:

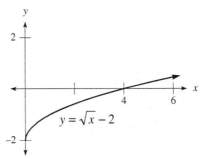

53. Graphing the equation:

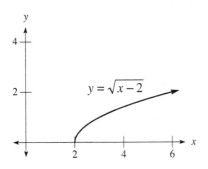

55. Graphing the equation:

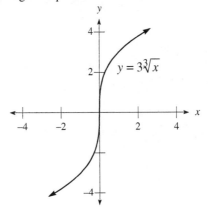

57.   Graphing the equation:

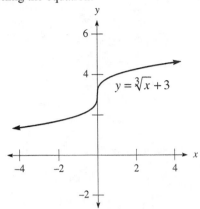

59.   Graphing the equation:

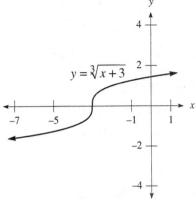

61.   Solving for h:

$$t = \frac{\sqrt{100 - h}}{4}$$
$$4t = \sqrt{100 - h}$$
$$16t^2 = 100 - h$$
$$h = 100 - 16t^2$$

63.   Solving for L:

$$2 = 2\left(\tfrac{22}{7}\right)\sqrt{\frac{L}{32}}$$
$$\tfrac{7}{22} = \sqrt{\frac{L}{32}}$$
$$\left(\tfrac{7}{22}\right)^2 = \frac{L}{32}$$
$$L = 32\left(\tfrac{7}{22}\right)^2 \approx 3.24 \text{ feet}$$

65.   The width is $\sqrt{25} = 5$ meters.

67.   Solving the equation:
$$\sqrt{x} = 50$$
$$x = 2500$$
The plume is 2,500 meters down river.

69.   Multiplying: $\sqrt{2}\left(\sqrt{3} - \sqrt{2}\right) = \sqrt{6} - \sqrt{4} = \sqrt{6} - 2$

71.   Multiplying: $\left(\sqrt{x} + 5\right)^2 = \left(\sqrt{x} + 5\right)\left(\sqrt{x} + 5\right) = x + 5\sqrt{x} + 5\sqrt{x} + 25 = x + 10\sqrt{x} + 25$

73.   Rationalizing the denominator: $\dfrac{\sqrt{x}}{\sqrt{x} + 3} = \dfrac{\sqrt{x}}{\sqrt{x} + 3} \cdot \dfrac{\sqrt{x} - 3}{\sqrt{x} - 3} = \dfrac{x - 3\sqrt{x}}{x - 9}$

75.   Solving the equation:
$$\frac{x}{3\sqrt{2x - 3}} - \frac{1}{\sqrt{2x - 3}} = \tfrac{1}{3}$$
$$3\sqrt{2x - 3}\left(\frac{x}{3\sqrt{2x - 3}} - \frac{1}{\sqrt{2x - 3}}\right) = 3\sqrt{2x - 3}\left(\tfrac{1}{3}\right)$$
$$x - 3 = \sqrt{2x - 3}$$
$$(x - 3)^2 = 2x - 3$$
$$x^2 - 6x + 9 = 2x - 3$$
$$x^2 - 8x + 12 = 0$$
$$(x - 6)(x - 2) = 0$$
$$x = 2, 6$$
The solution is 6 (2 does not check).

77. Solving the equation:

$$x+1 = \sqrt[3]{4x+4}$$
$$(x+1)^3 = \left(\sqrt[3]{4x+4}\right)^3$$
$$x^3 + 3x^2 + 3x + 1 = 4x + 4$$
$$x^3 + 3x^2 - x - 3 = 0$$
$$x^2(x+3) - 1(x+3) = 0$$
$$(x+3)\left(x^2 - 1\right) = 0$$
$$(x+3)(x+1)(x-1) = 0$$
$$x = -3, -1, 1$$

79. Solving for y:

$$y+2 = \sqrt{x^2 + (y-2)^2}$$
$$(y+2)^2 = x^2 + (y-2)^2$$
$$y^2 + 4y + 4 = x^2 + y^2 - 4y + 4$$
$$8y = x^2$$
$$y = \tfrac{1}{8}x^2$$

7.7   Complex Numbers

1. Writing in terms of i: $\sqrt{-36} = 6i$

3. Writing in terms of i: $-\sqrt{-25} = -5i$

5. Writing in terms of i: $\sqrt{-72} = 6i\sqrt{2}$

7. Writing in terms of i: $-\sqrt{-12} = -2i\sqrt{3}$

9. Rewriting the expression: $i^{28} = \left(i^4\right)^7 = (1)^7 = 1$

11. Rewriting the expression: $i^{26} = i^{24}i^2 = \left(i^4\right)^6 i^2 = (1)^6(-1) = -1$

13. Rewriting the expression: $i^{75} = i^{72}i^3 = \left(i^4\right)^{18} i^2 i = (1)^{18}(-1)i = -i$

15. Setting real and imaginary parts equal:

$$2x = 6 \qquad 3y = -3$$
$$x = 3 \qquad y = -1$$

17. Setting real and imaginary parts equal:

$$-x = 2 \qquad 10y = -5$$
$$x = -2 \qquad y = -\tfrac{1}{2}$$

19. Setting real and imaginary parts equal:

$$2x = -16 \qquad -2y = 10$$
$$x = -8 \qquad y = -5$$

21. Setting real and imaginary parts equal:

$$2x - 4 = 10 \qquad -6y = -3$$
$$2x = 14 \qquad y = \tfrac{1}{2}$$
$$x = 7$$

23. Setting real and imaginary parts equal:

$$7x - 1 = 2 \qquad 5y + 2 = 4$$
$$7x = 3 \qquad 5y = 2$$
$$x = \tfrac{3}{7} \qquad y = \tfrac{2}{5}$$

25. Combining the numbers: $(2+3i)+(3+6i) = 5 + 9i$

27. Combining the numbers: $(3-5i)+(2+4i) = 5 - i$

29. Combining the numbers: $(5+2i)-(3+6i) = 5 + 2i - 3 - 6i = 2 - 4i$

31. Combining the numbers: $(3-5i)-(2+i) = 3 - 5i - 2 - i = 1 - 6i$

33. Combining the numbers: $[(3+2i)-(6+i)]+(5+i) = 3 + 2i - 6 - i + 5 + i = 2 + 2i$

35. Combining the numbers: $[(7-i)-(2+4i)]-(6+2i) = 7 - i - 2 - 4i - 6 - 2i = -1 - 7i$

37. Combining the numbers:

$$(3+2i)-[(3-4i)-(6+2i)] = (3+2i)-(3-4i-6-2i) = (3+2i)-(-3-6i) = 3 + 2i + 3 + 6i = 6 + 8i$$

39. Combining the numbers: $(4-9i)+[(2-7i)-(4+8i)] = (4-9i)+(2-7i-4-8i) = (4-9i)+(-2-15i) = 2 - 24i$

41. Finding the product: $3i(4+5i) = 12i + 15i^2 = -15 + 12i$

43. Finding the product: $6i(4-3i) = 24i - 18i^2 = 18 + 24i$

45. Finding the product: $(3+2i)(4+i) = 12 + 8i + 3i + 2i^2 = 12 + 11i - 2 = 10 + 11i$

47. Finding the product: $(4+9i)(3-i) = 12 + 27i - 4i - 9i^2 = 12 + 23i + 9 = 21 + 23i$

49. Finding the product: $(1+i)^3 = (1+i)(1+i)^2 = (1+i)(1+2i-1) = (1+i)(2i) = -2 + 2i$

51. Finding the product: $(2-i)^3 = (2-i)(2-i)^2 = (2-i)(4-4i-1) = (2-i)(3-4i) = 6 - 11i - 4 = 2 - 11i$

53. Finding the product: $(2+5i)^2 = (2+5i)(2+5i) = 4 + 10i + 10i - 25 = -21 + 20i$

55. Finding the product: $(1-i)^2 = (1-i)(1-i) = 1-i-i-1 = -2i$

57. Finding the product: $(3-4i)^2 = (3-4i)(3-4i) = 9-12i-12i-16 = -7-24i$

59. Finding the product: $(2+i)(2-i) = 4-i^2 = 4+1 = 5$

61. Finding the product: $(6-2i)(6+2i) = 36-4i^2 = 36+4 = 40$

63. Finding the product: $(2+3i)(2-3i) = 4-9i^2 = 4+9 = 13$

65. Finding the product: $(10+8i)(10-8i) = 100-64i^2 = 100+64 = 164$

67. Finding the quotient: $\dfrac{2-3i}{i} = \dfrac{2-3i}{i} \cdot \dfrac{i}{i} = \dfrac{2i+3}{-1} = -3-2i$

69. Finding the quotient: $\dfrac{5+2i}{-i} = \dfrac{5+2i}{-i} \cdot \dfrac{i}{i} = \dfrac{5i-2}{1} = -2+5i$

71. Finding the quotient: $\dfrac{4}{2-3i} = \dfrac{4}{2-3i} \cdot \dfrac{2+3i}{2+3i} = \dfrac{8+12i}{4+9} = \dfrac{8+12i}{13} = \frac{8}{13} + \frac{12}{13}i$

73. Finding the quotient: $\dfrac{6}{-3+2i} = \dfrac{6}{-3+2i} \cdot \dfrac{-3-2i}{-3-2i} = \dfrac{-18-12i}{9+4} = \dfrac{-18-12i}{13} = -\frac{18}{13} - \frac{12}{13}i$

75. Finding the quotient: $\dfrac{2+3i}{2-3i} = \dfrac{2+3i}{2-3i} \cdot \dfrac{2+3i}{2+3i} = \dfrac{4+12i-9}{4+9} = \dfrac{-5+12i}{13} = -\frac{5}{13} + \frac{12}{13}i$

77. Finding the quotient: $\dfrac{5+4i}{3+6i} = \dfrac{5+4i}{3+6i} \cdot \dfrac{3-6i}{3-6i} = \dfrac{15-18i+24}{9+36} = \dfrac{39-18i}{45} = \frac{13}{15} - \frac{2}{5}i$

79. Dividing to find R: $R = \dfrac{80+20i}{-6+2i} = \dfrac{80+20i}{-6+2i} \cdot \dfrac{-6-2i}{-6-2i} = \dfrac{-480-280i+40}{36+4} = \dfrac{-440-280i}{40} = (-11-7i)$ ohms

81. Solving the equation:

$$\frac{t}{3} - \frac{1}{2} = -1$$
$$6\left(\frac{t}{3} - \frac{1}{2}\right) = 6(-1)$$
$$2t - 3 = -6$$
$$2t = -3$$
$$t = -\frac{3}{2}$$

83. Solving the equation:

$$2 + \frac{5}{y} = \frac{3}{y^2}$$
$$y^2\left(2 + \frac{5}{y}\right) = y^2\left(\frac{3}{y^2}\right)$$
$$2y^2 + 5y = 3$$
$$2y^2 + 5y - 3 = 0$$
$$(2y-1)(y+3) = 0$$
$$y = -3, \tfrac{1}{2}$$

85. Let x represent the number. The equation is:

$$x + \frac{1}{x} = \tfrac{41}{20}$$
$$20x\left(x + \frac{1}{x}\right) = 20x\left(\tfrac{41}{20}\right)$$
$$20x^2 + 20 = 41x$$
$$20x^2 - 41x + 20 = 0$$
$$(5x-4)(4x-5) = 0$$
$$x = \tfrac{4}{5}, \tfrac{5}{4}$$

The number is either $\frac{5}{4}$ or $\frac{4}{5}$.

87. Simplifying: $\dfrac{1}{i} = \dfrac{1}{i} \cdot \dfrac{i}{i} = \dfrac{i}{i^2} = \dfrac{i}{-1} = -i$

89. Substituting into the equation: $x^2 - 2x + 2 = (1+i)^2 - 2(1+i) + 2 = 1+2i-1-2-2i+2 = 0$

Thus $x = 1+i$ is a solution to the equation.

91. Substituting into the equation:
$$x^3 - 11x + 20 = (2+i)^3 - 11(2+i) + 20$$
$$= (2+i)(4+4i-1) - 22 - 11i + 20$$
$$= (2+i)(3+4i) - 2 - 11i$$
$$= 6 + 11i - 4 - 2 - 11i$$
$$= 0$$
Thus $x = 2+i$ is a solution to the equation.

Chapter 7 Review

1. Simplifying: $\sqrt{49} = 7$

3. Simplifying: $16^{1/4} = 2$

5. Simplifying: $\sqrt[5]{32x^{15}y^{10}} = 2x^3y^2$

7. Simplifying: $x^{2/3} \cdot x^{4/3} = x^{2/3+4/3} = x^2$

9. Simplifying: $\dfrac{a^{3/5}}{a^{1/4}} = a^{3/5-1/4} = a^{12/20-5/20} = a^{7/20}$

11. Multiplying: $\left(3x^{1/2} + 5y^{1/2}\right)\left(4x^{1/2} - 3y^{1/2}\right) = 12x - 9x^{1/2}y^{1/2} + 20x^{1/2}y^{1/2} - 15y = 12x + 11x^{1/2}y^{1/2} - 15y$

13. Dividing: $\dfrac{28x^{5/6} + 14x^{7/6}}{7x^{1/3}} = \dfrac{28x^{5/6}}{7x^{1/3}} + \dfrac{14x^{7/6}}{7x^{1/3}} = 4x^{5/6-1/3} + 2x^{7/6-1/3} = 4x^{1/2} + 2x^{5/6}$

15. Simplifying: $x^{3/4} + \dfrac{5}{x^{1/4}} = x^{3/4} \cdot \dfrac{x^{1/4}}{x^{1/4}} + \dfrac{5}{x^{1/4}} = \dfrac{x+5}{x^{1/4}}$

17. Simplifying: $\sqrt{50} = \sqrt{25 \cdot 2} = 5\sqrt{2}$

19. Simplifying: $\sqrt{18x^2} = \sqrt{9x^2 \cdot 2} = 3x\sqrt{2}$

21. Simplifying: $\sqrt[4]{32a^4b^5c^6} = \sqrt[4]{16a^4b^4c^4 \cdot 2bc^2} = 2abc\sqrt[4]{2bc^2}$

23. Rationalizing the denominator: $\dfrac{6}{\sqrt[3]{2}} = \dfrac{6}{\sqrt[3]{2}} \cdot \dfrac{\sqrt[3]{4}}{\sqrt[3]{4}} = \dfrac{6\sqrt[3]{4}}{2} = 3\sqrt[3]{4}$

25. Simplifying: $\sqrt[3]{\dfrac{40x^2y^3}{3z}} = \dfrac{2y\sqrt[3]{5x^2}}{\sqrt[3]{3z}} \cdot \dfrac{\sqrt[3]{9z^2}}{\sqrt[3]{9z^2}} = \dfrac{2y\sqrt[3]{45x^2z^2}}{3z}$

27. Combining the expressions: $\sqrt{12} + \sqrt{3} = 2\sqrt{3} + \sqrt{3} = 3\sqrt{3}$

29. Combining the expressions: $3\sqrt{8} - 4\sqrt{72} + 5\sqrt{50} = 6\sqrt{2} - 24\sqrt{2} + 25\sqrt{2} = 7\sqrt{2}$

31. Combining the expressions: $2x\sqrt[3]{xy^3z^2} - 6y\sqrt[3]{x^4z^2} = 2xy\sqrt[3]{xz^2} - 6xy\sqrt[3]{xz^2} = -4xy\sqrt[3]{xz^2}$

33. Multiplying: $\left(\sqrt{x} - 2\right)\left(\sqrt{x} - 3\right) = x - 2\sqrt{x} - 3\sqrt{x} + 6 = x - 5\sqrt{x} + 6$

35. Rationalizing the denominator: $\dfrac{\sqrt{7} + \sqrt{5}}{\sqrt{7} - \sqrt{5}} = \dfrac{\sqrt{7} + \sqrt{5}}{\sqrt{7} - \sqrt{5}} \cdot \dfrac{\sqrt{7} + \sqrt{5}}{\sqrt{7} + \sqrt{5}} = \dfrac{7 + 2\sqrt{35} + 5}{7 - 5} = \dfrac{12 + 2\sqrt{35}}{2} = 6 + \sqrt{35}$

37. Solving the equation:
$$\sqrt{4a+1} = 1$$
$$\left(\sqrt{4a+1}\right)^2 = (1)^2$$
$$4a + 1 = 1$$
$$4a = 0$$
$$a = 0$$

39. Solving the equation:
$$\sqrt{3x+1} - 3 = 1$$
$$\sqrt{3x+1} = 4$$
$$\left(\sqrt{3x+1}\right)^2 = (4)^2$$
$$3x + 1 = 16$$
$$3x = 15$$
$$x = 5$$

41.   Graphing the equation:

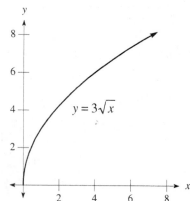

$$y = 3\sqrt{x}$$

43.   Writing in terms of i: $i^{24} = \left(i^4\right)^6 = (1)^6 = 1$

45.   Setting real and imaginary parts equal:
$$-2x = 3 \qquad 8y = -4$$
$$x = -\tfrac{3}{2} \qquad y = -\tfrac{1}{2}$$

47.   Combining the numbers: $(3 + 5i) + (6 - 2i) = 9 + 3i$

49.   Multiplying: $3i(4 + 2i) = 12i + 6i^2 = -6 + 12i$

51.   Multiplying: $(4 + 2i)^2 = (4 + 2i)(4 + 2i) = 16 + 8i + 8i + 4i^2 = 16 + 16i - 4 = 12 + 16i$

53.   Dividing: $\dfrac{3+i}{i} = \dfrac{3+i}{i} \cdot \dfrac{i}{i} = \dfrac{3i-1}{-1} = 1 - 3i$

55.   Let l represent the desired length. Using the Pythagorean theorem:
$$18^2 + 13.5^2 = l^2$$
$$l^2 = 506.25$$
$$l = \sqrt{506.25} = 22.5$$
The length of one side of the roof is 22.5 feet.

Chapters 1-7 Cumulative Review

1.   Simplifying: $33 - 22 - (-11) + 1 = 33 - 22 + 11 + 1 = 23$

3.   Simplifying: $-6 + 5\big[3 - 2(-4 - 1)\big] = -6 + 5\big[3 - 2(-5)\big] = -6 + 5(3 + 10) = -6 + 5(13) = -6 + 65 = 59$

5.   Simplifying: $\left(2y^{-3}\right)^{-1}\left(4y^{-3}\right)^2 = \tfrac{1}{2}y^3 \cdot 16y^{-6} = 8y^{-3} = \dfrac{8}{y^3}$

7.   Simplifying: $\sqrt{72y^5} = \sqrt{36y^4 \cdot 2y} = 6y^2\sqrt{2y}$

9.   The opposite is -12 and the reciprocal is $\tfrac{1}{12}$.

11.   Writing in scientific notation: $41,500 = 4.15 \times 10^4$

13.   Factoring completely: $625a^4 - 16b^4 = \left(25a^2 + 4b^2\right)\left(25a^2 - 4b^2\right) = (5a - 2b)(5a + 2b)\left(25a^2 + 4b^2\right)$

15.   Reducing to lowest terms: $\dfrac{246}{861} = \dfrac{2 \cdot 123}{7 \cdot 123} = \dfrac{2}{7}$

17.   Reducing to lowest terms: $\dfrac{x^2 - 9x + 20}{x^2 - 7x + 12} = \dfrac{(x - 4)(x - 5)}{(x - 4)(x - 3)} = \dfrac{x - 5}{x - 3}$

19.   Multiplying: $(2x - 5y)(3x - 2y) = 6x^2 - 4xy - 15xy + 10y^2 = 6x^2 - 19xy + 10y^2$

21.   Multiplying: $\left(\sqrt{x} - 2\right)^2 = \left(\sqrt{x} - 2\right)\left(\sqrt{x} - 2\right) = x - 2\sqrt{x} - 2\sqrt{x} + 4 = x - 4\sqrt{x} + 4$

23. Dividing: $\dfrac{27x^3y^2}{13x^2y^4} \div \dfrac{9xy}{26y} = \dfrac{27x^3y^2}{13x^2y^4} \cdot \dfrac{26y}{9xy} = \dfrac{27 \cdot 26x^3y^3}{9 \cdot 13x^3y^5} = \dfrac{6}{y^2}$

25. Subtracting:

$$\left(\tfrac{2}{3}x^3 + \tfrac{1}{6}x^2 + \tfrac{1}{2}\right) - \left(\tfrac{1}{4}x^2 - \tfrac{1}{3}x + \tfrac{1}{12}\right) = \tfrac{2}{3}x^3 + \tfrac{1}{6}x^2 + \tfrac{1}{2} - \tfrac{1}{4}x^2 + \tfrac{1}{3}x - \tfrac{1}{12}$$

$$= \tfrac{2}{3}x^3 + \left(\tfrac{1}{6} - \tfrac{1}{4}\right)x^2 + \tfrac{1}{3}x + \left(\tfrac{1}{2} - \tfrac{1}{12}\right)$$

$$= \tfrac{2}{3}x^3 - \tfrac{1}{12}x^2 + \tfrac{1}{3}x + \tfrac{5}{12}$$

27. Solving the equation:

$$3y - 8 = -4y + 6$$
$$7y = 14$$
$$y = 2$$

29. Solving the equation:

$$|3x - 1| - 2 = 6$$
$$|3x - 1| = 8$$
$$3x - 1 = -8, 8$$
$$3x = -7, 9$$
$$x = -\tfrac{7}{3}, 3$$

31. Solving the equation:

$$\dfrac{x+2}{x+1} - 2 = \dfrac{1}{x+1}$$
$$(x+1)\left(\dfrac{x+2}{x+1} - 2\right) = (x+1)\left(\dfrac{1}{x+1}\right)$$
$$x + 2 - 2(x+1) = 1$$
$$x + 2 - 2x - 2 = 1$$
$$-x = 1$$
$$x = -1$$

There is no solution (–1 does not check).

33. Solving the equation:

$$(x+3)^2 - 3(x+3) - 70 = 0$$
$$(x+3-10)(x+3+7) = 0$$
$$(x-7)(x+10) = 0$$
$$x = -10, 7$$

35. Solving for L:

$$P = 2L + 2W$$
$$18 = 2L + 2(3)$$
$$18 = 2L + 6$$
$$2L = 12$$
$$L = 6$$

37. Solving the inequality:

$$|5x - 4| \geq 6$$

$$5x - 4 \leq -6 \qquad \text{or} \qquad 5x - 4 \geq 6$$
$$5x \leq -2 \qquad\qquad\qquad 5x \geq 10$$
$$x \leq -\tfrac{2}{5} \qquad\qquad\qquad x \geq 2$$

The solution set is $\left\{x \mid x \leq -\tfrac{2}{5} \text{ or } x \geq 2\right\}$. Graphing the solution set:

$$-2/5 \qquad\qquad\qquad 2$$

39. Substituting into the first equation:

$$4x + 9(2x - 12) = 2$$
$$4x + 18x - 108 = 2$$
$$22x = 110$$
$$x = 5$$

The solution is $(5, -2)$.

41.   Multiply the first equation by –1 and add it to the first equation:
$$-x - y = 2$$
$$y + 10z = -1$$
Adding yields the equation $-x + 10z = 1$. So the system becomes:
$$-x + 10z = 1$$
$$2x - 13z = 5$$
Multiply the first equation by 2:
$$-2x + 20z = 2$$
$$2x - 13z = 5$$
Adding yields:
$$7z = 7$$
$$z = 1$$
Substituting to find x:
$$2x - 13 = 5$$
$$2x = 18$$
$$x = 9$$
Substituting to find y:
$$9 + y = -2$$
$$y = -11$$
The solution is $(9, -11, 1)$.

43.   Graphing the equation:

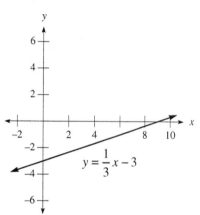

45.   Graphing the inequality:

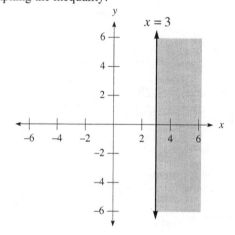

47.   Solving for y:
$$4x - 5y = 15$$
$$-5y = -4x + 15$$
$$y = \frac{4}{5}x - 3$$

The slope is $\frac{4}{5}$ and the y-intercept is –3.

49.   Using the point-slope formula:
$$y - 3 = \frac{2}{3}(x - 9)$$
$$y - 3 = \frac{2}{3}x - 6$$
$$y = \frac{2}{3}x - 3$$

Chapter 7 Test

1. Simplifying: $27^{-2/3} = \left(27^{1/3}\right)^{-2} = 3^{-2} = \dfrac{1}{3^2} = \dfrac{1}{9}$ **2.** Simplifying: $\left(\dfrac{25}{49}\right)^{-1/2} = \left(\dfrac{49}{25}\right)^{1/2} = \dfrac{7}{5}$

3. Simplifying: $a^{3/4} \cdot a^{-1/3} = a^{3/4-1/3} = a^{9/12-4/12} = a^{5/12}$

4. Simplifying: $\dfrac{\left(x^{2/3}y^{-3}\right)^{1/2}}{\left(x^{3/4}y^{1/2}\right)^{-1}} = \dfrac{x^{1/3}y^{-3/2}}{x^{-3/4}y^{-1/2}} = x^{1/3+3/4}y^{-3/2+1/2} = x^{13/12}y^{-1} = \dfrac{x^{13/12}}{y}$

5. Simplifying: $\sqrt{49x^8y^{10}} = 7x^4y^5$ **6.** Simplifying: $\sqrt[5]{32x^{10}y^{20}} = 2x^2y^4$

7. Simplifying: $\dfrac{\left(36a^8b^4\right)^{1/2}}{\left(27a^9b^6\right)^{1/3}} = \dfrac{6a^4b^2}{3a^3b^2} = 2a$ **8.** Simplifying: $\dfrac{\left(x^ny^{1/n}\right)^n}{\left(x^{1/n}y^n\right)^{n^2}} = \dfrac{x^{n^2}y}{x^ny^{n^3}} = x^{n^2-n}y^{1-n^3}$

9. Multiplying: $2a^{1/2}\left(3a^{3/2} - 5a^{1/2}\right) = 6a^2 - 10a$

10. Multiplying: $\left(4a^{3/2} - 5\right)^2 = \left(4a^{3/2} - 5\right)\left(4a^{3/2} - 5\right) = 16a^3 - 20a^{3/2} - 20a^{3/2} + 25 = 16a^3 - 40a^{3/2} + 25$

11. Factoring: $3x^{2/3} + 5x^{1/3} - 2 = \left(3x^{1/3} - 1\right)\left(x^{1/3} + 2\right)$ **12.** Factoring: $9x^{2/3} - 49 = \left(3x^{1/3} - 7\right)\left(3x^{1/3} + 7\right)$

13. Combining: $\dfrac{4}{x^{1/2}} + x^{1/2} = \dfrac{4}{x^{1/2}} + x^{1/2} \cdot \dfrac{x^{1/2}}{x^{1/2}} = \dfrac{x+4}{x^{1/2}}$

14. Combining: $\dfrac{x^2}{\left(x^2-3\right)^{1/2}} - \left(x^2-3\right)^{1/2} = \dfrac{x^2}{\left(x^2-3\right)^{1/2}} - \left(x^2-3\right)^{1/2} \cdot \dfrac{\left(x^2-3\right)^{1/2}}{\left(x^2-3\right)^{1/2}} = \dfrac{x^2 - x^2 + 3}{\left(x^2-3\right)^{1/2}} = \dfrac{3}{\left(x^2-3\right)^{1/2}}$

15. Simplifying: $\sqrt{125x^3y^5} = \sqrt{25x^2y^4 \cdot 5xy} = 5xy^2\sqrt{5xy}$

16. Simplifying: $\sqrt[3]{40x^7y^8} = \sqrt[3]{8x^6y^6 \cdot 5xy^2} = 2x^2y^2\sqrt[3]{5xy^2}$

17. Simplifying: $\sqrt{\dfrac{2}{3}} = \dfrac{\sqrt{2}}{\sqrt{3}} \cdot \dfrac{\sqrt{3}}{\sqrt{3}} = \dfrac{\sqrt{6}}{3}$

18. Simplifying: $\sqrt{\dfrac{12a^4b^3}{5c}} = \dfrac{2a^2b\sqrt{3b}}{\sqrt{5c}} \cdot \dfrac{\sqrt{5c}}{\sqrt{5c}} = \dfrac{2a^2b\sqrt{15bc}}{5c}$

19. Combining: $3\sqrt{12} - 4\sqrt{27} = 6\sqrt{3} - 12\sqrt{3} = -6\sqrt{3}$

20. Combining: $2\sqrt[3]{24a^3b^3} - 5a\sqrt[3]{3b^3} = 4ab\sqrt[3]{3} - 5ab\sqrt[3]{3} = -ab\sqrt[3]{3}$

21. Multiplying: $\left(\sqrt{x} + 7\right)\left(\sqrt{x} - 4\right) = x - 4\sqrt{x} + 7\sqrt{x} - 28 = x + 3\sqrt{x} - 28$

22. Multiplying: $\left(3\sqrt{2} - \sqrt{3}\right)^2 = \left(3\sqrt{2} - \sqrt{3}\right)\left(3\sqrt{2} - \sqrt{3}\right) = 18 - 3\sqrt{6} - 3\sqrt{6} + 3 = 21 - 6\sqrt{6}$

23. Rationalizing the denominator: $\dfrac{5}{\sqrt{3}-1} = \dfrac{5}{\sqrt{3}-1} \cdot \dfrac{\sqrt{3}+1}{\sqrt{3}+1} = \dfrac{5\sqrt{3}+5}{3-1} = \dfrac{5+5\sqrt{3}}{2}$

24. Rationalizing the denominator: $\dfrac{\sqrt{x}-\sqrt{2}}{\sqrt{x}+\sqrt{2}} = \dfrac{\sqrt{x}-\sqrt{2}}{\sqrt{x}+\sqrt{2}} \cdot \dfrac{\sqrt{x}-\sqrt{2}}{\sqrt{x}-\sqrt{2}} = \dfrac{x - \sqrt{2x} - \sqrt{2x} + 2}{x-2} = \dfrac{x - 2\sqrt{2x} + 2}{x-2}$

25. Solving the equation:
$$\sqrt{3x+1} = x-3$$
$$\left(\sqrt{3x+1}\right)^2 = (x-3)^2$$
$$3x+1 = x^2 - 6x + 9$$
$$0 = x^2 - 9x + 8$$
$$0 = (x-1)(x-8)$$
$$x = 1, 8$$
The solution is 8 (1 does not check).

26. Solving the equation:
$$\sqrt[3]{2x+7} = -1$$
$$\left(\sqrt[3]{2x+7}\right)^3 = (-1)^3$$
$$2x+7 = -1$$
$$2x = -8$$
$$x = -4$$

27. Solving the equation:
$$\sqrt{x+3} = \sqrt{x+4} - 1$$
$$\left(\sqrt{x+3}\right)^2 = \left(\sqrt{x+4} - 1\right)^2$$
$$x+3 = x+4 - 2\sqrt{x+4} + 1$$
$$-2 = -2\sqrt{x+4}$$
$$\sqrt{x+4} = 1$$
$$x+4 = 1$$
$$x = -3$$

28. Graphing the equation:

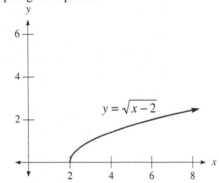

29. Graphing the equation:

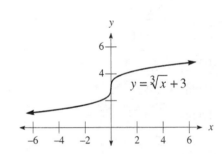

30. Setting the real and imaginary parts equal:
$$2x+5 = 6 \qquad -(y-3) = -4$$
$$2x = 1 \qquad\quad y-3 = 4$$
$$x = \tfrac{1}{2} \qquad\quad y = 7$$

31. Performing the operations:
$$(3+2i) - \left[(7-i) - (4+3i)\right] = (3+2i) - (7-i-4-3i) = (3+2i) - (3-4i) = 3+2i-3+4i = 6i$$

32. Performing the operations: $(2-3i)(4+3i) = 8+6i-12i+9 = 17-6i$

33. Performing the operations: $(5-4i)^2 = (5-4i)(5-4i) = 25-20i-20i-16 = 9-40i$

34. Performing the operations: $\dfrac{2-3i}{2+3i} = \dfrac{2-3i}{2+3i} \cdot \dfrac{2-3i}{2-3i} = \dfrac{4-12i-9}{4+9} = \dfrac{-5-12i}{13} = -\dfrac{5}{13} - \dfrac{12}{13}i$

35. Rewriting the exponent: $i^{38} = \left(i^2\right)^{19} = (-1)^{19} = -1$

Chapter 8
Quadratic Functions

8.1 Completing the Square

1. Solving the equation:

$$x^2 = 25$$
$$x = \pm\sqrt{25} = \pm5$$

3. Solving the equation:

$$y^2 = \tfrac{3}{4}$$
$$y = \pm\sqrt{\tfrac{3}{4}} = \pm\frac{\sqrt{3}}{2}$$

5. Solving the equation:

$$x^2 + 12 = 0$$
$$x^2 = -12$$
$$x = \pm\sqrt{-12} = \pm2i\sqrt{3}$$

7. Solving the equation:

$$4a^2 - 45 = 0$$
$$4a^2 = 45$$
$$a^2 = \tfrac{45}{4}$$
$$a = \pm\sqrt{\tfrac{45}{4}} = \pm\frac{3\sqrt{5}}{2}$$

9. Solving the equation:

$$(2y - 1)^2 = 25$$
$$2y - 1 = \pm\sqrt{25} = \pm5$$
$$2y - 1 = -5, 5$$
$$2y = -4, 6$$
$$y = -2, 3$$

11. Solving the equation:

$$(2a + 3)^2 = -9$$
$$2a + 3 = \pm\sqrt{-9} = \pm3i$$
$$2a = -3 \pm 3i$$
$$a = \frac{-3 \pm 3i}{2}$$

13. Solving the equation:

$$x^2 + 8x + 16 = -27$$
$$(x + 4)^2 = -27$$
$$x + 4 = \pm\sqrt{-27} = \pm3i\sqrt{3}$$
$$x = -4 \pm 3i\sqrt{3}$$

15. Solving the equation:

$$4a^2 - 12a + 9 = -4$$
$$(2a - 3)^2 = -4$$
$$2a - 3 = \pm\sqrt{-4} = \pm2i$$
$$2a = 3 \pm 2i$$
$$a = \frac{3 \pm 2i}{2}$$

17.   Solving the equation:

$$(x+5)^2 + (x-5)^2 = 52$$
$$x^2 + 10x + 25 + x^2 - 10x + 25 = 52$$
$$2x^2 + 50 = 52$$
$$2x^2 = 2$$
$$x^2 = 1$$
$$x = \pm\sqrt{1} = \pm 1$$

19.   Completing the square: $x^2 + 12x + 36 = (x+6)^2$

21.   Completing the square: $x^2 - 4x + 4 = (x-2)^2$

23.   Completing the square: $a^2 - 10a + 25 = (a-5)^2$

25.   Completing the square: $x^2 + 5x + \frac{25}{4} = \left(x + \frac{5}{2}\right)^2$

27.   Completing the square: $y^2 - 7y + \frac{49}{4} = \left(y - \frac{7}{2}\right)^2$

29.   Solving the equation:

$$x^2 + 4x = 12$$
$$x^2 + 4x + 4 = 12 + 4$$
$$(x+2)^2 = 16$$
$$x + 2 = \pm\sqrt{16} = \pm 4$$
$$x + 2 = -4, 4$$
$$x = -6, 2$$

31.   Solving the equation:

$$x^2 + 12x = -27$$
$$x^2 + 12x + 36 = -27 + 36$$
$$(x+6)^2 = 9$$
$$x + 6 = \pm\sqrt{9} = \pm 3$$
$$x + 6 = -3, 3$$
$$x = -9, -3$$

33.   Solving the equation:

$$a^2 - 2a + 5 = 0$$
$$a^2 - 2a + 1 = -5 + 1$$
$$(a-1)^2 = -4$$
$$a - 1 = \pm\sqrt{-4} = \pm 2i$$
$$a = 1 \pm 2i$$

35.   Solving the equation:

$$y^2 - 8y + 1 = 0$$
$$y^2 - 8y + 16 = -1 + 16$$
$$(y-4)^2 = 15$$
$$y - 4 = \pm\sqrt{15}$$
$$y = 4 \pm \sqrt{15}$$

37.   Solving the equation:

$$x^2 - 5x - 3 = 0$$
$$x^2 - 5x + \frac{25}{4} = 3 + \frac{25}{4}$$
$$\left(x - \frac{5}{2}\right)^2 = \frac{37}{4}$$
$$x - \frac{5}{2} = \pm\frac{\sqrt{37}}{2}$$
$$x = \frac{5 \pm \sqrt{37}}{2}$$

39.   Solving the equation:

$$2x^2 - 4x - 8 = 0$$
$$x^2 - 2x - 4 = 0$$
$$x^2 - 2x + 1 = 4 + 1$$
$$(x-1)^2 = 5$$
$$x - 1 = \pm\sqrt{5}$$
$$x = 1 \pm \sqrt{5}$$

41.   Solving the equation:

$$3t^2 - 8t + 1 = 0$$
$$t^2 - \frac{8}{3}t + \frac{1}{3} = 0$$
$$t^2 - \frac{8}{3}t + \frac{16}{9} = -\frac{1}{3} + \frac{16}{9}$$
$$\left(t - \frac{4}{3}\right)^2 = \frac{13}{9}$$
$$t - \frac{4}{3} = \pm\sqrt{\frac{13}{9}} = \pm\frac{\sqrt{13}}{3}$$
$$t = \frac{4 \pm \sqrt{13}}{3}$$

43.   Solving the equation:

$$4x^2 - 3x + 5 = 0$$
$$x^2 - \frac{3}{4}x + \frac{5}{4} = 0$$
$$x^2 - \frac{3}{4}x + \frac{9}{64} = -\frac{5}{4} + \frac{9}{64}$$
$$\left(x - \frac{3}{8}\right)^2 = -\frac{71}{64}$$
$$x - \frac{3}{8} = \pm\sqrt{-\frac{71}{64}} = \pm\frac{i\sqrt{71}}{8}$$
$$x = \frac{3 \pm i\sqrt{71}}{8}$$

45.   The other two sides are $\dfrac{\sqrt{3}}{2}$ inch, 1 inch.

47.   The other two sides are $3\sqrt{3}$ feet, 6 feet.

49. The hypotenuse is $x\sqrt{2}$.

51. Let x represent the horizontal distance. Using the Pythagorean theorem:
$$x^2 + 120^2 = 790^2$$
$$x^2 + 14400 = 624100$$
$$x^2 = 609700$$
$$x = \sqrt{609700} \approx 781 \text{ feet}$$

53. Solving for r:
$$3456 = 3000(1+r)^2$$
$$(1+r)^2 = 1.152$$
$$1+r = \sqrt{1.152}$$
$$r = \sqrt{1.152} - 1 \approx 0.073$$
The annual interest rate is 7.3%.

55. Its length is $20\sqrt{2} \approx 28$ feet.

57. Simplifying: $\sqrt{45} = \sqrt{9 \cdot 5} = 3\sqrt{5}$

59. Simplifying: $\sqrt{27y^5} = \sqrt{9y^4 \cdot 3y} = 3y^2\sqrt{3y}$

61. Simplifying: $\sqrt[3]{54x^6y^5} = \sqrt[3]{27x^6y^3 \cdot 2y^2} = 3x^2y\sqrt[3]{2y^2}$

63. Substituting values: $\sqrt{b^2 - 4ac} = \sqrt{7^2 - 4(6)(-5)} = \sqrt{49 + 120} = \sqrt{169} = 13$

65. Rationalizing the denominator: $\dfrac{3}{\sqrt{2}} = \dfrac{3}{\sqrt{2}} \cdot \dfrac{\sqrt{2}}{\sqrt{2}} = \dfrac{3\sqrt{2}}{2}$

67. Rationalizing the denominator: $\dfrac{2}{\sqrt[3]{4}} = \dfrac{2}{\sqrt[3]{4}} \cdot \dfrac{\sqrt[3]{2}}{\sqrt[3]{2}} = \dfrac{2\sqrt[3]{2}}{2} = \sqrt[3]{2}$

69. **a.** Using the formula:
$$z^n = x^n + y^n$$
$$z^2 = 6^2 + 8^2$$
$$z^2 = 36 + 64 = 100$$
$$z = \sqrt{100} = 10$$

   b. Using the formula:
$$x^n + y^n = z^n$$
$$5^2 + y^2 = 13^2$$
$$25 + y^2 = 169$$
$$y^2 = 144$$
$$y = \sqrt{144} = 12$$

8.2 The Quadratic Formula

1. Solving the equation:
$$x^2 + 5x + 6 = 0$$
$$(x+3)(x+2) = 0$$
$$x = -3, -2$$

3. Solving the equation:
$$a^2 - 4a + 1 = 0$$
$$a = \frac{4 \pm \sqrt{16-4}}{2} = \frac{4 \pm \sqrt{12}}{2} = \frac{4 \pm 2\sqrt{3}}{2} = 2 \pm \sqrt{3}$$

5. Solving the equation:
$$\tfrac{1}{6}x^2 - \tfrac{1}{2}x + \tfrac{1}{3} = 0$$
$$x^2 - 3x + 2 = 0$$
$$(x-1)(x-2) = 0$$
$$x = 1, 2$$

7. Solving the equation:

$$\frac{x^2}{2} + 1 = \frac{2x}{3}$$

$$3x^2 + 6 = 4x$$

$$3x^2 - 4x + 6 = 0$$

$$x = \frac{4 \pm \sqrt{16 - 72}}{6} = \frac{4 \pm \sqrt{-56}}{6} = \frac{4 \pm 2i\sqrt{14}}{6} = \frac{2 \pm i\sqrt{14}}{3}$$

9. Solving the equation:

$$y^2 - 5y = 0$$

$$y(y - 5) = 0$$

$$y = 0, 5$$

11. Solving the equation:

$$30x^2 + 40x = 0$$

$$10x(3x + 4) = 0$$

$$x = -\frac{4}{3}, 0$$

13. Solving the equation:

$$\frac{2t^2}{3} - t = -\frac{1}{6}$$

$$4t^2 - 6t = -1$$

$$4t^2 - 6t + 1 = 0$$

$$t = \frac{6 \pm \sqrt{36 - 16}}{8} = \frac{6 \pm \sqrt{20}}{8} = \frac{6 \pm 2\sqrt{5}}{8} = \frac{3 \pm \sqrt{5}}{4}$$

15. Solving the equation:

$$0.01x^2 + 0.06x - 0.08 = 0$$

$$x^2 + 6x - 8 = 0$$

$$x = \frac{-6 \pm \sqrt{36 + 32}}{2} = \frac{-6 \pm \sqrt{68}}{2} = \frac{-6 \pm 2\sqrt{17}}{2} = -3 \pm \sqrt{17}$$

17. Solving the equation:

$$2x + 3 = -2x^2$$

$$2x^2 + 2x + 3 = 0$$

$$x = \frac{-2 \pm \sqrt{4 - 24}}{4} = \frac{-2 \pm \sqrt{-20}}{4} = \frac{-2 \pm 2i\sqrt{5}}{4} = \frac{-1 \pm i\sqrt{5}}{2}$$

19. Solving the equation:

$$100x^2 - 200x + 100 = 0$$

$$100(x^2 - 2x + 1) = 0$$

$$100(x - 1)^2 = 0$$

$$x = 1$$

21. Solving the equation:

$$\frac{1}{2}r^2 = \frac{1}{6}r - \frac{2}{3}$$

$$3r^2 = r - 4$$

$$3r^2 - r + 4 = 0$$

$$r = \frac{1 \pm \sqrt{1 - 48}}{6} = \frac{1 \pm \sqrt{-47}}{6} = \frac{1 \pm i\sqrt{47}}{6}$$

23. Solving the equation:

$$(x - 3)(x - 5) = 1$$

$$x^2 - 8x + 15 = 1$$

$$x^2 - 8x + 14 = 0$$

$$x = \frac{8 \pm \sqrt{64 - 56}}{2} = \frac{8 \pm \sqrt{8}}{2} = \frac{8 \pm 2\sqrt{2}}{2} = 4 \pm \sqrt{2}$$

25. Solving the equation:

$$(x+3)^2 + (x-8)(x-1) = 16$$
$$x^2 + 6x + 9 + x^2 - 9x + 8 = 16$$
$$2x^2 - 3x + 1 = 0$$
$$(2x-1)(x-1) = 0$$
$$x = \tfrac{1}{2}, 1$$

27. Solving the equation:

$$\frac{x^2}{3} - \frac{5x}{6} = \tfrac{1}{2}$$
$$2x^2 - 5x = 3$$
$$2x^2 - 5x - 3 = 0$$
$$(2x+1)(x-3) = 0$$
$$x = -\tfrac{1}{2}, 3$$

29. Solving the equation:

$$\frac{1}{x+1} - \frac{1}{x} = \tfrac{1}{2}$$
$$2x(x+1)\left(\frac{1}{x+1} - \frac{1}{x}\right) = 2x(x+1) \cdot \tfrac{1}{2}$$
$$2x - (2x+2) = x^2 + x$$
$$2x - 2x - 2 = x^2 + x$$
$$x^2 + x + 2 = 0$$
$$x = \frac{-1 \pm \sqrt{1-8}}{2} = \frac{-1 \pm \sqrt{-7}}{2} = \frac{-1 \pm i\sqrt{7}}{2}$$

31. Solving the equation:

$$\frac{1}{y-1} + \frac{1}{y+1} = 1$$
$$(y+1)(y-1)\left(\frac{1}{y-1} + \frac{1}{y+1}\right) = (y+1)(y-1) \cdot 1$$
$$y + 1 + y - 1 = y^2 - 1$$
$$2y = y^2 - 1$$
$$y^2 - 2y - 1 = 0$$
$$y = \frac{2 \pm \sqrt{4+4}}{2} = \frac{2 \pm \sqrt{8}}{2} = \frac{2 \pm 2\sqrt{2}}{2} = 1 \pm \sqrt{2}$$

33. Solving the equation:

$$\frac{1}{x+2} + \frac{1}{x+3} = 1$$
$$(x+2)(x+3)\left(\frac{1}{x+2} + \frac{1}{x+3}\right) = (x+2)(x+3) \cdot 1$$
$$x + 3 + x + 2 = x^2 + 5x + 6$$
$$2x + 5 = x^2 + 5x + 6$$
$$x^2 + 3x + 1 = 0$$
$$x = \frac{-3 \pm \sqrt{9-4}}{2} = \frac{-3 \pm \sqrt{5}}{2}$$

35.   Solving the equation:

$$\frac{6}{r^2-1} - \frac{1}{2} = \frac{1}{r+1}$$

$$2(r+1)(r-1)\left(\frac{6}{(r+1)(r-1)} - \frac{1}{2}\right) = 2(r+1)(r-1)\cdot\frac{1}{r+1}$$

$$12-\left(r^2-1\right) = 2r-2$$

$$12-r^2+1 = 2r-2$$

$$r^2+2r-15 = 0$$

$$(r+5)(r-3) = 0$$

$$r = -5, 3$$

37.   Solving the equation:

$$x^3-8 = 0$$

$$(x-2)\left(x^2+2x+4\right) = 0$$

$$x = 2 \quad \text{or} \quad x = \frac{-2\pm\sqrt{4-16}}{2} = \frac{-2\pm\sqrt{-12}}{2} = \frac{-2\pm 2i\sqrt{3}}{2} = -1\pm i\sqrt{3}$$

$$x = 2, -1\pm i\sqrt{3}$$

39.   Solving the equation:

$$8a^3+27 = 0$$

$$(2a+3)\left(4a^2-6a+9\right) = 0$$

$$a = -\frac{3}{2} \quad \text{or} \quad a = \frac{6\pm\sqrt{36-144}}{8} = \frac{6\pm\sqrt{-108}}{8} = \frac{6\pm 6i\sqrt{3}}{8} = \frac{3\pm 3i\sqrt{3}}{4}$$

$$a = -\frac{3}{2}, \frac{3\pm 3i\sqrt{3}}{4}$$

41.   Solving the equation:

$$125t^3-1 = 0$$

$$(5t-1)\left(25t^2+5t+1\right) = 0$$

$$t = \frac{1}{5} \quad \text{or} \quad t = \frac{-5\pm\sqrt{25-100}}{50} = \frac{-5\pm\sqrt{-75}}{50} = \frac{-5\pm 5i\sqrt{3}}{50} = \frac{-1\pm i\sqrt{3}}{10}$$

$$t = \frac{1}{5}, \frac{-1\pm i\sqrt{3}}{10}$$

43.   Solving the equation:

$$2x^3+2x^2+3x = 0$$

$$x\left(2x^2+2x+3\right) = 0$$

$$x = 0 \quad \text{or} \quad x = \frac{-2\pm\sqrt{4-24}}{4} = \frac{-2\pm\sqrt{-20}}{4} = \frac{-2\pm 2i\sqrt{5}}{4} = \frac{-1\pm i\sqrt{5}}{2}$$

$$x = 0, \frac{-1\pm i\sqrt{5}}{2}$$

45.   Solving the equation:

$$3y^4 = 6y^3-6y^2$$

$$3y^4-6y^3+6y^2 = 0$$

$$3y^2\left(y^2-2y+2\right) = 0$$

$$y = 0 \quad \text{or} \quad y = \frac{2\pm\sqrt{4-8}}{2} = \frac{2\pm\sqrt{-4}}{2} = \frac{2\pm 2i}{2} = 1\pm i$$

$$y = 0, 1\pm i$$

47.   Solving the equation:
$$6t^5 + 4t^4 = -2t^3$$
$$6t^5 + 4t^4 + 2t^3 = 0$$
$$2t^3\left(3t^2 + 2t + 1\right) = 0$$

$$t = 0 \qquad \text{or} \qquad t = \frac{-2 \pm \sqrt{4-12}}{6} = \frac{-2 \pm \sqrt{-8}}{6} = \frac{-2 \pm 2i\sqrt{2}}{6} = \frac{-1 \pm i\sqrt{2}}{3}$$

$$t = 0, \frac{-1 \pm i\sqrt{2}}{3}$$

49.   Substituting $s = 74$:
$$5t + 16t^2 = 74$$
$$16t^2 + 5t - 74 = 0$$
$$(16t + 37)(t - 2) = 0$$
$$t = 2 \qquad \left(t = -\tfrac{37}{16} \text{ is impossible}\right)$$

It will take 2 seconds for the object to fall 74 feet.

51.   Since profit is revenue minus the cost, the equation is:
$$100x - 0.5x^2 - (60x + 300) = 300$$
$$100x - 0.5x^2 - 60x - 300 = 300$$
$$-0.5x^2 + 40x - 600 = 0$$
$$x^2 - 80x + 1200 = 0$$
$$(x - 20)(x - 60) = 0$$
$$x = 20, 60$$

The weekly profit is \$300 if 20 items or 60 items are sold.

53.   **a.**   The two equations are: $l + w = 10, lw = 15$

      b.   Since $l = 10 - w$, the equation is:
$$w(10 - w) = 15$$
$$10w - w^2 = 15$$
$$w^2 - 10w + 15 = 0$$
$$w = \frac{10 \pm \sqrt{100 - 60}}{2} = \frac{10 \pm \sqrt{40}}{2} = \frac{10 \pm 2\sqrt{10}}{2} = 5 \pm \sqrt{10} \approx 1.84, 8.16$$

The length and width are 8.16 yards and 1.84 yards.

      c.   Two answers are possible because either dimension (long or short) may be considered the length.

55.   Let x represent the width of strip being cut off. After removing the strip, the overall area is 80% of its original area. The equation is:
$$(10.5 - 2x)(8.2 - 2x) = 0.80(10.5 \times 8.2)$$
$$86.1 - 37.4x + 4x^2 = 68.88$$
$$4x^2 - 37.4x + 17.22 = 0$$

$$x = \frac{37.4 \pm \sqrt{(-37.4)^2 - 4(4)(17.22)}}{8} = \frac{37.4 \pm \sqrt{1123.24}}{8} \approx \frac{37.4 \pm 33.5}{8} \approx 0.49, 8.86$$

The width of strip is 0.49 centimeter (8.86 cm is impossible).

57. Dividing using long division:

$$
\begin{array}{r}
4y+1 \\
2y-7\overline{)8y^2-26y-9} \\
\underline{8y^2-28y} \\
2y-9 \\
\underline{2y-7} \\
-2
\end{array}
$$

The quotient is $4y+1-\dfrac{2}{2y-7}$.

59. Dividing using long division:

$$
\begin{array}{r}
x^2+7x+12 \\
x+2\overline{)x^3+9x^2+26x+24} \\
\underline{x^3+2x^2} \\
7x^2+26x \\
\underline{7x^2+14x} \\
12x+24 \\
\underline{12x+24} \\
0
\end{array}
$$

The quotient is $x^2+7x+12$.

61. Simplifying: $25^{1/2}=\sqrt{25}=5$

63. Simplifying: $\left(\frac{9}{25}\right)^{3/2}=\left(\frac{3}{5}\right)^3=\frac{27}{125}$

65. Simplifying: $8^{-2/3}=\left(8^{-1/3}\right)^2=\left(\frac{1}{2}\right)^2=\frac{1}{4}$

67. Simplifying: $\dfrac{\left(49x^8y^{-4}\right)^{1/2}}{\left(27x^{-3}y^9\right)^{-1/3}}=\dfrac{7x^4y^{-2}}{\frac{1}{3}xy^{-3}}=21x^{4-1}y^{-2+3}=21x^3y$

69. Solving the equation:
$$x^2+\sqrt{3}x-6=0$$
$$\left(x+2\sqrt{3}\right)\left(x-\sqrt{3}\right)=0$$
$$x=-2\sqrt{3},\sqrt{3}$$

71. Solving the equation:
$$\sqrt{2}x^2+2x-\sqrt{2}=0$$
$$x=\dfrac{-2\pm\sqrt{4+4(2)}}{2\sqrt{2}}=\dfrac{-2\pm\sqrt{12}}{2\sqrt{2}}=\dfrac{-2\pm2\sqrt{3}}{2\sqrt{2}}\cdot\dfrac{\sqrt{2}}{\sqrt{2}}=\dfrac{-2\sqrt{2}\pm2\sqrt{6}}{4}=\dfrac{-\sqrt{2}\pm\sqrt{6}}{2}$$

73. Solving the equation:
$$x^2+ix+2=0$$
$$x=\dfrac{-i\pm\sqrt{-1-8}}{2}=\dfrac{-i\pm3i}{2}=-2i,i$$

8.3 Additional Items Involving Solutions to Equations

1. Computing the discriminant: $D=(-6)^2-4(1)(5)=36-20=16$. The equation will have two rational solutions.

3. First write the equation as $4x^2-4x+1=0$. Computing the discriminant: $D=(-4)^2-4(4)(1)=16-16=0$
The equation will have one rational solution.

5. Computing the discriminant: $D=1^2-4(1)(-1)=1+4=5$. The equation will have two irrational solutions.

7. First write the equation as $2y^2-3y-1=0$. Computing the discriminant: $D=(-3)^2-4(2)(-1)=9+8=17$
The equation will have two irrational solutions.

9. Computing the discriminant: $D=0^2-4(1)(-9)=36$. The equation will have two rational solutions.

11. First write the equation as $5a^2-4a-5=0$. Computing the discriminant: $D=(-4)^2-4(5)(-5)=16+100=116$
The equation will have two irrational solutions.

13. Setting the discriminant equal to 0:
$$(-k)^2-4(1)(25)=0$$
$$k^2-100=0$$
$$k^2=100$$
$$k=\pm10$$

15. First write the equation as $x^2 - kx + 36 = 0$. Setting the discriminant equal to 0:

$$(-k)^2 - 4(1)(36) = 0$$
$$k^2 - 144 = 0$$
$$k^2 = 144$$
$$k = \pm 12$$

17. Setting the discriminant equal to 0:

$$(-12)^2 - 4(4)(k) = 0$$
$$144 - 16k = 0$$
$$16k = 144$$
$$k = 9$$

19. First write the equation as $kx^2 - 40x - 25 = 0$. Setting the discriminant equal to 0:

$$(-40)^2 - 4(k)(-25) = 0$$
$$1600 + 100k = 0$$
$$100k = -1600$$
$$k = -16$$

21. Setting the discriminant equal to 0:

$$(-k)^2 - 4(3)(2) = 0$$
$$k^2 - 24 = 0$$
$$k^2 = 24$$
$$k = \pm\sqrt{24} = \pm 2\sqrt{6}$$

23. Writing the equation:

$$(x-5)(x-2) = 0$$
$$x^2 - 7x + 10 = 0$$

25. Writing the equation:

$$(t+3)(t-6) = 0$$
$$t^2 - 3t - 18 = 0$$

27. Writing the equation:

$$(y-2)(y+2)(y-4) = 0$$
$$(y^2 - 4)(y-4) = 0$$
$$y^3 - 4y^2 - 4y + 16 = 0$$

29. Writing the equation:

$$(2x-1)(x-3) = 0$$
$$2x^2 - 7x + 3 = 0$$

31. Writing the equation:

$$(4t+3)(t-3) = 0$$
$$4t^2 - 9t - 9 = 0$$

33. Writing the equation:

$$(x-3)(x+3)(6x-5) = 0$$
$$(x^2 - 9)(6x-5) = 0$$
$$6x^3 - 5x^2 - 54x + 45 = 0$$

35. Writing the equation:

$$(2a+1)(5a-3) = 0$$
$$10a^2 - a - 3 = 0$$

37. Writing the equation:

$$(3x+2)(3x-2)(x-1) = 0$$
$$(9x^2 - 4)(x-1) = 0$$
$$9x^3 - 9x^2 - 4x + 4 = 0$$

39. Writing the equation:

$$(x-2)(x+2)(x-3)(x+3) = 0$$
$$(x^2 - 4)(x^2 - 9) = 0$$
$$x^4 - 13x^2 + 36 = 0$$

41. Multiplying: $a^4\left(a^{3/2} - a^{1/2}\right) = a^{11/2} - a^{9/2}$

43. Multiplying: $\left(x^{3/2} - 3\right)^2 = \left(x^{3/2} - 3\right)\left(x^{3/2} - 3\right) = x^3 - 6x^{3/2} + 9$

45. Dividing: $\dfrac{30x^{3/4} - 25x^{5/4}}{5x^{1/4}} = \dfrac{30x^{3/4}}{5x^{1/4}} - \dfrac{25x^{5/4}}{5x^{1/4}} = 6x^{1/2} - 5x$

47. Factoring: $10(x-3)^{3/2} - 15(x-3)^{1/2} = 5(x-3)^{1/2}\left[2(x-3) - 3\right] = 5(x-3)^{1/2}(2x-9)$

49. Factoring: $2x^{2/3} - 11x^{1/3} + 12 = \left(2x^{1/3} - 3\right)\left(x^{1/3} - 4\right)$

51. Solving the equation:
$$x^2 = 4x + 5$$
$$x^2 - 4x - 5 = 0$$
$$(x-5)(x+1) = 0$$
$$x = -1, 5$$

53. Solving the equation:
$$x^2 - 1 = 2x$$
$$x^2 - 2x - 1 = 0$$
$$x = \frac{2 \pm \sqrt{4+4}}{2} = \frac{2 \pm \sqrt{8}}{2} = \frac{2 \pm 2\sqrt{2}}{2} = 1 \pm \sqrt{2} \approx -0.41, 2.41$$

55. Solving the equation:
$$2x^3 - x^2 - 2x + 1 = 0$$
$$x^2(2x-1) - 1(2x-1) = 0$$
$$(2x-1)(x^2-1) = 0$$
$$(2x-1)(x+1)(x-1) = 0$$
$$x = -1, \tfrac{1}{2}, 1$$

57. Solving the equation:
$$2x^3 + 2 = x^2 + 4x$$
$$2x^3 - x^2 - 4x + 2 = 0$$
$$x^2(2x-1) - 2(2x-1) = 0$$
$$(2x-1)(x^2-2) = 0$$
$$x = \tfrac{1}{2}, \pm\sqrt{2}$$
$$x \approx -1.41, 0.5, 1.41$$

8.4 Equations Quadratic in Form

1. Solving the equation:
$$(x-3)^2 + 3(x-3) + 2 = 0$$
$$(x-3+2)(x-3+1) = 0$$
$$(x-1)(x-2) = 0$$
$$x = 1, 2$$

3. Solving the equation:
$$2(x+4)^2 + 5(x+4) - 12 = 0$$
$$[2(x+4)-3][(x+4)+4] = 0$$
$$(2x+5)(x+8) = 0$$
$$x = -8, -\tfrac{5}{2}$$

5. Solving the equation:
$$x^4 - 6x^2 - 27 = 0$$
$$(x^2-9)(x^2+3) = 0$$
$$x^2 = 9, -3$$
$$x = \pm 3, \pm i\sqrt{3}$$

7. Solving the equation:
$$x^4 + 9x^2 = -20$$
$$x^4 + 9x^2 + 20 = 0$$
$$(x^2+4)(x^2+5) = 0$$
$$x^2 = -4, -5$$
$$x = \pm 2i, \pm i\sqrt{5}$$

9. Solving the equation:
$$(2a-3)^2 - 9(2a-3) = -20$$
$$(2a-3)^2 - 9(2a-3) + 20 = 0$$
$$(2a-3-4)(2a-3-5) = 0$$
$$(2a-7)(2a-8) = 0$$
$$a = \tfrac{7}{2}, 4$$

11. Solving the equation:
$$2(4a+2)^2 = 3(4a+2) + 20$$
$$2(4a+2)^2 - 3(4a+2) - 20 = 0$$
$$[2(4a+2)+5][(4a+2)-4] = 0$$
$$(8a+9)(4a-2) = 0$$
$$a = -\tfrac{9}{8}, \tfrac{1}{2}$$

13. Solving the equation:
$$6t^4 = -t^2 + 5$$
$$6t^4 + t^2 - 5 = 0$$
$$\left(6t^2 - 5\right)\left(t^2 + 1\right) = 0$$
$$t^2 = \tfrac{5}{6}, -1$$
$$t = \pm\sqrt{\tfrac{5}{6}} = \pm\frac{\sqrt{30}}{6}, \pm i$$

15. Solving the equation:
$$9x^4 - 49 = 0$$
$$\left(3x^2 - 7\right)\left(3x^2 + 7\right) = 0$$
$$x^2 = \tfrac{7}{3}, -\tfrac{7}{3}$$
$$x = \pm\sqrt{\tfrac{7}{3}}, \pm\sqrt{-\tfrac{7}{3}}$$
$$t = \pm\frac{\sqrt{21}}{3}, \pm\frac{i\sqrt{21}}{3}$$

17. Solving the equation:
$$x - 7\sqrt{x} + 10 = 0$$
$$\left(\sqrt{x} - 5\right)\left(\sqrt{x} - 2\right) = 0$$
$$\sqrt{x} = 2, 5$$
$$x = 4, 25$$
Both values check in the original equation.

19. Solving the equation:
$$t - 2\sqrt{t} - 15 = 0$$
$$\left(\sqrt{t} - 5\right)\left(\sqrt{t} + 3\right) = 0$$
$$\sqrt{t} = -3, 5$$
$$t = 9, 25$$
Only $t = 25$ checks in the original equation.

21. Solving the equation:
$$6x + 11\sqrt{x} = 35$$
$$6x + 11\sqrt{x} - 35 = 0$$
$$\left(3\sqrt{x} - 5\right)\left(2\sqrt{x} + 7\right) = 0$$
$$\sqrt{x} = \tfrac{5}{3}, -\tfrac{7}{2}$$
$$x = \tfrac{25}{9}, \tfrac{49}{4}$$

Only $x = \tfrac{25}{9}$ checks in the original equation.

23. Solving the equation:
$$(a - 2) - 11\sqrt{a - 2} + 30 = 0$$
$$\left(\sqrt{a - 2} - 6\right)\left(\sqrt{a - 2} - 5\right) = 0$$
$$\sqrt{a - 2} = 5, 6$$
$$a - 2 = 25, 36$$
$$a = 27, 38$$

25. Solving the equation:
$$(2x + 1) - 8\sqrt{2x + 1} + 15 = 0$$
$$\left(\sqrt{2x + 1} - 3\right)\left(\sqrt{2x + 1} - 5\right) = 0$$
$$\sqrt{2x + 1} = 3, 5$$
$$2x + 1 = 9, 25$$
$$2x = 8, 24$$
$$x = 4, 12$$

27. Solving for t:
$$16t^2 - vt - h = 0$$
$$t = \frac{v \pm \sqrt{v^2 - 4(16)(-h)}}{32} = \frac{v \pm \sqrt{v^2 + 64h}}{32}$$

29. Solving for x:
$$kx^2 + 8x + 4 = 0$$
$$x = \frac{-8 \pm \sqrt{64 - 16k}}{2k} = \frac{-8 \pm 4\sqrt{4 - k}}{2k} = \frac{-4 \pm 2\sqrt{4 - k}}{k}$$

31. Solving for x:
$$x^2 + 2xy + y^2 = 0$$
$$x = \frac{-2y \pm \sqrt{4y^2 - 4y^2}}{2} = \frac{-2y}{2} = -y$$

33. Solving for t (note that $t > 0$):
$$16t^2 - 8t - h = 0$$
$$t = \frac{8 + \sqrt{64 + 64h}}{32} = \frac{8 + 8\sqrt{1 + h}}{32} = \frac{1 + \sqrt{1 + h}}{4}$$

35. **a.**   Sketching the graph:

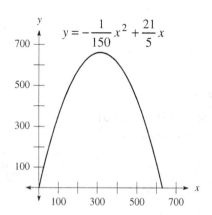

$$y = -\frac{1}{150}x^2 + \frac{21}{5}x$$

b.   Finding the x-intercepts:
$$-\frac{1}{150}x^2 + \frac{21}{5}x = 0$$
$$-\frac{1}{150}x(x - 630) = 0$$
$$x = 0, 630$$
The width is 630 feet.

37.   Let $x = BC$. Solving the proportion:
$$\frac{AB}{BC} = \frac{BC}{AC}$$
$$\frac{4}{x} = \frac{x}{4 + x}$$
$$16 + 4x = x^2$$
$$0 = x^2 - 4x - 16$$
$$x = \frac{4 \pm \sqrt{16 + 64}}{2} = \frac{4 \pm \sqrt{80}}{2} = \frac{4 \pm 4\sqrt{5}}{2} = 2 \pm 2\sqrt{5}$$

Thus $BC = 2 + 2\sqrt{5} = 4\left(\dfrac{1 + \sqrt{5}}{2}\right)$. **39**.   Combining: $5\sqrt{7} - 2\sqrt{7} = 3\sqrt{7}$

41.   Combining: $\sqrt{18} - \sqrt{8} + \sqrt{32} = 3\sqrt{2} - 2\sqrt{2} + 4\sqrt{2} = 5\sqrt{2}$

43.   Combining: $9x\sqrt{20x^3y^2} + 7y\sqrt{45x^5} = 9x \cdot 2xy\sqrt{5x} + 7y \cdot 3x^2\sqrt{5x} = 18x^2y\sqrt{5x} + 21x^2y\sqrt{5x} = 39x^2y\sqrt{5x}$

45.   Multiplying: $\left(\sqrt{5} - 2\right)\left(\sqrt{5} + 8\right) = 5 - 2\sqrt{5} + 8\sqrt{5} - 16 = -11 + 6\sqrt{5}$

47.   Multiplying: $\left(\sqrt{x} + 2\right)^2 = \left(\sqrt{x}\right)^2 + 4\sqrt{x} + 4 = x + 4\sqrt{x} + 4$

49.   Rationalizing the denominator: $\dfrac{\sqrt{7}}{\sqrt{7} - 2} \cdot \dfrac{\sqrt{7} + 2}{\sqrt{7} + 2} = \dfrac{7 + 2\sqrt{7}}{7 - 4} = \dfrac{7 + 2\sqrt{7}}{3}$

51.   To find the x-intercepts, set $y = 0$:
$$x^3 - 4x = 0$$
$$x\left(x^2 - 4\right) = 0$$
$$x(x + 2)(x - 2) = 0$$
$$x = -2, 0, 2$$
To find the y-intercept, set $x = 0$: $y = 0^3 - 4(0) = 0$. The x-intercepts are $-2, 0, 2$ and the y-intercept is 0.

53. To find the x-intercepts, set $y = 0$:
$$3x^3 + x^2 - 27x - 9 = 0$$
$$x^2(3x+1) - 9(3x+1) = 0$$
$$(3x+1)(x^2 - 9) = 0$$
$$(3x+1)(x+3)(x-3) = 0$$
$$x = -3, -\tfrac{1}{3}, 3$$

To find the y-intercept, set $x = 0$: $y = 3(0)^3 + (0)^2 - 27(0) - 9 = -9$

The x-intercepts are $-3, -\tfrac{1}{3}, 3$ and the y-intercept is -9.

55. Using long division:

$$
\begin{array}{r}
2x^2 + x - 1 \\
x-4 \overline{)2x^3 - 7x^2 - 5x + 4} \\
\underline{2x^3 - 8x^2} \\
x^2 - 5x \\
\underline{x^2 - 4x} \\
-x + 4 \\
\underline{-x + 4} \\
0
\end{array}
$$

Now solve the equation:
$$2x^2 + x - 1 = 0$$
$$(2x-1)(x+1) = 0$$
$$x = -1, \tfrac{1}{2}$$

It also crosses the x-axis at $\tfrac{1}{2}, -1$.

8.5   Graphing Parabolas

1. First complete the square: $y = x^2 + 2x - 3 = \left(x^2 + 2x + 1\right) - 1 - 3 = (x+1)^2 - 4$

The x-intercepts are $-3, 1$ and the vertex is $(-1, -4)$. Graphing the parabola:

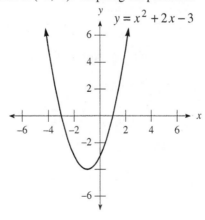

3.   First complete the square: $y = -x^2 - 4x + 5 = -\left(x^2 + 4x + 4\right) + 4 + 5 = -(x+2)^2 + 9$

The x-intercepts are $-5, 1$ and the vertex is $(-2, 9)$. Graphing the parabola:

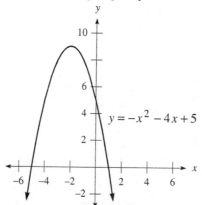

5.   The x-intercepts are $-1, 1$ and the vertex is $(0, -1)$. Graphing the parabola:

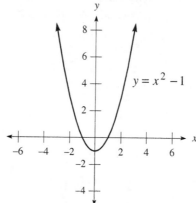

7.   The x-intercepts are $-3, 3$ and the vertex is $(0, 9)$. Graphing the parabola:

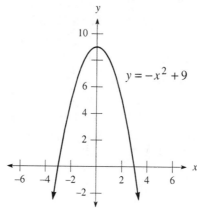

9.   First complete the square: $f(x) = 2x^2 - 4x - 6 = 2(x^2 - 2x + 1) - 2 - 6 = 2(x-1)^2 - 8$

The x-intercepts are $-1, 3$ and the vertex is $(1,-8)$. Graphing the parabola:

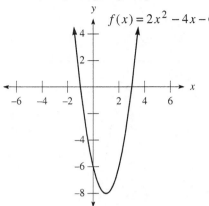

11.   First complete the square: $f(x) = x^2 - 2x - 4 = (x^2 - 2x + 1) - 1 - 4 = (x-1)^2 - 5$

The x-intercepts are $1 \pm \sqrt{5}$ and the vertex is $(1,-5)$. Graphing the parabola:

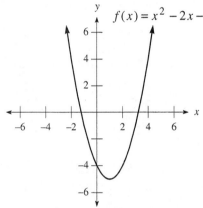

13.   First complete the square: $y = x^2 - 4x - 4 = (x^2 - 4x + 4) - 4 - 4 = (x-2)^2 - 8$

The vertex is $(2,-8)$. Graphing the parabola:

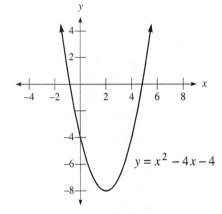

15. First complete the square: $y = -x^2 + 2x - 5 = -(x^2 - 2x + 1) + 1 - 5 = -(x-1)^2 - 4$

The vertex is $(1,-4)$. Graphing the parabola:

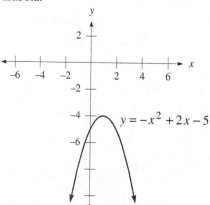

17. The vertex is $(0,1)$. Graphing the parabola:

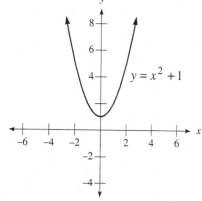

19. The vertex is $(0,-3)$. Graphing the parabola:

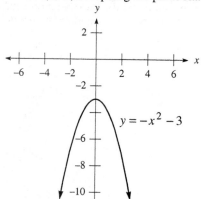

21. Completing the square: $y = x^2 - 6x + 5 = (x^2 - 6x + 9) - 9 + 5 = (x-3)^2 - 4$

The vertex is $(3,-4)$, which is the lowest point on the graph.

23. Completing the square: $y = -x^2 + 2x + 8 = -(x^2 - 2x + 1) + 1 + 8 = -(x-1)^2 + 9$

The vertex is $(1,9)$, which is the highest point on the graph.

25. Completing the square: $y = -x^2 + 4x + 12 = -(x^2 - 4x + 4) + 4 + 12 = -(x-2)^2 + 16$

The vertex is $(2,16)$, which is the highest point on the graph.

27. Completing the square: $y = -x^2 - 8x = -(x^2 + 8x + 16) + 16 = -(x+4)^2 + 16$

The vertex is $(-4,16)$, which is the highest point on the graph.

29. First complete the square:

$$P(x) = -0.002x^2 + 3.5x - 800 = -0.002(x^2 - 1750x + 765625) + 1531.25 - 800 = -0.002(x-875)^2 + 731.25$$

It must sell 875 patterns to obtain a maximum profit of \$731.25.

31. The ball is in her hand at times 0 sec and 2 sec.

Completing the square: $h(t) = -16t^2 + 32t = -16(t^2 - 2t + 1) + 16 = -16(t-1)^2 + 16$

The maximum height of the ball is 16 feet.

33. Let w represent the width, and $80 - 2w$ represent the length. Finding the area:

$$A(w) = w(80 - 2w) = -2w^2 + 80w = -2(w^2 - 40w + 400) + 800 = -2(w-20)^2 + 800$$

The dimensions are 40 feet by 20 feet.

35. Completing the square: $R = xp = 1200p - 100p^2 = -100\left(p^2 - 12p + 36\right) + 3600 = -100(p-6)^2 + 3600$

The price is $6.00 and the maximum revenue is $3,600. Sketching the graph:

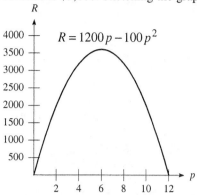

37. Completing the square: $R = xp = 1700p - 100p^2 = -100\left(p^2 - 17p + 72.25\right) + 7225 = -100(p-8.5)^2 + 7225$

The price is $8.50 and the maximum revenue is $7,225. Sketching the graph:

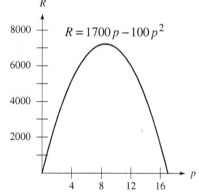

39. Let x represent the number of $10 increases in dues. Then the revenue is given by:

$$y = (10000 - 200x)(100 + 10x)$$
$$= -2000x^2 + 80000x + 1000000$$
$$= -2000\left(x^2 - 40x + 400\right) + 1000000 + 800000$$
$$= -2000(x-20)^2 + 1,800,000$$

The dues should be increased by $200, so the dues should be $300 to result in a maximum income of $1,800,000.

41. Performing the operations: $(3 - 5i) - (2 - 4i) = 3 - 5i - 2 + 4i = 1 - i$

43. Performing the operations: $(3 + 2i)(7 - 3i) = 21 + 5i - 6i^2 = 27 + 5i$

45. Performing the operations: $\dfrac{i}{3+i} = \dfrac{i}{3+i} \cdot \dfrac{3-i}{3-i} = \dfrac{3i - i^2}{9 - i^2} = \dfrac{3i+1}{9+1} = \dfrac{1}{10} + \dfrac{3}{10}i$

47. The equation is $y = (x-2)^2 - 4$.

49. The equation is given on the graph:

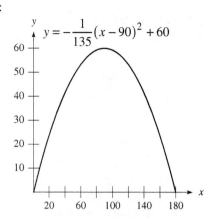

8.6   Quadratic Inequalities

1. Factoring the inequality:
$$x^2 + x - 6 > 0$$
$$(x+3)(x-2) > 0$$
Forming the sign chart:

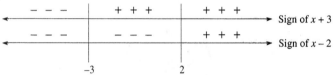

The solution set is $x < -3$ or $x > 2$. Graphing the solution set:

3. Factoring the inequality:
$$x^2 - x - 12 \leq 0$$
$$(x+3)(x-4) \leq 0$$
Forming the sign chart:

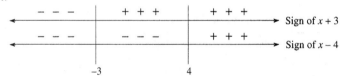

The solution set is $-3 \leq x \leq 4$. Graphing the solution set:

5. Factoring the inequality:
$$x^2 + 5x \geq -6$$
$$x^2 + 5x + 6 \geq 0$$
$$(x+2)(x+3) \geq 0$$
Forming the sign chart:

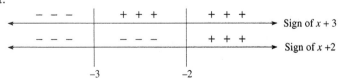

The solution set is $x \le -3$ or $x \ge -2$. Graphing the solution set:

7. Factoring the inequality:
$$6x^2 < 5x - 1$$
$$6x^2 - 5x + 1 < 0$$
$$(3x - 1)(2x - 1) < 0$$

Forming the sign chart:

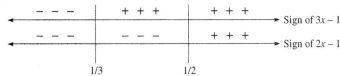

The solution set is $\frac{1}{3} < x < \frac{1}{2}$. Graphing the solution set:

9. Factoring the inequality:
$$x^2 - 9 < 0$$
$$(x + 3)(x - 3) < 0$$

Forming the sign chart:

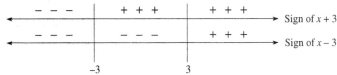

The solution set is $-3 < x < 3$. Graphing the solution set:

11. Factoring the inequality:
$$4x^2 - 9 \ge 0$$
$$(2x + 3)(2x - 3) \ge 0$$

Forming the sign chart:

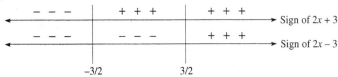

The solution set is $x \le -\frac{3}{2}$ or $x \ge \frac{3}{2}$. Graphing the solution set:

13. Factoring the inequality:
$$2x^2 - x - 3 < 0$$
$$(2x - 3)(x + 1) < 0$$
Forming the sign chart:

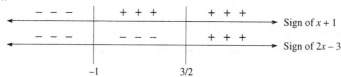

The solution set is $-1 < x < \frac{3}{2}$. Graphing the solution set:

15. Factoring the inequality:
$$x^2 - 4x + 4 \geq 0$$
$$(x - 2)^2 \geq 0$$
Since this inequality is always true, the solution set is all real numbers. Graphing the solution set:

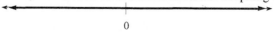

17. Factoring the inequality:
$$x^2 - 10x + 25 < 0$$
$$(x - 5)^2 < 0$$
Since this inequality is never true, there is no solution.

19. Forming the sign chart:

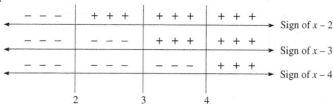

The solution set is $2 < x < 3$ or $x > 4$. Graphing the solution set:

21. Forming the sign chart:

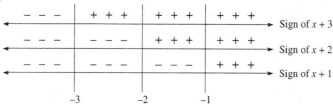

The solution set is $x \leq -3$ or $-2 \leq x \leq -1$. Graphing the solution set:

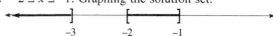

23.   Forming the sign chart:

The solution set is $-4 < x \le 1$. Graphing the solution set:

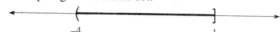

25.   Forming the sign chart:

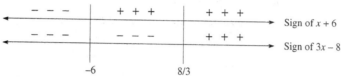

The solution set is $x < -6$ or $x > 8\,/\,3$. Graphing the solution set:

27.   Forming the sign chart:

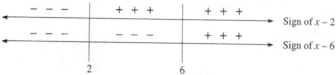

The solution set is $x < 2$ or $x > 6$. Graphing the solution set:

29.   Forming the sign chart:

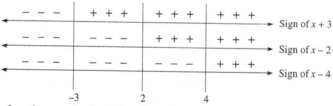

The solution set is $x < -3$ or $2 < x < 4$. Graphing the solution set:

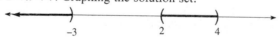

31. First simplify the inequality:
$$\frac{2}{x-4} - \frac{1}{x-3} > 0$$
$$\frac{2(x-3)-(x-4)}{(x-3)(x-4)} > 0$$
$$\frac{2x-6-x+4}{(x-3)(x-4)} > 0$$
$$\frac{x-2}{(x-3)(x-4)} > 0$$

Forming the sign chart:

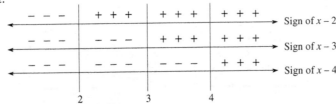

The solution set is $2 < x < 3$ or $x > 4$. Graphing the solution set:

33. a. The solution set is $-2 < x < 2$.
    b. The solution set is $x < -2$ or $x > 2$.
    c. The solution set is $x = -2, 2$.

35. a. The solution set is $-2 < x < 5$.
    b. The solution set is $x < -2$ or $x > 5$.
    c. The solution set is $x = -2, 5$.

37. a. The solution set is $x < -1$ or $1 < x < 3$.
    b. The solution set is $-1 < x < 1$ or $x > 3$.
    c. The solution set is $x = -1, 1, 3$.

39. Let w represent the width and $2w + 3$ represent the length. Using the area formula:
$$w(2w+3) \geq 44$$
$$2w^2 + 3w \geq 44$$
$$2w^2 + 3w - 44 \geq 0$$
$$(2w+11)(w-4) \geq 0$$

Forming the sign chart:

The width is at least 4 inches.

41. Solving the inequality:
$$1300p - 100p^2 \geq 4000$$
$$-100p^2 + 1300p - 4000 \geq 0$$
$$p^2 - 13p + 40 \leq 0$$
$$(p-8)(p-5) \leq 0$$

Forming the sign chart:

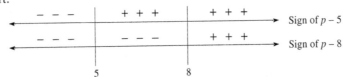

Charge at least $5 but no more than $8 per radio.

43. Solving the equation:
$$\sqrt{3t-1}=2$$
$$\left(\sqrt{3t-1}\right)^2=(2)^2$$
$$3t-1=4$$
$$3t=5$$
$$t=\tfrac{5}{3}$$

The solution is $\tfrac{5}{3}$.

45. Solving the equation:
$$\sqrt{x+3}=x-3$$
$$\left(\sqrt{x+3}\right)^2=(x-3)^2$$
$$x+3=x^2-6x+9$$
$$0=x^2-7x+6$$
$$0=(x-6)(x-1)$$
$$x=1,6 \qquad (x=1 \text{ does not check})$$

The solution is 6.

47. Graphing the equation:

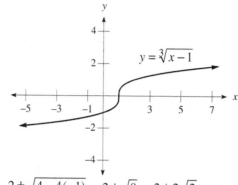

49. Using the quadratic formula: $x=\dfrac{2\pm\sqrt{4-4(-1)}}{2(1)}=\dfrac{2\pm\sqrt{8}}{2}=\dfrac{2\pm 2\sqrt{2}}{2}=1\pm\sqrt{2}$

The inequality is satisfied when $1-\sqrt{2}<x<1+\sqrt{2}$.

51. Using the quadratic formula: $x=\dfrac{8\pm\sqrt{64-4(13)}}{2(1)}=\dfrac{8\pm\sqrt{12}}{2}=\dfrac{8\pm 2\sqrt{3}}{2}=4\pm\sqrt{3}$

The inequality is satisfied when $x<4-\sqrt{3}$ or $x>4+\sqrt{3}$.

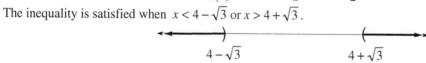

Chapter 8 Review

1. Solving the equation:
$$(2t-5)^2 = 25$$
$$2t-5 = \pm 5$$
$$2t-5 = -5, 5$$
$$2t = 0, 10$$
$$t = 0, 5$$

3. Solving the equation:
$$(3y-4)^2 = -49$$
$$3y-4 = \pm\sqrt{-49}$$
$$3y-4 = \pm 7i$$
$$3y = 4 \pm 7i$$
$$y = \frac{4 \pm 7i}{3}$$

5. Solving by completing the square:
$$2x^2 + 6x - 20 = 0$$
$$x^2 + 3x = 10$$
$$x^2 + 3x + \frac{9}{4} = 10 + \frac{9}{4}$$
$$\left(x + \frac{3}{2}\right)^2 = \frac{49}{4}$$
$$x + \frac{3}{2} = -\frac{7}{2}, \frac{7}{2}$$
$$x = -5, 2$$

7. Solving by completing the square:
$$a^2 + 9 = 6a$$
$$a^2 - 6a = -9$$
$$a^2 - 6a + 9 = -9 + 9$$
$$(a-3)^2 = 0$$
$$a - 3 = 0$$
$$a = 3$$

9. Solving by completing the square:
$$2y^2 + 6y = -3$$
$$y^2 + 3y = -\frac{3}{2}$$
$$y^2 + 3y + \frac{9}{4} = -\frac{3}{2} + \frac{9}{4}$$
$$\left(y + \frac{3}{2}\right)^2 = \frac{3}{4}$$
$$y + \frac{3}{2} = \pm\frac{\sqrt{3}}{2}$$
$$y = \frac{-3 \pm \sqrt{3}}{2}$$

11. Solving by completing the square:
$$\frac{1}{6}x^2 + \frac{1}{2}x - \frac{5}{3} = 0$$
$$x^2 + 3x - 10 = 0$$
$$(x+5)(x-2) = 0$$
$$x = -5, 2$$

13. Solving the equation:
$$4t^2 - 8t + 19 = 0$$
$$t = \frac{8 \pm \sqrt{64 - 304}}{8} = \frac{8 \pm \sqrt{-240}}{8} = \frac{8 \pm 4i\sqrt{15}}{8} = \frac{2 \pm i\sqrt{15}}{2}$$

15. Solving the equation:
$$0.06a^2 + 0.05a = 0.04$$
$$0.06a^2 + 0.05a - 0.04 = 0$$
$$6a^2 + 5a - 4 = 0$$
$$(2a-1)(3a+4) = 0$$
$$a = -\frac{4}{3}, \frac{1}{2}$$

17. Solving the equation:
$$(2x+1)(x-5) - (x+3)(x-2) = -17$$
$$2x^2 - 9x - 5 - x^2 - x + 6 = -17$$
$$x^2 - 10x + 1 = -17$$
$$x^2 - 10x + 18 = 0$$
$$x = \frac{10 \pm \sqrt{100 - 72}}{2} = \frac{10 \pm \sqrt{28}}{2} = \frac{10 \pm 2\sqrt{7}}{2} = 5 \pm \sqrt{7}$$

19. Solving the equation:
$$5x^2 = -2x + 3$$
$$5x^2 + 2x - 3 = 0$$
$$(x+1)(5x-3) = 0$$
$$x = -1, \tfrac{3}{5}$$

21. Solving the equation:
$$3 - \frac{2}{x} + \frac{1}{x^2} = 0$$
$$3x^2 - 2x + 1 = 0$$
$$x = \frac{2 \pm \sqrt{4 - 12}}{6} = \frac{2 \pm \sqrt{-8}}{6} = \frac{2 \pm 2i\sqrt{2}}{6} = \frac{1 \pm i\sqrt{2}}{3}$$

23. The profit equation is given by:
$$34x - 0.1x^2 - 7x - 400 = 1300$$
$$-0.1x^2 + 27x - 1700 = 0$$
$$x^2 - 270x + 17000 = 0$$
$$(x - 100)(x - 170) = 0$$
$$x = 100, 170$$
The company must sell either 100 or 170 items for its weekly profit to be \$1,300.

25. First write the equation as $2x^2 - 8x + 8 = 0$. Computing the discriminant: $D = (-8)^2 - 4(2)(8) = 64 - 64 = 0$
The equation will have one rational solution.

27. Computing the discriminant: $D = (1)^2 - 4(2)(-3) = 1 + 24 = 25$. The equation will have two rational solutions.

29. First write the equation as $x^2 - x - 1 = 0$. Computing the discriminant: $D = (-1)^2 - 4(1)(-1) = 1 + 4 = 5$
The equation will have two irrational solutions.

31. First write the equation as $3x^2 + 5x + 4 = 0$. Computing the discriminant: $D = (5)^2 - 4(3)(4) = 25 - 48 = -23$
The equation will have two complex solutions.

33. Setting the discriminant equal to 0:
$$(-k)^2 - 4(25)(4) = 0$$
$$k^2 - 400 = 0$$
$$k^2 = 400$$
$$k = \pm 20$$

35. Setting the discriminant equal to 0:
$$(12)^2 - 4(k)(9) = 0$$
$$144 - 36k = 0$$
$$36k = 144$$
$$k = 4$$

37. Setting the discriminant equal to 0:
$$30^2 - 4(9)(k) = 0$$
$$900 - 36k = 0$$
$$36k = 900$$
$$k = 25$$

39. Writing the equation:
$$(x - 3)(x - 5) = 0$$
$$x^2 - 8x + 15 = 0$$

41. Writing the equation:
$$(2y - 1)(y + 4) = 0$$
$$2y^2 + 7y - 4 = 0$$

43. Solving the equation:
$$(x - 2)^2 - 4(x - 2) - 60 = 0$$
$$(x - 2 - 10)(x - 2 + 6) = 0$$
$$(x - 12)(x + 4) = 0$$
$$x = -4, 12$$

45. Solving the equation:
$$x^4 - x^2 = 12$$
$$x^4 - x^2 - 12 = 0$$
$$(x^2 - 4)(x^2 + 3) = 0$$
$$x^2 = 4, -3$$
$$x = \pm 2, \pm i\sqrt{3}$$

47. Solving the equation:
$$2x - 11\sqrt{x} = -12$$
$$2x - 11\sqrt{x} + 12 = 0$$
$$(2\sqrt{x} - 3)(\sqrt{x} - 4) = 0$$
$$\sqrt{x} = \tfrac{3}{2}, 4$$
$$x = \tfrac{9}{4}, 16$$

49.   Solving the equation:
$$\sqrt{y+21} + \sqrt{y} = 7$$
$$\left(\sqrt{y+21}\right)^2 = \left(7 - \sqrt{y}\right)^2$$
$$y + 21 = 49 - 14\sqrt{y} + y$$
$$14\sqrt{y} = 28$$
$$\sqrt{y} = 2$$
$$y = 4$$

51.   Solving for t:
$$16t^2 - 10t - h = 0$$
$$t = \frac{10 \pm \sqrt{100 + 64h}}{32} = \frac{10 \pm 2\sqrt{25 + 16h}}{32} = \frac{5 \pm \sqrt{25 + 16h}}{16}$$

53.   Factoring the inequality:
$$x^2 - x - 2 < 0$$
$$(x-2)(x+1) < 0$$
Forming a sign chart:

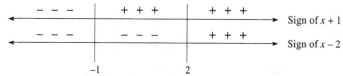

The solution set is $-1 < x < 2$. Graphing the solution set:

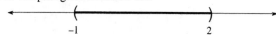

55.   Factoring the inequality:
$$2x^2 + 5x - 12 \geq 0$$
$$(2x-3)(x+4) \geq 0$$
Forming a sign chart:

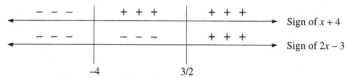

The solution set is $x \leq -4$ or $x \geq \frac{3}{2}$. Graphing the solution set:

57. First complete the square: $y = x^2 - 6x + 8 = \left(x^2 - 6x + 9\right) + 8 - 9 = (x-3)^2 - 1$

The x-intercepts are 2,4, and the vertex is $(3,-1)$. Graphing the parabola:

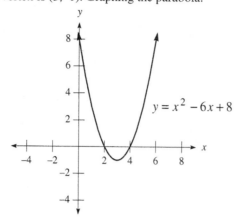

$y = x^2 - 6x + 8$

Chapters 1-8 Cumulative Review

1. Simplifying: $11 + 20 \div 5 - 3 \cdot 5 = 11 + 4 - 15 = 0$
3. Simplifying: $4(15-19)^2 - 3(17-19)^3 = 4(-4)^2 - 3(-2)^3 = 4(16) - 3(-8) = 64 + 24 = 88$
5. Simplifying: $3 - 5[2x - 4(x-2)] = 3 - 5[2x - 4x + 8] = 3 - 5(-2x + 8) = 3 + 10x - 40 = 10x - 37$
7. Simplifying: $\sqrt[3]{32} = \sqrt[3]{8 \cdot 4} = 2\sqrt[3]{4}$
9. Simplifying: $\dfrac{1 - \frac{3}{4}}{1 + \frac{3}{4}} = \dfrac{1 - \frac{3}{4}}{1 + \frac{3}{4}} \cdot \dfrac{4}{4} = \dfrac{4-3}{4+3} = \dfrac{1}{7}$

11. Reducing the fraction: $\dfrac{5x^2 - 26xy - 24y^2}{5x + 4y} = \dfrac{(5x + 4y)(x - 6y)}{5x + 4y} = x - 6y$

13. Multiplying: $(3x - 2)\left(x^2 - 3x - 2\right) = 3x^3 - 9x^2 - 6x - 2x^2 + 6x + 4 = 3x^3 - 11x^2 + 4$

15. Dividing: $\dfrac{7 - i}{3 - 2i} = \dfrac{7 - i}{3 - 2i} \cdot \dfrac{3 + 2i}{3 + 2i} = \dfrac{21 + 11i - 2i^2}{9 + 4} = \dfrac{23 + 11i}{13} = \dfrac{23}{13} + \dfrac{11}{13}i$

17. Solving the equation:
$$\frac{7}{5}a - 6 = 15$$
$$\frac{7}{5}a = 21$$
$$7a = 105$$
$$a = 15$$

19. Solving the equation:
$$\frac{a}{2} + \frac{3}{a-3} = \frac{a}{a-3}$$
$$2(a-3)\left(\frac{a}{2} + \frac{3}{a-3}\right) = 2(a-3)\left(\frac{a}{a-3}\right)$$
$$a(a-3) + 6 = 2a$$
$$a^2 - 3a + 6 = 2a$$
$$a^2 - 5a + 6 = 0$$
$$(a-2)(a-3) = 0$$
$$a = 2,3 \qquad (a = 3 \text{ does not check})$$

21.   Solving the equation:
$$(3x-4)^2 = 18$$
$$3x-4 = \pm\sqrt{18}$$
$$3x-4 = \pm3\sqrt{2}$$
$$3x = 4 \pm 3\sqrt{2}$$
$$x = \frac{4 \pm 3\sqrt{2}}{3}$$

23.   Solving the equation:
$$3y^3 - y = 5y^2$$
$$3y^3 - 5y^2 - y = 0$$
$$y\left(3y^2 - 5y - 1\right) = 0$$
$$y = 0, \frac{5 \pm \sqrt{25+12}}{6} = \frac{5 \pm \sqrt{37}}{6}$$

25.   Solving the equation:
$$\sqrt{x-2} = 2 - \sqrt{x}$$
$$\left(\sqrt{x-2}\right)^2 = \left(2-\sqrt{x}\right)^2$$
$$x - 2 = 4 - 4\sqrt{x} + x$$
$$-4\sqrt{x} = -6$$
$$\sqrt{x} = \tfrac{3}{2}$$
$$x = \tfrac{9}{4}$$

27.   Solving the inequality:
$$5 \le \tfrac{1}{4}x + 3 \le 8$$
$$2 \le \tfrac{1}{4}x \le 5$$
$$8 \le x \le 20$$
Graphing the solution set:

8   20

29.   Multiply the first equation by 2:
$$6x - 2y = 4$$
$$-6x + 2y = -4$$
Adding yields $0 = 0$, which is true. The two lines coincide.

31.   Substituting into the first equation:
$$3x - 2(3x-7) = 5$$
$$3x - 6x + 14 = 5$$
$$-3x + 14 = 5$$
$$-3x = -9$$
$$x = 3$$
$$y = 3 \cdot 3 - 7 = 2$$
The solution is $(3,2)$.

33.   Graphing the line:

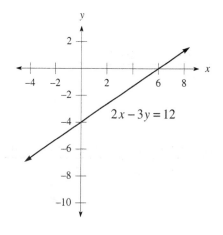

35. First find the slope: $m = \dfrac{-\frac{1}{3} - \frac{4}{3}}{\frac{1}{4} - \frac{3}{2}} = \dfrac{-\frac{5}{3}}{-\frac{5}{4}} = \frac{4}{3}$. Using the point-slope formula:

$$y - \tfrac{4}{3} = \tfrac{4}{3}\left(x - \tfrac{3}{2}\right)$$
$$y - \tfrac{4}{3} = \tfrac{4}{3}x - 2$$
$$y = \tfrac{4}{3}x - \tfrac{2}{3}$$

37. Rationalizing the denominator: $\dfrac{7}{\sqrt[3]{9}} = \dfrac{7}{\sqrt[3]{9}} \cdot \dfrac{\sqrt[3]{3}}{\sqrt[3]{3}} = \dfrac{7\sqrt[3]{3}}{3}$

39. Let x represent the largest angle, $\frac{1}{4}x$ represent the smallest angle, and $\frac{1}{4}x + 30$ represent the remaining angle. The equation is:

$$x + \tfrac{1}{4}x + \tfrac{1}{4}x + 30 = 180$$
$$\tfrac{3}{2}x + 30 = 180$$
$$\tfrac{3}{2}x = 150$$
$$3x = 300$$
$$x = 100$$

The angles are $25°, 55°,$ and $100°$.

Chapter 8 Test

1. Solving the equation:

$$(2x + 4)^2 = 25$$
$$2x + 4 = \pm 5$$
$$2x + 4 = -5, 5$$
$$2x = -9, 1$$
$$x = -\tfrac{9}{2}, \tfrac{1}{2}$$

2. Solving the equation:

$$(2x - 6)^2 = -8$$
$$2x - 6 = \pm\sqrt{-8}$$
$$2x - 6 = \pm 2i\sqrt{2}$$
$$2x = 6 \pm 2i\sqrt{2}$$
$$x = 3 \pm i\sqrt{2}$$

3. Solving the equation:

$$y^2 - 10y + 25 = -4$$
$$(y - 5)^2 = -4$$
$$y - 5 = \pm 2i$$
$$y = 5 \pm 2i$$

4. Solving the equation:

$$(y + 1)(y - 3) = -6$$
$$y^2 - 2y - 3 = -6$$
$$y^2 - 2y + 3 = 0$$
$$y = \frac{2 \pm \sqrt{4 - 12}}{2} = \frac{2 \pm \sqrt{-8}}{2} = \frac{2 \pm 2i\sqrt{2}}{2} = 1 \pm i\sqrt{2}$$

5. Solving the equation:

$$8t^3 - 125 = 0$$
$$(2t - 5)(4t^2 + 10t + 25) = 0$$
$$t = \tfrac{5}{2}, \frac{-10 \pm \sqrt{100 - 400}}{8} = \frac{-10 \pm \sqrt{-300}}{8} = \frac{-10 \pm 10i\sqrt{3}}{8} = \frac{-5 \pm 5i\sqrt{3}}{4}$$

6.   Solving the equation:
$$\frac{1}{a+2} - \frac{1}{3} = \frac{1}{a}$$
$$3a(a+2)\left(\frac{1}{a+2} - \frac{1}{3}\right) = 3a(a+2)\left(\frac{1}{a}\right)$$
$$3a - a(a+2) = 3(a+2)$$
$$3a - a^2 - 2a = 3a + 6$$
$$a^2 + 2a + 6 = 0$$
$$a = \frac{-2 \pm \sqrt{4-24}}{2} = \frac{-2 \pm \sqrt{-20}}{2} = \frac{-2 \pm 2i\sqrt{5}}{2} = -1 \pm i\sqrt{5}$$

7.   Solving for r:
$$64(1+r)^2 = A$$
$$(1+r)^2 = \frac{A}{64}$$
$$1 + r = \pm\frac{\sqrt{A}}{8}$$
$$r = \pm\frac{\sqrt{A}}{8} - 1$$

8.   Solving by completing the square:
$$x^2 - 4x = -2$$
$$x^2 - 4x + 4 = -2 + 4$$
$$(x-2)^2 = 2$$
$$x - 2 = \pm\sqrt{2}$$
$$x = 2 \pm \sqrt{2}$$

9.   Solving the equation:
$$32t - 16t^2 = 12$$
$$-16t^2 + 32t - 12 = 0$$
$$4t^2 - 8t + 3 = 0$$
$$(2t-1)(2t-3) = 0$$
$$t = \tfrac{1}{2}, \tfrac{3}{2}$$
The object will be 12 feet above the ground after $\frac{1}{2}$ or $\frac{3}{2}$ sec.

10.   Setting the profit equal to $200:
$$25x - 0.2x^2 - 2x - 100 = 200$$
$$-0.2x^2 + 23x - 300 = 0$$
$$x^2 - 115x + 1500 = 0$$
$$(x-15)(x-100) = 0$$
$$x = 15, 100$$
The company must sell 15 or 100 cups to make a weekly profit of $200.

11.   First write the equation as $kx^2 - 12x + 4 = 0$. Setting the discriminant equal to 0:
$$(-12)^2 - 4(k)(4) = 0$$
$$144 - 16k = 0$$
$$16k = 144$$
$$k = 9$$

12.   First write the equation as $2x^2 - 5x - 7 = 0$. Finding the discriminant: $D = (-5)^2 - 4(2)(-7) = 25 + 56 = 81$
The equation has two rational solutions.

13.   Finding the equation:
$$(x-5)(3x+2) = 0$$
$$3x^2 - 13x - 10 = 0$$

14.   Finding the equation:
$$(x-2)(x+2)(x-7) = 0$$
$$(x^2 - 4)(x-7) = 0$$
$$x^3 - 7x^2 - 4x + 28 = 0$$

15. Solving the equation:
$$4x^4 - 7x^2 - 2 = 0$$
$$\left(x^2 - 2\right)\left(4x^2 + 1\right) = 0$$
$$x^2 = 2, -\tfrac{1}{4}$$
$$x = \pm\sqrt{2}, \pm\tfrac{1}{2}i$$

16. Solving the equation:
$$(2t+1)^2 - 5(2t+1) + 6 = 0$$
$$(2t+1-3)(2t+1-2) = 0$$
$$(2t-2)(2t-1) = 0$$
$$t = \tfrac{1}{2}, 1$$

17. Solving the equation:
$$2t - 7\sqrt{t} + 3 = 0$$
$$\left(2\sqrt{t} - 1\right)\left(\sqrt{t} - 3\right) = 0$$
$$\sqrt{t} = \tfrac{1}{2}, 3$$
$$t = \tfrac{1}{4}, 9$$

18. Solving for t:
$$16t^2 - 14t - h = 0$$
$$t = \frac{14 + \sqrt{196 - 4(16)(-h)}}{32} = \frac{14 + \sqrt{196 + 64h}}{32} = \frac{14 + 4\sqrt{49 + 16h}}{32} = \frac{7 + \sqrt{49 + 16h}}{16}$$

19. Completing the square: $y = x^2 - 2x - 3 = \left(x^2 - 2x + 1\right) - 1 - 3 = (x-1)^2 - 4$

The vertex is $(1,-4)$. Graphing the parabola:

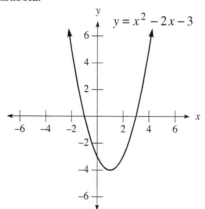

20. Completing the square: $y = -x^2 + 2x + 8 = -\left(x^2 - 2x + 1\right) + 1 + 8 = -(x-1)^2 + 9$

The vertex is $(1,9)$. Graphing the parabola:

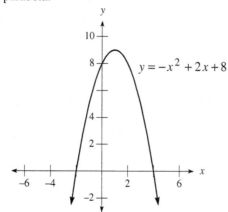

21.   Factoring the inequality:
$$x^2 - x - 6 \le 0$$
$$(x-3)(x+2) \le 0$$
Forming a sign chart:

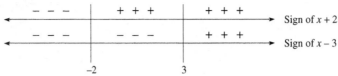

The solution set is $-2 \le x \le 3$. Graphing the solution set:

22.   Factoring the inequality:
$$2x^2 + 5x > 3$$
$$2x^2 + 5x - 3 > 0$$
$$(x+3)(2x-1) > 0$$
Forming a sign chart:

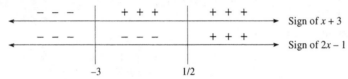

The solution set is $x < -3$ or $x > \frac{1}{2}$. Graphing the solution set:

23.   Finding the profit and completing the square:
$$P = 25x - 0.1x^2 - 5x - 100$$
$$= -0.1x^2 + 20x - 100$$
$$= -0.1\left(x^2 - 200x + 10000\right) + 1000 - 100$$
$$= -0.1(x-100)^2 + 900$$
The maximum weekly profit is \$900, obtained by selling 100 items per week.

Chapter 9
Exponential and Logarithmic Functions

9.1   Exponential Functions

1. Evaluating: $g(0) = \left(\frac{1}{2}\right)^0 = 1$

3. Evaluating: $g(-1) = \left(\frac{1}{2}\right)^{-1} = 2$

5. Evaluating: $f(-3) = 3^{-3} = \frac{1}{27}$

7. Evaluating: $f(2) + g(-2) = 3^2 + \left(\frac{1}{2}\right)^{-2} = 9 + 4 = 13$

9. Graphing the function:

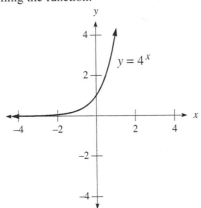

11. Graphing the function:

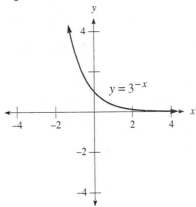

13. Graphing the function:

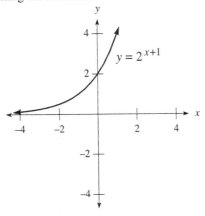

15. Graphing the function:

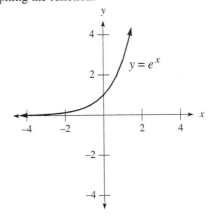

17. Graphing the function:

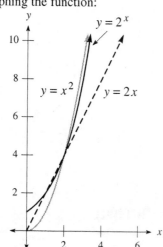

19. Graphing the function:

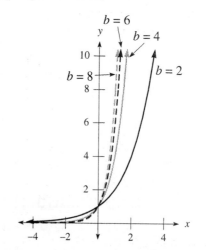

21. The equation is: $h(n) = 6\left(\frac{2}{3}\right)^n$

Substituting $n = 5$: $h(5) = 6\left(\frac{2}{3}\right)^5 \approx 0.79$ feet

23. After 8 days, there will be: $1400 \cdot 2^{-8/8} = 1400 \cdot \frac{1}{2} = 700$ micrograms

After 11 days, there will be: $1400 \cdot 2^{-11/8} \approx 539.8$ micrograms

25. **a.** The equation is $A(t) = 1200\left(1 + \frac{0.06}{4}\right)^{4t}$.   **b.** Substitute $t = 8$: $A(8) = 1200\left(1 + \frac{0.06}{4}\right)^{32} \approx \$1,932.39$

c. Using a graphing calculator, the time is approximately 11.6 years.

d. Substitute $t = 8$ into the compound interest formula: $A(8) = 1200e^{0.06 \times 8} \approx \$1,939.29$

27. **a.** Substitute $t = 20$: $E(20) = 78.16(1.11)^{20} \approx \$630$ billion. The estimate is \$69 billion too low.

b. For 2005, substitute $t = 35$: $E(35) = 78.16(1.11)^{35} \approx \$3,015$ billion

For 2006, substitute $t = 36$: $E(36) = 78.16(1.11)^{36} \approx \$3,347$ billion

For 2007, substitute $t = 37$: $E(37) = 78.16(1.11)^{37} \approx \$3,715$ billion

29. Graphing the function:

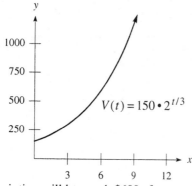

31. The painting will be worth \$600 after approximately 6 years.

33. **a.** Substitute $t = 3.5$: $V(5) = 450{,}000(1 - 0.30)^5 \approx \$129{,}138.48$

 b. The domain is $\{t \mid 0 \le t \le 6\}$.

 c. Sketching the graph:

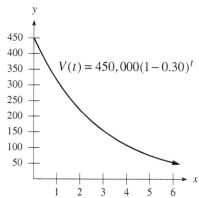

 d. The range is $\{V(t) \mid 52{,}942.05 \le V(t) \le 450{,}000\}$.

 e. From the graph, the crane will be worth \$85,000 after approximately 4.7 years, or 4 years 8 months.

35. The domain is $\{1, 3, 4\}$ and the range is $\{1, 2, 4\}$. This is a function.

37. Find where the quantity inside the radical is non-negative:

$$3x + 1 \ge 0$$
$$3x \ge -1$$
$$x \ge -\frac{1}{3}$$

The domain is $\left\{ x \mid x \ge -\frac{1}{3} \right\}$.

39. Evaluating the function: $f(0) = 2(0)^2 - 18 = 0 - 18 = -18$

41. Simplifying the function: $\dfrac{g(x+h) - g(x)}{h} = \dfrac{2(x+h) - 6 - (2x - 6)}{h} = \dfrac{2x + 2h - 6 - 2x + 6}{h} = \dfrac{2h}{h} = 2$

43. Graphing the function:

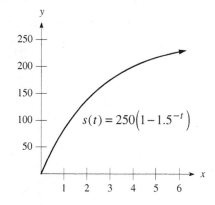

9.2   The Inverse of a Function

1.   Let $y = f(x)$. Switch x and y and solve for y:

$$3y - 1 = x$$
$$3y = x + 1$$
$$y = \frac{x+1}{3}$$

The inverse is $f^{-1}(x) = \frac{x+1}{3}$.

3.   Let $y = f(x)$. Switch x and y and solve for y:

$$y^3 = x$$
$$y = \sqrt[3]{x}$$

The inverse is $f^{-1}(x) = \sqrt[3]{x}$.

5.   Let $y = f(x)$. Switch x and y and solve for y:

$$\frac{y-3}{y-1} = x$$
$$y - 3 = xy - x$$
$$y - xy = 3 - x$$
$$y(1-x) = 3 - x$$
$$y = \frac{3-x}{1-x} = \frac{x-3}{x-1}$$

The inverse is $f^{-1}(x) = \frac{x-3}{x-1}$.

7.   Let $y = f(x)$. Switch x and y and solve for y:

$$\frac{y-3}{4} = x$$
$$y - 3 = 4x$$
$$y = 4x + 3$$

The inverse is $f^{-1}(x) = 4x + 3$.

9.   Let $y = f(x)$. Switch x and y and solve for y:

$$\tfrac{1}{2}y - 3 = x$$
$$y - 6 = 2x$$
$$y = 2x + 6$$

The inverse is $f^{-1}(x) = 2x + 6$.

11.   Let $y = f(x)$. Switch x and y and solve for y:

$$\frac{2y+1}{3y+1} = x$$
$$2y + 1 = 3xy + x$$
$$2y - 3xy = x - 1$$
$$y(2 - 3x) = x - 1$$
$$y = \frac{x-1}{2-3x} = \frac{1-x}{3x-2}$$

The inverse is $f^{-1}(x) = \frac{1-x}{3x-2}$.

13.   Finding the inverse:

$$2y - 1 = x$$
$$2y = x + 1$$
$$y = \frac{x+1}{2}$$

The inverse is $y^{-1} = \frac{x+1}{2}$. Graphing each curve:

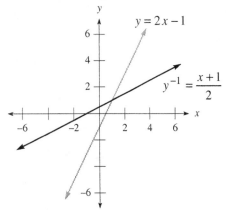

15.   Finding the inverse:

$$y^2 - 3 = x$$
$$y^2 = x + 3$$
$$y = \pm\sqrt{x+3}$$

The inverse is $y^{-1} = \pm\sqrt{x+3}$. Graphing each curve:

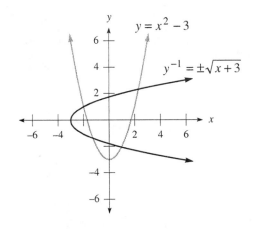

17.   Finding the inverse:
$$y^2 - 2y - 3 = x$$
$$y^2 - 2y + 1 = x + 3 + 1$$
$$(y-1)^2 = x + 4$$
$$y - 1 = \pm\sqrt{x+4}$$
$$y = 1 \pm \sqrt{x+4}$$

The inverse is $y^{-1} = 1 \pm \sqrt{x+4}$. Graphing each curve:

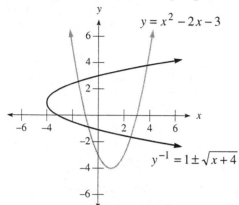

19.   The inverse is $x = 3^y$. Graphing each curve:

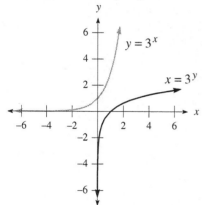

21.   The inverse is $x = 4$. Graphing each curve:

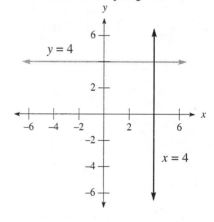

23.   Finding the inverse:
$$\tfrac{1}{2}y^3 = x$$
$$y^3 = 2x$$
$$y = \sqrt[3]{2x}$$
The inverse is $y^{-1} = \sqrt[3]{2x}$. Graphing each curve:

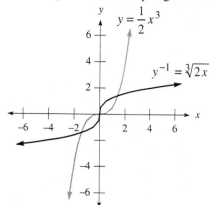

25.   Finding the inverse:
$$\tfrac{1}{2}y + 2 = x$$
$$y + 4 = 2x$$
$$y = 2x - 4$$
The inverse is $y^{-1} = 2x - 4$. Graphing each curve:

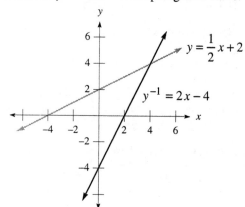

27.   Finding the inverse:
$$\sqrt{y+2} = x$$
$$y + 2 = x^2$$
$$y = x^2 - 2$$
The inverse is $y^{-1} = x^2 - 2, x \geq 0$. Graphing each curve:

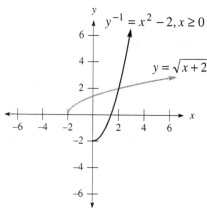

29.   **a.**   Yes, this function is one-to-one.   **b.**   No, this function is not one-to-one.
          c.   Yes, this function is one-to-one.

31.   **a.**   Evaluating the function: $f(2) = 3(2) - 2 = 6 - 2 = 4$

          b.   Evaluating the function: $f^{-1}(2) = \dfrac{2+2}{3} = \tfrac{4}{3}$

          c.   Evaluating the function: $f\left[f^{-1}(2)\right] = f\left(\tfrac{4}{3}\right) = 3\left(\tfrac{4}{3}\right) - 2 = 4 - 2 = 2$

          d.   Evaluating the function: $f^{-1}\left[f(2)\right] = f^{-1}(4) = \dfrac{4+2}{3} = \dfrac{6}{3} = 2$

33. Let $y = f(x)$. Switch x and y and solve for y:

$$\frac{1}{y} = x$$

$$y = \frac{1}{x}$$

The inverse is $f^{-1}(x) = \dfrac{1}{x}$.

35. The inverse is $f^{-1}(x) = 7(x+2)$.

37. **a.** The value is -3.

     c. The value is 2.

     e. The value is -2.

     g. Each is an inverse of the other.

     b. The value is -6.

     d. The value is 3.

     f. The value is 3.

39. Solving the equation:

$$(2x-1)^2 = 25$$

$$2x - 1 = \pm\sqrt{25}$$

$$2x - 1 = -5, 5$$

$$2x = -4, 6$$

$$x = -2, 3$$

41. The number is 25, since $x^2 - 10x + 25 = (x-5)^2$.

43. Solving the equation:

$$x^2 - 10x + 8 = 0$$

$$x^2 - 10x + 25 = -8 + 25$$

$$(x-5)^2 = 17$$

$$x - 5 = \pm\sqrt{17}$$

$$x = 5 \pm \sqrt{17}$$

45. Solving the equation:

$$3x^2 - 6x + 6 = 0$$

$$x^2 - 2x + 2 = 0$$

$$x^2 - 2x + 1 = -2 + 1$$

$$(x-1)^2 = -1$$

$$x - 1 = \pm i$$

$$x = 1 \pm i$$

47. Finding the inverse:

$$3y + 5 = x$$

$$3y = x - 5$$

$$y = \frac{x-5}{3}$$

So $f^{-1}(x) = \dfrac{x-5}{3}$. Now verifying the inverse: $f\left[f^{-1}(x)\right] = f\left(\dfrac{x-5}{3}\right) = 3\left(\dfrac{x-5}{3}\right) + 5 = x - 5 + 5 = x$

49. Finding the inverse:

$$y^3 + 1 = x$$

$$y^3 = x - 1$$

$$y = \sqrt[3]{x-1}$$

So $f^{-1}(x) = \sqrt[3]{x-1}$. Now verifying the inverse: $f\left[f^{-1}(x)\right] = f\left(\sqrt[3]{x-1}\right) = \left(\sqrt[3]{x-1}\right)^3 + 1 = x - 1 + 1 = x$

51.   Finding the inverse:
$$\frac{y-4}{y-2} = x$$
$$y - 4 = xy - 2x$$
$$y - xy = 4 - 2x$$
$$y(1-x) = 4 - 2x$$
$$y = \frac{4-2x}{1-x} = \frac{2x-4}{x-1}$$

So $f^{-1}(x) = \frac{2x-4}{x-1}$. Now verifying the inverse:

$$f\left[f^{-1}(x)\right] = f\left(\frac{2x-4}{x-1}\right) = \frac{\frac{2x-4}{x-1} - 4}{\frac{2x-4}{x-1} - 2} = \frac{2x-4-4(x-1)}{2x-4-2(x-1)} = \frac{2x-4-4x+4}{2x-4-2x+2} = \frac{-2x}{-2} = x$$

53.   **a.**   From the graph: $f(0) = 1$ **b.**   From the graph: $f(1) = 2$

   c.   From the graph: $f(2) = 5$ **d.**   From the graph: $f^{-1}(1) = 0$

   e.   From the graph: $f^{-1}(2) = 1$ **f.**   From the graph: $f^{-1}(5) = 2$

   g.   From the graph: $f^{-1}\left[f(2)\right] = 2$ **h.**   From the graph: $f\left[f^{-1}(5)\right] = 5$

9.3   Logarithms Are Exponents

1.   Writing in logarithmic form: $\log_2 16 = 4$

3.   Writing in logarithmic form: $\log_5 125 = 3$

5.   Writing in logarithmic form: $\log_{10} 0.01 = -2$

7.   Writing in logarithmic form: $\log_2 \frac{1}{32} = -5$

9.   Writing in logarithmic form: $\log_{1/2} 8 = -3$

11.   Writing in logarithmic form: $\log_3 27 = 3$

13.   Writing in exponential form: $10^2 = 100$

15.   Writing in exponential form: $2^6 = 64$

17.   Writing in exponential form: $8^0 = 1$

19.   Writing in exponential form: $10^{-3} = 0.001$

21.   Writing in exponential form: $6^2 = 36$

23.   Writing in exponential form: $5^{-2} = \frac{1}{25}$

25.   Solving the equation:
$$\log_3 x = 2$$
$$x = 3^2 = 9$$

27.   Solving the equation:
$$\log_5 x = -3$$
$$x = 5^{-3} = \frac{1}{125}$$

29.   Solving the equation:
$$\log_2 16 = x$$
$$2^x = 16$$
$$x = 4$$

31.   Solving the equation:
$$\log_8 2 = x$$
$$8^x = 2$$
$$x = \frac{1}{3}$$

33.   Solving the equation:
$$\log_x 4 = 2$$
$$x^2 = 4$$
$$x = 2$$

35.   Solving the equation:
$$\log_x 5 = 3$$
$$x^3 = 5$$
$$x = \sqrt[3]{5}$$

37. Sketching the graph:

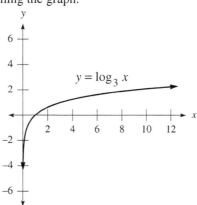

39. Sketching the graph:

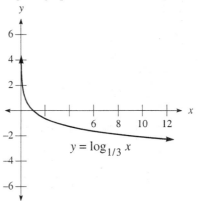

41. Sketching the graph:

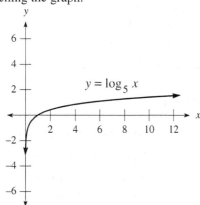

43. Sketching the graph:

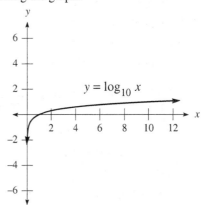

45. Simplifying the logarithm:

$$x = \log_2 16$$
$$2^x = 16$$
$$x = 4$$

47. Simplifying the logarithm:
$$x = \log_{25} 125$$
$$25^x = 125$$
$$5^{2x} = 5^3$$
$$2x = 3$$
$$x = \frac{3}{2}$$

49. Simplifying the logarithm:
$$x = \log_{10} 1000$$
$$10^x = 1000$$
$$x = 3$$

51. Simplifying the logarithm:
$$x = \log_3 3$$
$$3^x = 3$$
$$x = 1$$

53. Simplifying the logarithm:
$$x = \log_5 1$$
$$5^x = 1$$
$$x = 0$$

55. First find $\log_6 6$:

$$x = \log_6 6$$
$$6^x = 6$$
$$x = 1$$

Now find $\log_3 1$:

$$x = \log_3 1$$
$$3^x = 1$$
$$x = 0$$

57.   First find $\log_2 16$:

$$x = \log_2 16$$
$$2^x = 16$$
$$x = 4$$

Now find $\log_4 2$:

$$x = \log_4 2$$
$$4^x = 2$$
$$2^{2x} = 2$$
$$2x = 1$$
$$x = \tfrac{1}{2}$$

Now find $\log_2 4$:

$$x = \log_2 4$$
$$2^x = 4$$
$$x = 2$$

59.   The pH is given by: $\text{pH} = -\log_{10}\left(10^{-7}\right) = -(-7) = 7$

61.   The $\left[\text{H}^+\right]$ is given by:

$$-\log_{10}\left[\text{H}^+\right] = 6$$
$$\log_{10}\left[\text{H}^+\right] = -6$$
$$\left[\text{H}^+\right] = 10^{-6}$$

63.   Using the relationship $M = \log_{10} T$:

$$M = \log_{10} 100$$
$$10^M = 100$$
$$M = 2$$

65.   It is 10^8 times as large.

67.   Solving the equation:

$$2x^2 + 4x - 3 = 0$$
$$x = \frac{-4 \pm \sqrt{16+24}}{4} = \frac{-4 \pm \sqrt{40}}{4} = \frac{-4 \pm 2\sqrt{10}}{4} = \frac{-2 \pm \sqrt{10}}{2}$$

69.   Solving the equation:

$$(2y-3)(2y-1) = -4$$
$$4y^2 - 8y + 3 = -4$$
$$4y^2 - 8y + 7 = 0$$
$$y = \frac{8 \pm \sqrt{64-112}}{8} = \frac{8 \pm \sqrt{-48}}{8} = \frac{8 \pm 4i\sqrt{3}}{8} = \frac{2 \pm i\sqrt{3}}{2}$$

71.   Solving the equation:

$$t^3 - 125 = 0$$
$$(t-5)\left(t^2 + 5t + 25\right) = 0$$
$$t = 5, \frac{-5 \pm \sqrt{25-100}}{2} = \frac{-5 \pm \sqrt{-75}}{2} = \frac{-5 \pm 5i\sqrt{3}}{2}$$

73.   Solving the equation:

$$4x^5 - 16x^4 = 20x^3$$
$$4x^5 - 16x^4 - 20x^3 = 0$$
$$4x^3\left(x^2 - 4x - 5\right) = 0$$
$$4x^3(x-5)(x+1) = 0$$
$$x = -1, 0, 5$$

75. Solving the equation:

$$\frac{1}{x-3}+\frac{1}{x+2}=1$$
$$x+2+x-3=(x-3)(x+2)$$
$$2x-1=x^2-x-6$$
$$x^2-3x-5=0$$
$$x=\frac{3\pm\sqrt{9+20}}{2}=\frac{3\pm\sqrt{29}}{2}$$

77. **a.** Completing the table:

x	-1	0	1	2
$f(x)$	$\frac{1}{8}$	1	8	64

b. Completing the table:

x	$\frac{1}{8}$	1	8	64
$f^{-1}(x)$	-1	0	1	2

c. The equation is $f(x)=8^x$.

d. The equation is $f^{-1}(x)=\log_8 x$.

9.4   Properties of Logarithms

1. Using properties of logarithms: $\log_3 4x=\log_3 4+\log_3 x$

3. Using properties of logarithms: $\log_6 \dfrac{5}{x}=\log_6 5-\log_6 x$

5. Using properties of logarithms: $\log_2 y^5=5\log_2 y$

7. Using properties of logarithms: $\log_9 \sqrt[3]{z}=\log_9 z^{1/3}=\frac{1}{3}\log_9 z$

9. Using properties of logarithms: $\log_6 x^2 y^4=\log_6 x^2+\log_6 y^4=2\log_6 x+4\log_6 y$

11. Using properties of logarithms: $\log_5 \sqrt{x}\cdot y^4=\log_5 x^{1/2}+\log_5 y^4=\frac{1}{2}\log_5 x+4\log_5 y$

13. Using properties of logarithms: $\log_b \dfrac{xy}{z}=\log_b xy-\log_b z=\log_b x+\log_b y-\log_b z$

15. Using properties of logarithms: $\log_{10} \dfrac{4}{xy}=\log_{10} 4-\log_{10} xy=\log_{10} 4-\log_{10} x-\log_{10} y$

17. Using properties of logarithms: $\log_{10} \dfrac{x^2 y}{\sqrt{z}}=\log_{10} x^2+\log_{10} y-\log_{10} z^{1/2}=2\log_{10} x+\log_{10} y-\frac{1}{2}\log_{10} z$

19. Using properties of logarithms: $\log_{10} \dfrac{x^3\sqrt{y}}{z^4}=\log_{10} x^3+\log_{10} y^{1/2}-\log_{10} z^4=3\log_{10} x+\frac{1}{2}\log_{10} y-4\log_{10} z$

21. Using properties of logarithms:

$$\log_b \sqrt[3]{\frac{x^2 y}{z^4}}=\log_b \frac{x^{2/3}y^{1/3}}{z^{4/3}}=\log_b x^{2/3}+\log_b y^{1/3}-\log_b z^{4/3}=\frac{2}{3}\log_b x+\frac{1}{3}\log_b y-\frac{4}{3}\log_b z$$

23. Writing as a single logarithm: $\log_b x+\log_b z=\log_b xz$

25. Writing as a single logarithm: $2\log_3 x-3\log_3 y=\log_3 x^2-\log_3 y^3=\log_3 \dfrac{x^2}{y^3}$

27. Writing as a single logarithm: $\frac{1}{2}\log_{10} x+\frac{1}{3}\log_{10} y=\log_{10} x^{1/2}+\log_{10} y^{1/3}=\log_{10} \sqrt{x}\sqrt[3]{y}$

29. Writing as a single logarithm: $3\log_2 x+\frac{1}{2}\log_2 y-\log_2 z=\log_2 x^3+\log_2 y^{1/2}-\log_2 z=\log_2 \dfrac{x^3\sqrt{y}}{z}$

31. Writing as a single logarithm: $\frac{1}{2}\log_2 x-3\log_2 y-4\log_2 z=\log_2 x^{1/2}-\log_2 y^3-\log_2 z^4=\log_2 \dfrac{\sqrt{x}}{y^3 z^4}$

33. Writing as a single logarithm:

$$\frac{3}{2}\log_{10} x-\frac{3}{4}\log_{10} y-\frac{4}{5}\log_{10} z=\log_{10} x^{3/2}-\log_{10} y^{3/4}-\log_{10} z^{4/5}=\log_{10} \frac{x^{3/2}}{y^{3/4}z^{4/5}}$$

35. Solving the equation:

$$\log_2 x + \log_2 3 = 1$$
$$\log_2 3x = 1$$
$$3x = 2^1$$
$$3x = 2$$
$$x = \tfrac{2}{3}$$

37. Solving the equation:

$$\log_3 x - \log_3 2 = 2$$
$$\log_3 \frac{x}{2} = 2$$
$$\frac{x}{2} = 3^2$$
$$\frac{x}{2} = 9$$
$$x = 18$$

39. Solving the equation:

$$\log_3 x + \log_3 (x-2) = 1$$
$$\log_3 \left(x^2 - 2x\right) = 1$$
$$x^2 - 2x = 3^1$$
$$x^2 - 2x - 3 = 0$$
$$(x-3)(x+1) = 0$$
$$x = 3, -1$$

The solution is 3 (−1 does not check).

41. Solving the equation:

$$\log_3 (x+3) - \log_3 (x-1) = 1$$
$$\log_3 \frac{x+3}{x-1} = 1$$
$$\frac{x+3}{x-1} = 3^1$$
$$x+3 = 3x - 3$$
$$-2x = -6$$
$$x = 3$$

43. Solving the equation:

$$\log_2 x + \log_2 (x-2) = 3$$
$$\log_2 \left(x^2 - 2x\right) = 3$$
$$x^2 - 2x = 2^3$$
$$x^2 - 2x - 8 = 0$$
$$(x-4)(x+2) = 0$$
$$x = 4, -2$$

The solution is 4 (−2 does not check).

45. Solving the equation:

$$\log_8 x + \log_8 (x-3) = \tfrac{2}{3}$$
$$\log_8 \left(x^2 - 3x\right) = \tfrac{2}{3}$$
$$x^2 - 3x = 8^{2/3}$$
$$x^2 - 3x - 4 = 0$$
$$(x-4)(x+1) = 0$$
$$x = 4, -1$$

The solution is 4 (−1 does not check).

47. Solving the equation:

$$\log_5 \sqrt{x} + \log_5 \sqrt{6x+5} = 1$$
$$\log_5 \sqrt{6x^2 + 5x} = 1$$
$$\tfrac{1}{2}\log_5 \left(6x^2 + 5x\right) = 1$$
$$\log_5 \left(6x^2 + 5x\right) = 2$$
$$6x^2 + 5x = 5^2$$
$$6x^2 + 5x - 25 = 0$$
$$(3x-5)(2x+5) = 0$$
$$x = \tfrac{5}{3}, -\tfrac{5}{2}$$

The solution is $\tfrac{5}{3}$ ($-\tfrac{5}{2}$ does not check).

49. Solving for M: $M = 0.21\log_{10} \dfrac{1}{10^{-12}} = 0.21\log_{10} 10^{12} = 0.21(12) = 2.52$

51. Rewriting the expression: $\text{pH} = 6.1 + \log_{10}\left(\dfrac{x}{y}\right) = 6.1 + \log_{10} x - \log_{10} y$

53. Rewriting the formula:
$$D = 10\log_{10}\left(\frac{I}{I_0}\right)$$
$$D = 10\left(\log_{10} I - \log_{10} I_0\right)$$

55. Computing the discriminant: $D = (-5)^2 - 4(2)(4) = 25 - 32 = -7$. There are two complex solutions.

57. Writing the equation:
$$(x+3)(x-5) = 0$$
$$x^2 - 2x - 15 = 0$$

59. Writing the equation:
$$(3y-2)(y-3) = 0$$
$$3y^2 - 11y + 6 = 0$$

9.5   Common Logarithms and Natural Logarithms

1. Evaluating the logarithm: $\log 378 \approx 2.5775$

3. Evaluating the logarithm: $\log 37.8 \approx 1.5775$

5. Evaluating the logarithm: $\log 3,780 \approx 3.5775$

7. Evaluating the logarithm: $\log 0.0378 \approx -1.4225$

9. Evaluating the logarithm: $\log 37,800 \approx 4.5775$

11. Evaluating the logarithm: $\log 600 \approx 2.7782$

13. Evaluating the logarithm: $\log 2,010 \approx 3.3032$

15. Evaluating the logarithm: $\log 0.00971 \approx -2.0128$

17. Evaluating the logarithm: $\log 0.0314 \approx -1.5031$

19. Evaluating the logarithm: $\log 0.399 \approx -0.3990$

21. Solving for x:
$$\log x = 2.8802$$
$$x = 10^{2.8802} \approx 759$$

23. Solving for x:
$$\log x = -2.1198$$
$$x = 10^{-2.1198} \approx 0.00759$$

25. Solving for x:
$$\log x = 3.1553$$
$$x = 10^{3.1553} \approx 1,430$$

27. Solving for x:
$$\log x = -5.3497$$
$$x = 10^{-5.3497} \approx 0.00000447$$

29. Solving for x:
$$\log x = -7.0372$$
$$x = 10^{-7.0372} \approx 0.0000000918$$

31. Solving for x:
$$\log x = 10$$
$$x = 10^{10}$$

33. Solving for x:
$$\log x = -10$$
$$x = 10^{-10}$$

35. Solving for x:
$$\log x = 20$$
$$x = 10^{20}$$

37. Solving for x:
$$\log x = -2$$
$$x = 10^{-2} = \tfrac{1}{100}$$

39. Solving for x:
$$\log x = \log_2 8$$
$$\log x = 3$$
$$x = 10^3 = 1,000$$

41. Simplifying the logarithm: $\ln e = \ln e^1 = 1$

43. Simplifying the logarithm: $\ln e^5 = 5$

45. Simplifying the logarithm: $\ln e^x = x$

47. Using properties of logarithms: $\ln 10e^{3t} = \ln 10 + \ln e^{3t} = \ln 10 + 3t$

49. Using properties of logarithms: $\ln Ae^{-2t} = \ln A + \ln e^{-2t} = \ln A - 2t$

51. Evaluating the logarithm: $\ln 15 = \ln(3 \bullet 5) = \ln 3 + \ln 5 = 1.0986 + 1.6094 = 2.7080$

53. Evaluating the logarithm: $\ln \tfrac{1}{3} = \ln 3^{-1} = -\ln 3 = -1.0986$

55. Evaluating the logarithm: $\ln 9 = \ln 3^2 = 2\ln 3 = 2(1.0986) = 2.1972$

57. Evaluating the logarithm: $\ln 16 = \ln 2^4 = 4\ln 2 = 4(0.6931) = 2.7724$

59. Computing the pH: $\text{pH} = -\log(6.50 \times 10^{-4}) \approx 3.19$

61. Finding the concentration:
$$4.75 = -\log[H^+]$$
$$-4.75 = \log[H^+]$$
$$[H^+] = 10^{-4.75} \approx 1.78 \times 10^{-5}$$

63. Finding the magnitude:
$$5.5 = \log T$$
$$T = 10^{5.5} \approx 3.16 \times 10^5$$

65. Finding the magnitude:
$$8.3 = \log T$$
$$T = 10^{8.3} \approx 2.00 \times 10^8$$

66. Finding the magnitude:
$$8.7 = \log T$$
$$T = 10^{8.7} \approx 5.01 \times 10^8$$

67. For the first earthquake:
$$\log T_1 = 6.5$$
$$T_1 = 10^{6.5}$$

For the second earthquake:
$$\log T_2 = 5.5$$
$$T_2 = 10^{5.5}$$

The ratio is $\dfrac{T_1}{T_2} = \dfrac{10^{6.5}}{10^{5.5}} = 10$ times stronger.

69. Completing the table:

Location	Date	Magnitude (M)	Shockwave (T)
Moresby Island	January 23	4.0	1.00×10^4
Vancouver Island	April 30	5.3	1.99×10^5
Quebec City	June 29	3.2	1.58×10^3
Mould Bay	November 13	5.2	1.58×10^5
St. Lawrence	December 14	3.7	5.01×10^3

71. Finding the rate of depreciation:
$$\log(1-r) = \tfrac{1}{5}\log\frac{4500}{9000}$$
$$\log(1-r) \approx -0.0602$$
$$1-r \approx 10^{-0.0602}$$
$$r = 1 - 10^{-0.0602}$$
$$r \approx 0.129 = 12.9\%$$

73. Finding the rate of depreciation:
$$\log(1-r) = \tfrac{1}{5}\log\frac{5750}{7550}$$
$$\log(1-r) \approx -0.0237$$
$$1-r \approx 10^{-0.0237}$$
$$r = 1 - 10^{-0.0237}$$
$$r \approx 0.053 = 5.3\%$$

75. It appears to approach e. Completing the table:

x	$(1+x)^{1/x}$
1	2
0.5	2.25
0.1	2.5937
0.01	2.7048
0.001	2.7169
0.0001	2.7181
0.00001	2.7183

77. Solving the equation:
$$x^4 - 2x^2 - 8 = 0$$
$$\left(x^2 - 4\right)\left(x^2 + 2\right) = 0$$
$$x^2 = 4, -2$$
$$x = \pm 2, \pm i\sqrt{2}$$

79. Solving the equation:
$$2x - 5\sqrt{x} + 3 = 0$$
$$\left(2\sqrt{x} - 3\right)\left(\sqrt{x} - 1\right) = 0$$
$$\sqrt{x} = 1, \tfrac{3}{2}$$
$$x = 1, \tfrac{9}{4}$$

9.6 Exponential Equations and Change of Base

1. Solving the equation:
$$3^x = 5$$
$$\ln 3^x = \ln 5$$
$$x \ln 3 = \ln 5$$
$$x = \frac{\ln 5}{\ln 3} \approx 1.4650$$

3. Solving the equation:
$$5^x = 3$$
$$\ln 5^x = \ln 3$$
$$x \ln 5 = \ln 3$$
$$x = \frac{\ln 3}{\ln 5} \approx 0.6826$$

5. Solving the equation:
$$5^{-x} = 12$$
$$\ln 5^{-x} = \ln 12$$
$$-x \ln 5 = \ln 12$$
$$x = -\frac{\ln 12}{\ln 5} \approx -1.5440$$

7. Solving the equation:
$$12^{-x} = 5$$
$$\ln 12^{-x} = \ln 5$$
$$-x \ln 12 = \ln 5$$
$$x = -\frac{\ln 5}{\ln 12} \approx -0.6477$$

9. Solving the equation:
$$8^{x+1} = 4$$
$$2^{3x+3} = 2^2$$
$$3x + 3 = 2$$
$$3x = -1$$
$$x = -\tfrac{1}{3}$$

11. Solving the equation:
$$4^{x-1} = 4$$
$$4^{x-1} = 4^1$$
$$x - 1 = 1$$
$$x = 2$$

13. Solving the equation:
$$3^{2x+1} = 2$$
$$\ln 3^{2x+1} = \ln 2$$
$$(2x + 1) \ln 3 = \ln 2$$
$$2x + 1 = \frac{\ln 2}{\ln 3}$$
$$2x = \frac{\ln 2}{\ln 3} - 1$$
$$x = \tfrac{1}{2}\left(\frac{\ln 2}{\ln 3} - 1\right) \approx -0.1845$$

15. Solving the equation:
$$3^{1-2x} = 2$$
$$\ln 3^{1-2x} = \ln 2$$
$$(1 - 2x) \ln 3 = \ln 2$$
$$1 - 2x = \frac{\ln 2}{\ln 3}$$
$$-2x = \frac{\ln 2}{\ln 3} - 1$$
$$x = \tfrac{1}{2}\left(1 - \frac{\ln 2}{\ln 3}\right) \approx 0.1845$$

17. Solving the equation:
$$15^{3x-4} = 10$$
$$\ln 15^{3x-4} = \ln 10$$
$$(3x - 4) \ln 15 = \ln 10$$
$$3x - 4 = \frac{\ln 10}{\ln 15}$$
$$3x = \frac{\ln 10}{\ln 15} + 4$$
$$x = \tfrac{1}{3}\left(\frac{\ln 10}{\ln 15} + 4\right) \approx 1.6168$$

19. Solving the equation:
$$6^{5-2x} = 4$$
$$\ln 6^{5-2x} = \ln 4$$
$$(5 - 2x) \ln 6 = \ln 4$$
$$5 - 2x = \frac{\ln 4}{\ln 6}$$
$$-2x = \frac{\ln 4}{\ln 6} - 5$$
$$x = \tfrac{1}{2}\left(5 - \frac{\ln 4}{\ln 6}\right) \approx 2.1131$$

21. Evaluating the logarithm: $\log_8 16 = \dfrac{\log 16}{\log 8} \approx 1.3333$

23. Evaluating the logarithm: $\log_{16} 8 = \dfrac{\log 8}{\log 16} = 0.7500$

25. Evaluating the logarithm: $\log_7 15 = \dfrac{\log 15}{\log 7} \approx 1.3917$

27. Evaluating the logarithm: $\log_{15} 7 = \dfrac{\log 7}{\log 15} \approx 0.7186$

29. Evaluating the logarithm: $\log_8 240 = \dfrac{\log 240}{\log 8} \approx 2.6356$

31. Evaluating the logarithm: $\log_4 321 = \dfrac{\log 321}{\log 4} \approx 4.1632$

33. Evaluating the logarithm: $\ln 345 \approx 5.8435$

35. Evaluating the logarithm: $\ln 0.345 \approx -1.0642$

37. Evaluating the logarithm: $\ln 10 \approx 2.3026$

39. Evaluating the logarithm: $\ln 45,000 \approx 10.7144$

41. Using the compound interest formula:

$$500\left(1+\frac{0.06}{2}\right)^{2t} = 1000$$

$$\left(1+\frac{0.06}{2}\right)^{2t} = 2$$

$$\ln\left(1+\frac{0.06}{2}\right)^{2t} = \ln 2$$

$$2t\ln\left(1+\frac{0.06}{2}\right) = \ln 2$$

$$t = \frac{\ln 2}{2\ln\left(1+\frac{0.06}{2}\right)} \approx 11.72$$

It will take 11.72 years.

45. Using the compound interest formula:

$$P\left(1+\frac{0.08}{4}\right)^{4t} = 2P$$

$$\left(1+\frac{0.08}{4}\right)^{4t} = 2$$

$$\ln\left(1+\frac{0.08}{4}\right)^{4t} = \ln 2$$

$$4t\ln\left(1+\frac{0.08}{4}\right) = \ln 2$$

$$t = \frac{\ln 2}{4\ln\left(1+\frac{0.08}{4}\right)} \approx 8.75$$

It will take 8.75 years.

49. Using the continuous interest formula:

$$500e^{0.06t} = 1000$$

$$e^{0.06t} = 2$$

$$0.06t = \ln 2$$

$$t = \frac{\ln 2}{0.06} \approx 11.55$$

It will take 11.55 years.

43. Using the compound interest formula:

$$1000\left(1+\frac{0.12}{6}\right)^{6t} = 3000$$

$$\left(1+\frac{0.12}{6}\right)^{6t} = 3$$

$$\ln\left(1+\frac{0.12}{6}\right)^{6t} = \ln 3$$

$$6t\ln\left(1+\frac{0.12}{6}\right) = \ln 3$$

$$t = \frac{\ln 3}{6\ln\left(1+\frac{0.12}{6}\right)} \approx 9.25$$

It will take 9.25 years.

47. Using the compound interest formula:

$$25\left(1+\frac{0.06}{2}\right)^{2t} = 75$$

$$\left(1+\frac{0.06}{2}\right)^{2t} = 3$$

$$\ln\left(1+\frac{0.06}{2}\right)^{2t} = \ln 3$$

$$2t\ln\left(1+\frac{0.06}{2}\right) = \ln 3$$

$$t = \frac{\ln 3}{2\ln\left(1+\frac{0.06}{2}\right)} \approx 18.58$$

It was invested 18.58 years ago.

51. Using the continuous interest formula:

$$500e^{0.06t} = 1500$$

$$e^{0.06t} = 3$$

$$0.06t = \ln 3$$

$$t = \frac{\ln 3}{0.06} \approx 18.31$$

It will take 18.31 years.

53. Completing the square: $y = 2x^2 + 8x - 15 = 2\left(x^2 + 4x + 4\right) - 8 - 15 = 2(x+2)^2 - 23$. The lowest point is $(-2,-23)$.

55. Completing the square: $y = 12x - 4x^2 = -4\left(x^2 - 3x + \frac{9}{4}\right) + 9 = -4\left(x - \frac{3}{2}\right)^2 + 9$. The highest point is $\left(\frac{3}{2}, 9\right)$.

57. Completing the square: $y = 64t - 16t^2 = -16\left(t^2 - 4t + 4\right) + 64 = -16(t-2)^2 + 64$

The object reaches a maximum height after 2 seconds, and the maximum height is 64 feet.

59. Using the population model:

$$32000e^{0.05t} = 64000$$

$$e^{0.05t} = 2$$

$$0.05t = \ln 2$$

$$t = \frac{\ln 2}{0.05} \approx 13.9$$

The city will reach 64,000 toward the end of the year 2007.

61. Finding when $P(t) = 45,000$:

$$15,000e^{0.04t} = 45,000$$
$$e^{0.04t} = 3$$
$$0.04t = \ln 3$$
$$t = \frac{\ln 3}{0.04} \approx 27.5$$

It will take approximately 27.5 years.

65. Solving for t:

$$A = P2^{-kt}$$
$$2^{-kt} = \frac{A}{P}$$
$$\ln 2^{-kt} = \ln \frac{A}{P}$$
$$-kt \ln 2 = \ln A - \ln P$$
$$t = \frac{\ln A - \ln P}{-k \ln 2} = \frac{\ln P - \ln A}{k \ln 2}$$

63. Solving for t:

$$A = Pe^{rt}$$
$$e^{rt} = \frac{A}{P}$$
$$rt = \ln \frac{A}{P}$$
$$t = \frac{\ln A - \ln P}{r} = \frac{1}{r} \ln \frac{A}{P}$$

67. Solving for t:

$$A = P(1-r)^t$$
$$(1-r)^t = \frac{A}{P}$$
$$\ln(1-r)^t = \ln \frac{A}{P}$$
$$t \ln(1-r) = \ln A - \ln P$$
$$t = \frac{\ln A - \ln P}{\ln(1-r)}$$

Chapter 9 Review

1. Evaluating the function: $f(4) = 2^4 = 16$

3. Evaluating the function: $g(2) = \left(\frac{1}{3}\right)^2 = \frac{1}{9}$

5. Evaluating the function: $f(-1) + g(1) = 2^{-1} + \left(\frac{1}{3}\right)^1 = \frac{1}{2} + \frac{1}{3} = \frac{5}{6}$

7. Graphing the function:

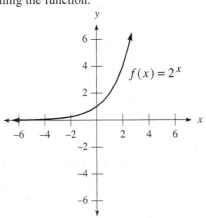

9.   Finding the inverse:
$$2y + 1 = x$$
$$2y = x - 1$$
$$y = \frac{x-1}{2}$$

The inverse is $y^{-1} = \frac{x-1}{2}$. Sketching the graph:

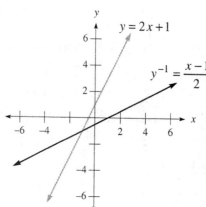

11.   Finding the inverse:
$$2y + 3 = x$$
$$2y = x - 3$$
$$y = \frac{x-3}{2}$$

The inverse is $f^{-1}(x) = \frac{x-3}{2}$.

13.   Finding the inverse:
$$\tfrac{1}{2}y + 2 = x$$
$$y + 4 = 2x$$
$$y = 2x - 4$$

The inverse is $f^{-1}(x) = 2x - 4$.

15.   Writing in logarithmic form: $\log_3 81 = 4$

17.   Writing in logarithmic form: $\log_{10} 0.01 = -2$

19.   Writing in exponential form: $2^3 = 8$

21.   Writing in exponential form: $4^{1/2} = 2$

23.   Solving for x:

$$\log_5 x = 2$$
$$x = 5^2 = 25$$

25.   Solving for x:
$$\log_x 0.01 = -2$$
$$x^{-2} = 0.01$$
$$x^2 = 100$$
$$x = 10$$

27.   Graphing the equation:

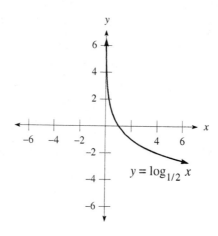

29. Simplifying the logarithm:
$$\log_{27} 9 = x$$
$$27^x = 9$$
$$3^{3x} = 3^2$$
$$3x = 2$$
$$x = \tfrac{2}{3}$$

31. Expanding the logarithm: $\log_2 5x = \log_2 5 + \log_2 x$

33. Expanding the logarithm: $\log_a \dfrac{y^3 \sqrt{x}}{z} = \log_a y^3 + \log_a x^{1/2} - \log_a z = 3\log_a y + \tfrac{1}{2}\log_a x - \log_a z$

35. Writing as a single logarithm: $\log_2 x + \log_2 y = \log_2 xy$

37. Writing as a single logarithm: $2\log_a 5 - \tfrac{1}{2}\log_a 9 = \log_a 5^2 - \log_a 9^{1/2} = \log_a 25 - \log_a 3 = \log_a \tfrac{25}{3}$

39. Solving the equation:
$$\log_2 x + \log_2 4 = 3$$
$$\log_2 4x = 3$$
$$4x = 2^3$$
$$4x = 8$$
$$x = 2$$

41. Solving the equation:
$$\log_3 x + \log_3 (x - 2) = 1$$
$$\log_3 (x^2 - 2x) = 1$$
$$x^2 - 2x = 3^1$$
$$x^2 - 2x - 3 = 0$$
$$(x - 3)(x + 1) = 0$$
$$x = 3, -1$$
The solution is 3 (–1 does not check).

43. Solving the equation:
$$\log_6 (x - 1) + \log_6 x = 1$$
$$\log_6 (x^2 - x) = 1$$
$$x^2 - x = 6^1$$
$$x^2 - x - 6 = 0$$
$$(x - 3)(x + 2) = 0$$
$$x = 3, -2$$
The solution is 3 (–2 does not check).

45. Evaluating: $\log 346 \approx 2.5391$

47. Solving for x:
$$\log x = 3.9652$$
$$x = 10^{3.9652} \approx 9,230$$

49. Simplifying: $\ln e = \ln e^1 = 1$

51. Simplifying: $\ln e^2 = 2$

53. Finding the pH: $pH = -\log(7.9 \times 10^{-3}) \approx 2.1$

55. Finding $[H^+]$:
$$2.7 = -\log[H^+]$$
$$-2.7 = \log[H^+]$$
$$[H^+] = 10^{-2.7} \approx 2.0 \times 10^{-3}$$

57. Solving the equation:
$$4^x = 8$$
$$2^{2x} = 2^3$$
$$2x = 3$$
$$x = \tfrac{3}{2}$$

59. Using a calculator: $\log_{16} 8 = \dfrac{\ln 8}{\ln 16} = 0.75$

61. Using the compound interest formula:

$$5000(1+0.16)^t = 10000$$

$$1.16^t = 2$$

$$\ln 1.16^t = \ln 2$$

$$t \ln 1.16 = \ln 2$$

$$t = \frac{\ln 2}{\ln 1.16} \approx 4.67$$

It will take approximately 4.67 years for the amount to double.

Chapters 1-9 Cumulative Review

1. Simplifying: $-8 + 2\big[5 - 3(-2-3)\big] = -8 + 2\big[5 - 3(-5)\big] = -8 + 2(5+15) = -8 + 2(20) = -8 + 40 = 32$

3. Simplifying: $\left(\frac{3}{5}\right)^{-2} - \left(\frac{3}{13}\right)^{-2} = \left(\frac{5}{3}\right)^2 - \left(\frac{13}{3}\right)^2 = \frac{25}{9} - \frac{169}{9} = -\frac{144}{9} = -16$

5. Simplifying: $\sqrt[3]{27x^4 y^3} = \sqrt[3]{27x^3 y^3 \cdot x} = 3xy\sqrt[3]{x}$

7. Simplifying: $\big[(6+2i) - (3-4i)\big] - (5-i) = (6+2i-3+4i) - (5-i) = 3 + 6i - 5 + i = -2 + 7i$

9. Simplifying: $1 + \dfrac{x}{1+\dfrac{1}{x}} = 1 + \dfrac{x}{1+\dfrac{1}{x}} \cdot \dfrac{x}{x} = 1 + \dfrac{x^2}{x+1} = \dfrac{x+1}{x+1} + \dfrac{x^2}{x+1} = \dfrac{x^2+x+1}{x+1}$

11. Multiplying: $\left(3t^2 + \frac{1}{4}\right)\left(4t^2 - \frac{1}{3}\right) = 12t^4 + t^2 - t^2 - \frac{1}{12} = 12t^4 - \frac{1}{12}$

13. Using long division:

$$
\begin{array}{r}
3x + 7 \\
3x-4 \overline{\smash{\big)}\, 9x^2 + 9x - 18} \\
\underline{9x^2 - 12x} \\
21x - 18 \\
\underline{21x - 28} \\
10
\end{array}
$$

The quotient is $3x + 7 + \dfrac{10}{3x-4}$.

15. Subtracting:

$$\frac{7}{4x^2 - x - 3} - \frac{1}{4x^2 - 7x + 3} = \frac{7}{(4x+3)(x-1)} \cdot \frac{4x-3}{4x-3} - \frac{1}{(4x-3)(x-1)} \cdot \frac{4x+3}{4x+3}$$

$$= \frac{28x - 21}{(4x+3)(4x-3)(x-1)} - \frac{4x+3}{(4x+3)(4x-3)(x-1)}$$

$$= \frac{24x - 24}{(4x+3)(4x-3)(x-1)}$$

$$= \frac{24(x-1)}{(4x+3)(4x-3)(x-1)}$$

$$= \frac{24}{(4x+3)(4x-3)}$$

17. Solving the equation:
$$\tfrac{2}{3}(6x-5)+\tfrac{1}{3}=13$$
$$2(6x-5)+1=39$$
$$12x-10+1=39$$
$$12x-9=39$$
$$12x=48$$
$$x=4$$

19. Solving the equation:
$$|3x-5|+6=2$$
$$|3x-5|=-4$$
Since this statement is impossible, there is no solution.

21. Solving the equation:
$$\frac{1}{x+3}+\frac{1}{x-2}=1$$
$$(x+3)(x-2)\left(\frac{1}{x+3}+\frac{1}{x-2}\right)=(x+3)(x-2)\cdot 1$$
$$x-2+x+3=x^2+x-6$$
$$2x+1=x^2+x-6$$
$$x^2-x-7=0$$
$$x=\frac{1\pm\sqrt{1+28}}{2}=\frac{1\pm\sqrt{29}}{2}$$

23. Solving the equation:
$$2x-1=x^2$$
$$x^2-2x+1=0$$
$$(x-1)^2=0$$
$$x=1$$

25. Solving the equation:
$$\sqrt{7x-4}=-2$$
$$7x-4=4$$
$$7x=8$$
$$x=\tfrac{8}{7}$$
Since $x=\tfrac{8}{7}$ does not check in the original equation, there is no solution.

27. Solving the equation:
$$x-3\sqrt{x}+2=0$$
$$(\sqrt{x}-2)(\sqrt{x}-1)=0$$
$$\sqrt{x}=1,2$$
$$x=1,4$$

29. Solving the equation:
$$\log_3 x=3$$
$$x=3^3=27$$

31. Solving the equation:
$$\log_3(x-3)-\log_3(x+2)=1$$
$$\log_3\frac{x-3}{x+2}=1$$
$$\frac{x-3}{x+2}=3^1$$
$$x-3=3x+6$$
$$-2x=9$$
$$x=-\tfrac{9}{2}\ \text{(impossible)}$$
There is no solution.

33. Substituting into the first equation:
$$4(-2y-2)+7y=-3$$
$$-8y-8+7y=-3$$
$$-y=5$$
$$y=-5$$
$$x=-2(-5)-2=8$$
The solution is $(8,-5)$.

35. Solving the inequality:
$$3y - 6 \geq 3 \qquad \text{or} \qquad 3y - 6 \leq -3$$
$$3y \geq 9 \qquad\qquad\qquad 3y \leq 3$$
$$y \geq 3 \qquad\qquad\qquad y \leq 1$$

The solution set is $y \leq 1$ or $y \geq 3$. Graphing the solution set:

37. Factoring the inequality:
$$x^3 + 2x^2 - 9x - 18 < 0$$
$$x^2(x + 2) - 9(x + 2) < 0$$
$$(x + 2)(x^2 - 9) < 0$$
$$(x + 2)(x + 3)(x - 3) < 0$$

Forming the sign chart:

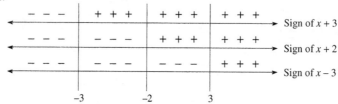

The solution set is $x < -3$ or $-2 < x < 3$. Graphing the solution set:

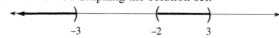

39. Using the point-slope formula:
$$y - (-3) = -\tfrac{5}{3}(x - 3)$$
$$y + 3 = -\tfrac{5}{3}x + 5$$
$$y = -\tfrac{5}{3}x + 2$$

41. Writing in scientific notation: $0.0000972 = 9.72 \times 10^{-5}$

43. Finding the inverse:
$$\tfrac{1}{2}y + 3 = x$$
$$y + 6 = 2x$$
$$y = 2x - 6$$

The inverse function is $f^{-1}(x) = 2x - 6$.

45. The domain is $\{-3, 2\}$ and the range is $\{-3, -1, 3\}$. This is not a function.

47. The pattern is to multiply by -3, so the next number is $18(-3) = -54$. This is a geometric sequence.

49. Let x represent the gallons of 25% alcohol and y represent the gallons of 50% alcohol. The system of equations is:
$$x + y = 20$$
$$0.25x + 0.50y = 0.425(20)$$

Multiply the first equation by -0.25:
$$-0.25x - 0.25y = -5$$
$$0.25x + 0.50y = 8.5$$

Adding yields:
$$0.25y = 3.5$$
$$y = 14$$
$$x = 6$$

The mixture contains 6 gallons of 25% alcohol and 14 gallons of 50% alcohol.

Chapter 9 Test

1.   Graphing the function:

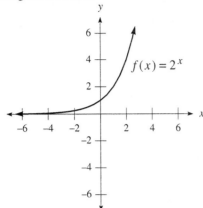

2.   Graphing the function:

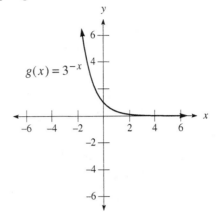

3.   Finding the inverse:
$$2y - 3 = x$$
$$2y = x + 3$$
$$y = \frac{x + 3}{2}$$

The inverse is $f^{-1}(x) = \dfrac{x + 3}{2}$. Sketching the graph:

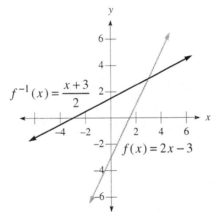

4.   Graphing the function and its inverse:

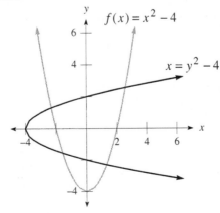

5.   Solving for x:

$$\log_4 x = 3$$
$$x = 4^3 = 64$$

6.   Solving for x:

$$\log_x 5 = 2$$
$$x^2 = 5$$
$$x = \sqrt{5}$$

7.   Graphing the function:

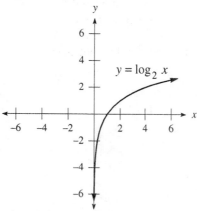

8.   Graphing the function:

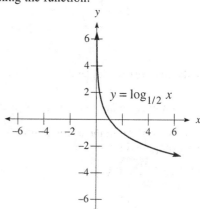

9.   Evaluating the logarithm:

$$x = \log_8 4$$
$$8^x = 4$$
$$2^{3x} = 2^2$$
$$3x = 2$$
$$x = \tfrac{2}{3}$$

10.   Evaluating the logarithm: $\log_7 21 = \dfrac{\ln 21}{\ln 7} \approx 1.5646$

11.   Evaluating the logarithm: $\log 23,400 \approx 4.3692$

12.   Evaluating the logarithm: $\log 0.0123 \approx -1.9101$

13.   Evaluating the logarithm: $\ln 46.2 \approx 3.8330$

14.   Evaluating the logarithm: $\ln 0.0462 \approx -3.0748$

15.   Expanding the logarithm: $\log_2 \dfrac{8x^2}{y} = \log_2 2^3 + \log_2 x^2 - \log_2 y = 3 + 2\log_2 x - \log_2 y$

16.   Expanding the logarithm: $\log \dfrac{\sqrt{x}}{y^4 \sqrt[5]{z}} = \log \dfrac{x^{1/2}}{y^4 z^{1/5}} = \log x^{1/2} - \log y^4 - \log z^{1/5} = \tfrac{1}{2}\log x - 4\log y - \tfrac{1}{5}\log z$

17.   Writing as a single logarithm: $2\log_3 x - \tfrac{1}{2}\log_3 y = \log_3 x^2 - \log_3 y^{1/2} = \log_3 \dfrac{x^2}{\sqrt{y}}$

18.   Writing as a single logarithm: $\tfrac{1}{3}\log x - \log y - 2\log z = \log x^{1/3} - \log y - \log z^2 = \log \dfrac{\sqrt[3]{x}}{yz^2}$

19.   Solving for x:
$$\log x = 4.8476$$
$$x = 10^{4.8476} \approx 70,404$$

20.   Solving for x:
$$\log x = -2.6478$$
$$x = 10^{-2.6478} \approx 0.00225$$

21.   Solving for x:

$$3^x = 5$$
$$\ln 3^x = \ln 5$$
$$x \ln 3 = \ln 5$$
$$x = \dfrac{\ln 5}{\ln 3} \approx 1.4650$$

22.   Solving for x:

$$4^{2x-1} = 8$$
$$2^{4x-2} = 2^3$$
$$4x - 2 = 3$$
$$4x = 5$$
$$x = \tfrac{5}{4}$$

23. Solving for x:

$$\log_5 x - \log_5 3 = 1$$
$$\log_5 \frac{x}{3} = 1$$
$$\frac{x}{3} = 5^1$$
$$x = 15$$

24. Solving for x:

$$\log_2 x + \log_2 (x-7) = 3$$
$$\log_2 \left(x^2 - 7x \right) = 3$$
$$x^2 - 7x = 2^3$$
$$x^2 - 7x - 8 = 0$$
$$(x-8)(x+1) = 0$$
$$x = 8, -1$$

The solution is 8 (–1 does not check).

25. Finding the pH: $\text{pH} = -\log\left(6.6 \times 10^{-7}\right) \approx 6.18$

26. Using the compound interest formula: $A = 400\left(1 + \dfrac{0.10}{2}\right)^{2 \cdot 5} = 400(1.05)^{10} \approx 651.56$

There will be \$651.56 in the account after 5 years.

27. Using the compound interest formula:

$$600\left(1 + \frac{0.08}{4}\right)^{4t} = 1800$$
$$\left(1 + \frac{0.08}{4}\right)^{4t} = 3$$
$$\ln(1.02)^{4t} = \ln 3$$
$$4t \ln 1.02 = \ln 3$$
$$t = \frac{\ln 3}{4 \ln 1.02} \approx 13.87$$

It will take 13.87 years for the account to reach \$1,800.

Chapter 10
Conic Sections

10.1 The Circle

1. Using the distance formula: $d = \sqrt{(6-3)^2 + (3-7)^2} = \sqrt{9+16} = \sqrt{25} = 5$

3. Using the distance formula: $d = \sqrt{(5-0)^2 + (0-9)^2} = \sqrt{25+81} = \sqrt{106}$

5. Using the distance formula: $d = \sqrt{(-2-3)^2 + (1+5)^2} = \sqrt{25+36} = \sqrt{61}$

7. Using the distance formula: $d = \sqrt{(-10+1)^2 + (5+2)^2} = \sqrt{81+49} = \sqrt{130}$

9. Solving the equation:
$$\sqrt{(x-1)^2 + (2-5)^2} = \sqrt{13}$$
$$(x-1)^2 + 9 = 13$$
$$(x-1)^2 = 4$$
$$x - 1 = \pm 2$$
$$x - 1 = -2, 2$$
$$x = -1, 3$$

11. Solving the equation:
$$\sqrt{(7-8)^2 + (y-3)^2} = 1$$
$$(y-3)^2 + 1 = 1$$
$$(y-3)^2 = 0$$
$$y - 3 = 0$$
$$y = 3$$

13. The equation is $(x-2)^2 + (y-3)^2 = 16$.

15. The equation is $(x-3)^2 + (y+2)^2 = 9$.

17. The equation is $(x+5)^2 + (y+1)^2 = 5$.

19. The equation is $x^2 + (y+5)^2 = 1$.

21. The equation is $x^2 + y^2 = 4$.

23. The center is (0,0) and the radius is 2.

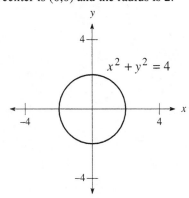

25. The center is (1,3) and the radius is 5.

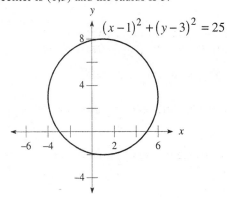

27. The center is $(-2,4)$ and the radius is $2\sqrt{2}$.

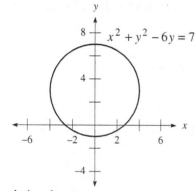

$$(x+2)^2 + (y-4)^2 = 8$$

29. The center is $(-1,-1)$ and the radius is 1.

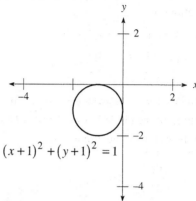

$$(x+1)^2 + (y+1)^2 = 1$$

31. Completing the square:
$$x^2 + y^2 - 6y = 7$$
$$x^2 + \left(y^2 - 6y + 9\right) = 7 + 9$$
$$x^2 + (y-3)^2 = 16$$
The center is $(0,3)$ and the radius is 4.

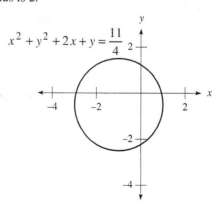

$$x^2 + y^2 - 6y = 7$$

33. Completing the square:
$$x^2 + y^2 - 4x - 6y = -4$$
$$\left(x^2 - 4x + 4\right) + \left(y^2 - 6y + 9\right) = -4 + 4 + 9$$
$$(x-2)^2 + (y-3)^2 = 9$$
The center is $(2,3)$ and the radius is 3.

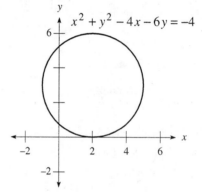

$$x^2 + y^2 - 4x - 6y = -4$$

35. Completing the square:
$$x^2 + y^2 + 2x + y = \tfrac{11}{4}$$
$$\left(x^2 + 2x + 1\right) + \left(y^2 + y + \tfrac{1}{4}\right) = \tfrac{11}{4} + 1 + \tfrac{1}{4}$$
$$(x+1)^2 + \left(y + \tfrac{1}{2}\right)^2 = 4$$
The center is $\left(-1, -\tfrac{1}{2}\right)$ and the radius is 2.

$$x^2 + y^2 + 2x + y = \frac{11}{4}$$

37. The equation is $(x-3)^2 + (y-4)^2 = 25$.

39. The equations are:

     A: $\left(x - \frac{1}{2}\right)^2 + (y-1)^2 = \frac{1}{4}$

     B: $(x-1)^2 + (y-1)^2 = 1$

     C: $(x-2)^2 + (y-1)^2 = 4$

41. The equations are:

     A: $(x+8)^2 + y^2 = 64$

     B: $x^2 + y^2 = 64$

     C: $(x-8)^2 + y^2 = 64$

43. The x-coordinate of the center is $x = 500$, the y-coordinate of the center is $12 + 120 = 132$, and the radius is 120. Thus the equation of the circle is $(x-500)^2 + (y-132)^2 = 120^2 = 14,400$.

45. Finding the radius of the circular path:

$$2\pi r = 1005$$
$$r = \frac{1005}{2\pi} \approx 160$$

The equation of the circular path is $(x-160)^2 + (y-160)^2 = 160^2 = 25,600$.

47. Since the radius is 5 meters, the circumference is: $C = 2\pi(5) = 10\pi$ meters

49.   **a.**   Evaluating: $f(-2) = 3^{-2} = \frac{1}{9}$ **b.**   Evaluating: $f(-1) = 3^{-1} = \frac{1}{3}$

     c.   Evaluating: $f(0) = 3^0 = 1$ **d.**   Evaluating: $f(1) = 3^1 = 3$

     e.   Evaluating: $f(2) = 3^2 = 9$

51. Let $y = f(x)$. Switch x and y and solve for y:

$$2y + 3 = x$$
$$2y = x - 3$$
$$y = \frac{x-3}{2}$$

The inverse is $f^{-1}(x) = \frac{x-3}{2}$.

53. Let $y = f(x)$. Switch x and y and solve for y:

$$y^2 - 4 = x$$
$$y^2 = x + 4$$
$$y = \pm\sqrt{x+4}$$

55. The radius is 2, so the equation is $(x-2)^2 + (y-3)^2 = 4$.

57. The radius is 2, so the equation is $(x-2)^2 + (y-3)^2 = 4$.

59. Completing the square:

$$x^2 + y^2 - 6x + 8y = 144$$
$$\left(x^2 - 6x + 9\right) + \left(y^2 + 8y + 16\right) = 144 + 9 + 16$$
$$(x-3)^2 + (y+4)^2 = 169$$

The center is (3,–4), so the distance is: $d = \sqrt{(-3)^2 + 4^2} = \sqrt{9+16} = 5$

61. Completing the square:

$$x^2 + y^2 - 6x - 8y = 144$$
$$\left(x^2 - 6x + 9\right) + \left(y^2 - 8y + 16\right) = 144 + 9 + 16$$
$$(x-3)^2 + (y-4)^2 = 169$$

The center is (3,4), so the distance is: $d = \sqrt{3^2 + 4^2} = \sqrt{9+16} = 5$

63. The equation $y = \sqrt{9-x^2}$ corresponds to the top half of the circle, and the equation $y = -\sqrt{9-x^2}$ corresponds to the bottom half of the circle.

10.2   Ellipses and Hyperbolas

1.   Graphing the ellipse:

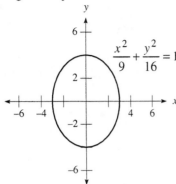

3.   Graphing the ellipse:

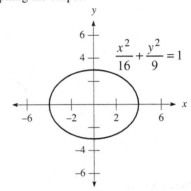

5.   Graphing the ellipse:

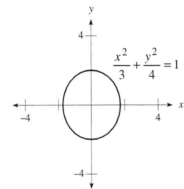

7.   The standard form is $\dfrac{x^2}{25} + \dfrac{y^2}{4} = 1$. Graphing the ellipse:

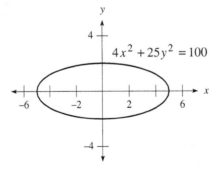

9. The standard form is $\dfrac{x^2}{16} + \dfrac{y^2}{2} = 1$. Graphing the ellipse:

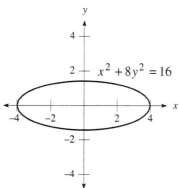

$x^2 + 8y^2 = 16$

11. Graphing the hyperbola:

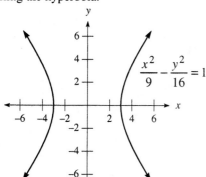

$\dfrac{x^2}{9} - \dfrac{y^2}{16} = 1$

13. Graphing the hyperbola:

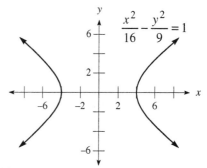

$\dfrac{x^2}{16} - \dfrac{y^2}{9} = 1$

15. Graphing the hyperbola:

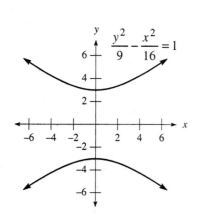

$\dfrac{y^2}{9} - \dfrac{x^2}{16} = 1$

17. Graphing the hyperbola:

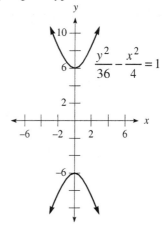

$\dfrac{y^2}{36} - \dfrac{x^2}{4} = 1$

19.   The standard form is $\dfrac{x^2}{4} - \dfrac{y^2}{1} = 1$. Graphing the hyperbola:

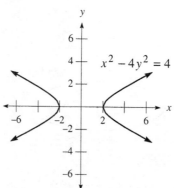

21.   The standard form is $\dfrac{y^2}{9} - \dfrac{x^2}{16} = 1$. Graphing the hyperbola:

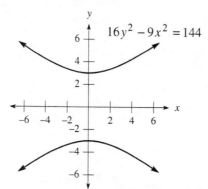

23.   For the x-intercepts, set $y = 0$:
$$0.4x^2 = 3.6$$
$$x^2 = 9$$
$$x = \pm 3$$

For the y-intercepts, set $x = 0$:
$$0.9y^2 = 3.6$$
$$y^2 = 4$$
$$y = \pm 2$$

25.   For the x-intercepts, set $y = 0$:
$$\dfrac{x^2}{0.04} = 1$$
$$x^2 = 0.04$$
$$x = \pm 0.2$$

For the y-intercepts, set $x = 0$:
$$-\dfrac{y^2}{0.09} = 1$$
$$y^2 = -0.09$$
There are no y-intercepts.

27.   For the x-intercepts, set $y = 0$:
$$\dfrac{25x^2}{9} = 1$$
$$x^2 = \dfrac{9}{25}$$
$$x = \pm \dfrac{3}{5}$$

For the y-intercepts, set $x = 0$:
$$\dfrac{25y^2}{4} = 1$$
$$y^2 = \dfrac{4}{25}$$
$$y = \pm \dfrac{2}{5}$$

29. Substituting $x = 4$:

$$\frac{4^2}{25} + \frac{y^2}{9} = 1$$

$$\frac{y^2}{9} + \frac{16}{25} = 1$$

$$\frac{y^2}{9} = \frac{9}{25}$$

$$\frac{y}{3} = \pm\frac{3}{5}$$

$$y = \pm\frac{9}{5}$$

31. Substituting $x = 1.8$:

$$16(1.8)^2 + 9y^2 = 144$$

$$51.84 + 9y^2 = 144$$

$$9y^2 = 92.16$$

$$3y = \pm 9.6$$

$$y = \pm 3.2$$

33. The asymptotes are $y = \frac{3}{4}x$ and $y = -\frac{3}{4}x$.

35. The equation has the form $\frac{x^2}{16} + \frac{y^2}{b^2} = 1$. Substituting the point $\left(2, \sqrt{3}\right)$:

$$\frac{(2)^2}{16} + \frac{\left(\sqrt{3}\right)^2}{b^2} = 1$$

$$\frac{1}{4} + \frac{3}{b^2} = 1$$

$$\frac{3}{b^2} = \frac{3}{4}$$

$$b^2 = 4$$

The equation of the ellipse is $\frac{x^2}{16} + \frac{y^2}{4} = 1$.

37. The equation of the ellipse is $\frac{x^2}{20^2} + \frac{y^2}{10^2} = 1$. Substituting $y = 6$:

$$\frac{x^2}{20^2} + \frac{6^2}{10^2} = 1$$

$$\frac{x^2}{400} + \frac{9}{25} = 1$$

$$\frac{x^2}{400} = \frac{16}{25}$$

$$x^2 = 256$$

$$x = 16$$

The man can walk within 16 feet of the center.

39. Substituting $a = 4$ and $c = 3$:

$$4^2 = b^2 + 3^2$$

$$16 = b^2 + 9$$

$$b^2 = 7$$

$$b = \sqrt{7} \approx 2.65$$

The width should be approximately $2(2.65) = 5.3$ feet wide.

41. Writing in logarithmic form: $\log_{10} 100 = 2$

43. Writing in exponential form: $3^4 = 81$

45. Solving for x:

$$\log_9 x = \frac{3}{2}$$

$$x = 9^{3/2} = 27$$

47. Simplifying: $\log_2 32 = \log_2 2^5 = 5$

49. Simplifying: $\log_3\left[\log_2 8\right] = \log_3\left[\log_2 2^3\right] = \log_3 3 = 1$

51. Finding the logarithm: $\log 576 \approx 2.7604$

55. Solving for x:
$$\log x = 2.6484$$
$$x = 10^{2.6484} \approx 445$$

53. Finding the logarithm: $\log 0.0576 \approx -1.2396$

57. Solving for x:
$$\log x = -7.3516$$
$$x = 10^{-7.3516} \approx 4.45 \times 10^{-8}$$

10.3   Second-Degree Inequalities and Nonlinear Systems

1. Graphing the inequality:

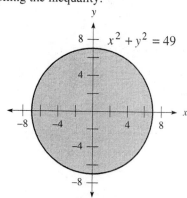

3. Graphing the inequality:

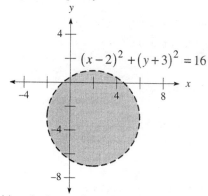

5. Graphing the inequality:

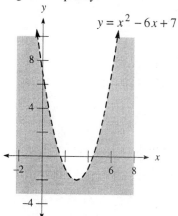

7. Graphing the inequality:

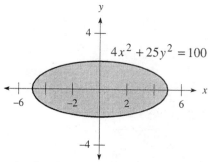

9. Solving the second equation for y yields $y = 3 - 2x$. Substituting into the first equation:
$$x^2 + (3 - 2x)^2 = 9$$
$$x^2 + 9 - 12x + 4x^2 = 9$$
$$5x^2 - 12x = 0$$
$$x(5x - 12) = 0$$
$$x = 0, \tfrac{12}{5}$$
$$y = 3, -\tfrac{9}{5}$$
The solutions are $(0, 3), \left(\tfrac{12}{5}, -\tfrac{9}{5}\right)$.

11. Solving the second equation for x yields $x = 8 - 2y$. Substituting into the first equation:
$$(8-2y)^2 + y^2 = 16$$
$$64 - 32y + 4y^2 + y^2 = 16$$
$$5y^2 - 32y + 48 = 0$$
$$(y-4)(5y-12) = 0$$
$$y = 4, \tfrac{12}{5}$$
$$x = 0, \tfrac{16}{5}$$
The solutions are $\left(0,4\right), \left(\tfrac{16}{5}, \tfrac{12}{5}\right)$.

13. Adding the two equations yields:
$$2x^2 = 50$$
$$x^2 = 25$$
$$x = -5, 5$$
$$y = 0$$

The solutions are $(-5,0), (5,0)$.

15. Substituting into the first equation:
$$x^2 + \left(x^2 - 3\right)^2 = 9$$
$$x^2 + x^4 - 6x^2 + 9 = 9$$
$$x^4 - 5x^2 = 0$$
$$x^2\left(x^2 - 5\right) = 0$$
$$x = 0, -\sqrt{5}, \sqrt{5}$$
$$y = -3, 2, 2$$
The solutions are $\left(0,-3\right), \left(-\sqrt{5}, 2\right), \left(\sqrt{5}, 2\right)$.

17. Substituting into the first equation:
$$x^2 + \left(x^2 - 4\right)^2 = 16$$
$$x^2 + x^4 - 8x^2 + 16 = 16$$
$$x^4 - 7x^2 = 0$$
$$x^2\left(x^2 - 7\right) = 0$$
$$x = 0, -\sqrt{7}, \sqrt{7}$$
$$y = -4, 3, 3$$
The solutions are $\left(0,-4\right), \left(-\sqrt{7}, 3\right), \left(\sqrt{7}, 3\right)$.

19. Substituting into the first equation:
$$3x + 2\left(x^2 - 5\right) = 10$$
$$3x + 2x^2 - 10 = 10$$
$$2x^2 + 3x - 20 = 0$$
$$(x+4)(2x-5) = 0$$
$$x = -4, \tfrac{5}{2}$$
$$y = 11, \tfrac{5}{4}$$
The solutions are $\left(-4,11\right), \left(\tfrac{5}{2}, \tfrac{5}{4}\right)$.

21. Substituting into the first equation:
$$-x + 1 = x^2 + 2x - 3$$
$$x^2 + 3x - 4 = 0$$
$$(x+4)(x-1) = 0$$
$$x = -4, 1$$
$$y = 5, 0$$
The solutions are $(-4,5), (1,0)$.

23. Substituting into the first equation:
$$x - 5 = x^2 - 6x + 5$$
$$x^2 - 7x + 10 = 0$$
$$(x-2)(x-5) = 0$$
$$x = 2, 5$$
$$y = -3, 0$$
The solutions are $(2,-3), (5,0)$.

25. Adding the two equations yields:
$$8x^2 = 72$$
$$x^2 = 9$$
$$x = \pm 3$$
$$y = 0$$
The solutions are $(-3,0), (3,0)$.

27.   Solving the first equation for x yields $x = y + 4$. Substituting into the second equation:

$$(y+4)^2 + y^2 = 16$$
$$y^2 + 8y + 16 + y^2 = 16$$
$$2y^2 + 8y = 0$$
$$2y(y+4) = 0$$
$$y = 0, -4$$
$$x = 4, 0$$

The solutions are $(0,-4)$, $(4,0)$.

29.   **a.**   Subtracting the two equations yields:

$$(x+8)^2 - x^2 = 0$$
$$x^2 + 16x + 64 - x^2 = 0$$
$$16x = -64$$
$$x = -4$$

Substituting to find y:

$$(-4)^2 + y^2 = 64$$
$$y^2 + 16 = 64$$
$$y^2 = 48$$
$$y = \pm\sqrt{48} = \pm 4\sqrt{3}$$

The intersection points are $\left(-4, -4\sqrt{3}\right)$ and $\left(-4, 4\sqrt{3}\right)$.

b.   Subtracting the two equations yields:

$$x^2 - (x-8)^2 = 0$$
$$x^2 - x^2 + 16x - 64 = 0$$
$$16x = 64$$
$$x = 4$$

Substituting to find y:

$$4^2 + y^2 = 64$$
$$y^2 + 16 = 64$$
$$y^2 = 48$$
$$y = \pm\sqrt{48} = \pm 4\sqrt{3}$$

The intersection points are $\left(4, -4\sqrt{3}\right)$ and $\left(4, 4\sqrt{3}\right)$.

31. Graphing the inequality:

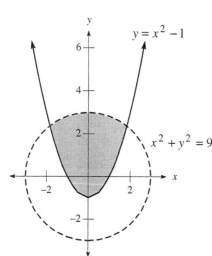

33. Graphing the inequality:

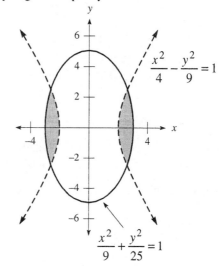

35. There is no intersection.

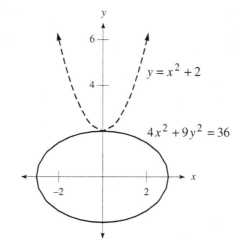

37. The system of inequalities is:
$$x^2 + y^2 < 16$$
$$y > 4 - \tfrac{1}{4}x^2$$

39. The system of equations is:
$$x^2 + y^2 = 89$$
$$x^2 - y^2 = 39$$
Adding the two equations yields:
$$2x^2 = 128$$
$$x^2 = 64$$
$$x = \pm 8$$
$$y = \pm 5$$
The numbers are either 8 and 5, 8 and −5, −8 and 5, or −8 and −5.

41. The system of equations is:
$$y = x^2 - 3$$
$$x + y = 9$$
Substituting into the second equation:
$$x + x^2 - 3 = 9$$
$$x^2 + x - 12 = 0$$
$$(x+4)(x-3) = 0$$
$$x = -4, 3$$
$$y = 13, 6$$
The numbers are either –4 and 13, or 3 and 6.

43. Expanding the logarithm: $\log_2 x^3 y = \log_2 x^3 + \log_2 y = 3\log_2 x + \log_2 y$

45. Writing as a single logarithm: $\log_{10} x - \log_{10} y^2 = \log_{10} \dfrac{x}{y^2}$

47. Solving the equation:
$$\log_4 x - \log_4 5 = 2$$
$$\log_4 \frac{x}{5} = 2$$
$$\frac{x}{5} = 4^2$$
$$\frac{x}{5} = 16$$
$$x = 80$$
The solution is 80.

49. Solving the equation:
$$5^x = 7$$
$$\log 5^x = \log 7$$
$$x \log 5 = \log 7$$
$$x = \frac{\log 7}{\log 5} \approx 1.21$$

51. Using the compound interest formula:
$$400\left(1 + \frac{0.10}{4}\right)^{4t} = 800$$
$$(1.025)^{4t} = 2$$
$$\ln(1.025)^{4t} = \ln 2$$
$$4t \ln 1.025 = \ln 2$$
$$t = \frac{\ln 2}{4 \ln 1.025} \approx 7.02$$
It will take 7.02 years.

53. Evaluating the logarithm: $\log_4 20 = \dfrac{\log 20}{\log 4} \approx 2.16$

55. Evaluating the logarithm: $\ln 576 \approx 6.36$

Chapter 10 Review

1. Using the distance formula: $d = \sqrt{(-1-2)^2 + (5-6)^2} = \sqrt{9+1} = \sqrt{10}$

3. Using the distance formula: $d = \sqrt{(-4-0)^2 + (0-3)^2} = \sqrt{16+9} = \sqrt{25} = 5$

5. Solving the equation:
$$\sqrt{(x-2)^2 + (-1+4)^2} = 5$$
$$(x-2)^2 + 9 = 25$$
$$(x-2)^2 = 16$$
$$x - 2 = \pm\sqrt{16}$$
$$x - 2 = -4, 4$$
$$x = -2, 6$$

7.   The equation is $(x-3)^2 + (y-1)^2 = 4$.

9.   The equation is $(x+5)^2 + y^2 = 9$.

11.   The equation is $x^2 + y^2 = 25$.

13.   Finding the radius: $r = \sqrt{(-2-2)^2 + (3-0)^2} = \sqrt{16+9} = \sqrt{25} = 5$. The equation is $(x+2)^2 + (y-3)^2 = 25$.

15.   The center is $(0,0)$ and the radius is 2.

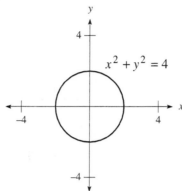

17.   Completing the square:
$$x^2 + y^2 - 6x + 4y = -4$$
$$\left(x^2 - 6x + 9\right) + \left(y^2 + 4y + 4\right) = -4 + 9 + 4$$
$$(x-3)^2 + (y+2)^2 = 9$$
The center is $(3,-2)$ and the radius is 3.

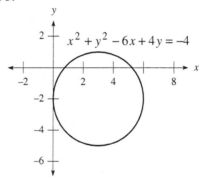

19.   Graphing the ellipse:

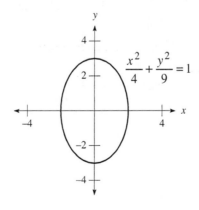

21.   Graphing the hyperbola:

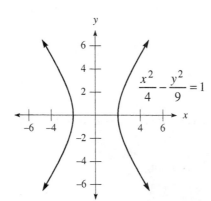

$$\frac{x^2}{4} - \frac{y^2}{9} = 1$$

23.   Graphing the inequality:

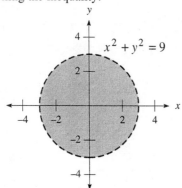

$$x^2 + y^2 = 9$$

25.   Graphing the inequality:

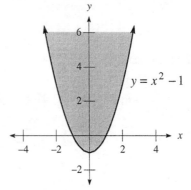

$$y = x^2 - 1$$

27.   Graphing the solution set:

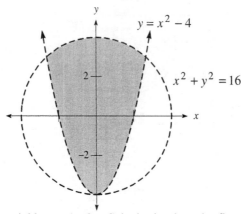

$$y = x^2 - 4$$

$$x^2 + y^2 = 16$$

29.   Solving the second equation for y yields $y = 4 - 2x$. Substituting into the first equation:

$$x^2 + (4 - 2x)^2 = 16$$
$$x^2 + 16 - 16x + 4x^2 = 16$$
$$5x^2 - 16x = 0$$
$$x(5x - 16) = 0$$
$$x = 0, \frac{16}{5}$$
$$y = 4, -\frac{12}{5}$$

The solutions are $(0, 4), \left(\frac{16}{5}, -\frac{12}{5}\right)$.

31. Adding the two equations yields:
$$18x^2 = 72$$
$$x^2 = 4$$
$$x = \pm 2$$
$$y = 0$$
The solutions are $(-2,0)$, $(2,0)$.

Chapters 1-10 Cumulative Review

1. Simplifying: $2^3 + 3(2 + 20 \div 4) = 8 + 3(2 + 5) = 8 + 3(7) = 8 + 21 = 29$

3. Simplifying: $-5(2x + 3) + 8x = -10x - 15 + 8x = -2x - 15$

5. Simplifying: $(3y + 2)^2 - (3y - 2)^2 = 9y^2 + 12y + 4 - 9y^2 + 12y - 4 = 24y$

7. Simplifying: $x^{2/3} \cdot x^{1/5} = x^{2/3+1/5} = x^{10/15+3/15} = x^{13/15}$

9. Solving the equation:
$$5y - 2 = -3y + 6$$
$$8y = 8$$
$$y = 1$$

11. Solving the equation:
$$|3x - 1| - 2 = 6$$
$$|3x - 1| = 8$$
$$3x - 1 = -8, 8$$
$$3x = -7, 9$$
$$x = -\tfrac{7}{3}, 3$$

13. Solving the equation:
$$x^3 - 3x^2 - 4x + 12 = 0$$
$$x^2(x - 3) - 4(x - 3) = 0$$
$$(x - 3)(x^2 - 4) = 0$$
$$(x - 3)(x + 2)(x - 2) = 0$$
$$x = -2, 2, 3$$

15. Solving the equation:
$$x - 2 = \sqrt{3x + 4}$$
$$(x - 2)^2 = 3x + 4$$
$$x^2 - 4x + 4 = 3x + 4$$
$$x^2 - 7x = 0$$
$$x(x - 7) = 0$$
$$x = 0, 7$$
The solution is 7 (0 does not check).

17. Solving the equation:
$$4x^2 + 6x = -5$$
$$4x^2 + 6x + 5 = 0$$
$$x = \frac{-6 \pm \sqrt{36 - 80}}{8} = \frac{-6 \pm \sqrt{-44}}{8} = \frac{-6 \pm 2i\sqrt{11}}{8} = -\frac{3}{4} \pm \frac{i\sqrt{11}}{4}$$

19. Solving the equation:
$$\log_2 x + \log_2 5 = 1$$
$$\log_2 5x = 1$$
$$5x = 2^1$$
$$x = \tfrac{2}{5}$$

21. Substituting into the first equation:
$$4x + 2(-3x + 1) = 4$$
$$4x - 6x + 2 = 4$$
$$-2x = 2$$
$$x = -1$$
$$y = 4$$
The solution is $(-1, 4)$.

23.   Adding the first and third equations yields the equation $4y + z = 7$.
Multiply the third equation by 2 and add it to the second equation:
$$2x - y - 3z = -1$$
$$-2x + 4y + 4z = 6$$
Adding yields the equation $3y + z = 5$. So the system of equations becomes:
$$4y + z = 7$$
$$3y + z = 5$$
Multiply the second equation by -1:
$$4y + z = 7$$
$$-3y - z = -5$$
Adding yields $y = 2$. Substituting to find z:
$$3(2) + z = 5$$
$$6 + z = 5$$
$$z = -1$$
Substituting into the original first equation:
$$x + 2(2) - (-1) = 4$$
$$x + 5 = 4$$
$$x = -1$$
The solution is $(-1, 2, -1)$.

25.   Multiplying: $\left(x^{1/5} + 3\right)\left(x^{1/5} - 3\right) = \left(x^{1/5}\right)^2 - (3)^2 = x^{2/5} - 9$

27.   Dividing: $\dfrac{3 - 2i}{1 + 2i} = \dfrac{3 - 2i}{1 + 2i} \cdot \dfrac{1 - 2i}{1 - 2i} = \dfrac{3 - 8i + 4i^2}{1 - 4i^2} = \dfrac{3 - 8i - 4}{1 + 4} = \dfrac{-1 - 8i}{5} = -\dfrac{1}{5} - \dfrac{8}{5}i$

29.   Graphing the inequality: **31.**   Graphing the curve:

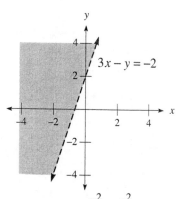

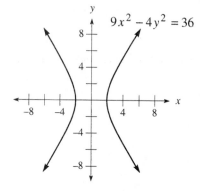

33.   The standard form is $\dfrac{x^2}{4} - \dfrac{y^2}{9} = 1$. Graphing the hyperbola:

35. Graphing the parabola:

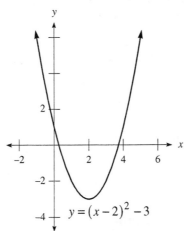

$$y = (x-2)^2 - 3$$

37. Factoring as a sum of cubes: $x^3 + 8y^3 = (x+2y)(x^2 - 2xy + 4y^2)$

39. Finding the value: $f(4) = -\frac{3}{2}(4) + 1 = -6 + 1 = -5$

41. First find the slope: $m = \dfrac{-5+1}{-3+6} = -\frac{4}{3}$. Using the point-slope formula:
$$y + 5 = -\frac{4}{3}(x+3)$$
$$y + 5 = -\frac{4}{3}x - 4$$
$$y = -\frac{4}{3}x - 9$$

43. Using the distance formula: $d = \sqrt{(4+3)^2 + (5-1)^2} = \sqrt{49+16} = \sqrt{65}$

45. Solving the equation:
$$\begin{vmatrix} 2x & -4 \\ 4 & x \end{vmatrix} = 18x$$
$$2x^2 + 16 = 18x$$
$$2x^2 - 18x + 16 = 0$$
$$x^2 - 9x + 8 = 0$$
$$(x-8)(x-1) = 0$$
$$x = 1, 8$$

47. Let $s = 20$ and solve for t:
$$48t - 16t^2 = 20$$
$$16t^2 - 48t + 20 = 0$$
$$4t^2 - 12t + 5 = 0$$
$$(2t-1)(2t-5) = 0$$
$$t = \frac{1}{2}, \frac{5}{2}$$

The object will be 20 feet above the ground after $\frac{1}{2}$ second and $\frac{5}{2}$ seconds.

49.   Factoring the inequality:
$$x^2 + x - 6 > 0$$
$$(x+3)(x-2) > 0$$
Forming the sign chart:

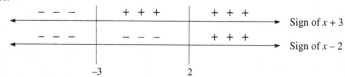

The solution set is $x < -3$ or $x > 2$. Graphing the solution set:

50.   The domain is $\{x \mid x \neq 2\}$.

Chapter 10 Test

1.   Solving the equation:
$$\sqrt{(x+1)^2 + (2-4)^2} = \left(2\sqrt{5}\right)^2$$
$$(x+1)^2 + 4 = 20$$
$$(x+1)^2 = 16$$
$$x + 1 = \pm\sqrt{16}$$
$$x + 1 = -4, 4$$
$$x = -5, 3$$

2.   The equation is $(x+2)^2 + (y-4)^2 = 9$.

3.   Finding the radius: $r = \sqrt{(-3)^2 + (-4)^2} = \sqrt{9+16} = \sqrt{25} = 5$. The equation is $x^2 + y^2 = 25$.

4.   Completing the square:
$$x^2 + y^2 - 10x + 6y = 5$$
$$\left(x^2 - 10x + 25\right) + \left(y^2 + 6y + 9\right) = 5 + 25 + 9$$
$$(x-5)^2 + (y+3)^2 = 39$$
The center is (5,–3) and the radius is $\sqrt{39}$.

5.   The standard form is $\dfrac{x^2}{4} - \dfrac{y^2}{16} = 1$. Graphing the hyperbola:

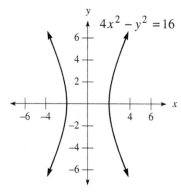

6.   Graphing the ellipse:

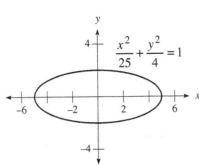

7.   Graphing the inequality:

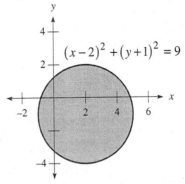

8.   Graphing the inequality:

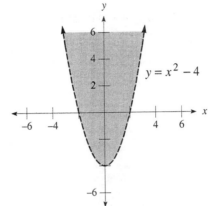

9.   Solving the first equation for y yields $y = 5 - 2x$. Substituting into the first equation:

$$x^2 + (5 - 2x)^2 = 25$$
$$x^2 + 25 - 20x + 4x^2 = 25$$
$$5x^2 - 20x = 0$$
$$5x(x - 4) = 0$$
$$x = 0, 4$$
$$y = 5, -3$$

The solutions are $(0,5), (4,-3)$.

10.   Substituting into the first equation:

$$x^2 + (x^2 - 4)^2 = 16$$
$$x^2 + x^4 - 8x^2 + 16 = 16$$
$$x^4 - 7x^2 = 0$$
$$x^2(x^2 - 7) = 0$$
$$x = 0, \pm\sqrt{7}$$
$$y = -4, 3$$

The solutions are $(0,-4), (\sqrt{7}, 3), (-\sqrt{7}, 3)$.

Chapter 11
Sequences and Series

11.1 Sequences

1. The first five terms are: $4, 7, 10, 13, 16$

3. The first five terms are: $3, 7, 11, 15, 19$

5. The first five terms are: $1, 2, 3, 4, 5$

7. The first five terms are: $4, 7, 12, 19, 28$

9. The first five terms are: $\frac{1}{4}, \frac{2}{5}, \frac{1}{2}, \frac{4}{7}, \frac{4}{8}$

11. The first five terms are: $1, \frac{1}{4}, \frac{1}{9}, \frac{1}{16}, \frac{1}{25}$

13. The first five terms are: $2, 4, 8, 16, 32$

15. The first five terms are: $2, \frac{3}{2}, \frac{4}{3}, \frac{5}{4}, \frac{6}{5}$

17. The first five terms are: $-2, 4, -8, 16, -32$

19. The first five terms are: $3, 5, 3, 5, 3$

21. The first five terms are: $1, -\frac{2}{3}, \frac{3}{5}, -\frac{4}{7}, \frac{5}{9}$

23. The first five terms are: $\frac{1}{2}, 1, \frac{9}{8}, 1, \frac{25}{32}$

25. The first five terms are: $3, -9, 27, -81, 243$

27. The first five terms are: $1, 5, 13, 29, 61$

29. The first five terms are: $2, 3, 5, 9, 17$

31. The first five terms are: $5, 11, 29, 83, 245$

33. The first five terms are: $4, 4, 4, 4, 4$

35. The general term is: $a_n = 4n$

37. The general term is: $a_n = n^2$

39. The general term is: $a_n = 2^{n+1}$

41. The general term is: $a_n = \dfrac{1}{2^{n+1}}$

43. The general term is: $a_n = 3n + 2$

45. The general term is: $a_n = -4n + 2$

47. The general term is: $a_n = (-2)^{n-1}$

49. The general term is: $a_n = \log_{n+1}(n+2)$

51. **a.** The sequence of salaries is: \$28000, \$29120, \$30284.80, \$31,496.19, \$32756.04

    b. The general term is: $a_n = 28000(1.04)^{n-1}$

53. **a.** The sequence of values is: 16 ft, 48 ft, 80 ft, 112 ft, 144 ft

    b. The sum of the values is 400 feet.   **c.** No, since the sum is less than 420 feet.

55. **a.** The sequence of angles is: $180°, 360°, 540°, 720°$ **b.** Substituting $n = 20$: $a_{20} = 180°(20 - 2) = 3240°$

    c. The sum of the interior angles would be $0°$.

57. Using the distance formula: $d = \sqrt{(7-0)^2 + (0-2)^2} = \sqrt{49 + 4} = \sqrt{53}$

59. Using the distance formula: $d = \sqrt{(-1-0)^2 + (-10+5)^2} = \sqrt{1 + 25} = \sqrt{26}$

61. Solving the equation:

$$\sqrt{(5+6)^2 + (x-2)^2} = \sqrt{130}$$
$$121 + (x-2)^2 = 130$$
$$(x-2)^2 = 9$$
$$x - 2 = \pm 3$$
$$x = -1, 5$$

63. Computing the values: $a_{100} \approx 2.7048, a_{1,000} \approx 2.7169, a_{10,000} \approx 2.7181, a_{100,000} \approx 2.7183$

65. The first ten terms are: 1,1,2,3,5,8,13,21,34,55

11.2 Series

1. Expanding the sum: $\displaystyle\sum_{i=1}^{4}(2i+4) = 6+8+10+12 = 36$

3. Expanding the sum: $\displaystyle\sum_{i=2}^{3}\left(i^2 - 1\right) = 3+8 = 11$

5. Expanding the sum: $\displaystyle\sum_{i=1}^{4}\left(i^2 - 3\right) = -2+1+6+13 = 18$

7. Expanding the sum: $\displaystyle\sum_{i=1}^{4}\frac{i}{1+i} = \frac{1}{2}+\frac{2}{3}+\frac{3}{4}+\frac{4}{5} = \frac{30}{60}+\frac{40}{60}+\frac{45}{60}+\frac{48}{60} = \frac{163}{60}$

9. Expanding the sum: $\displaystyle\sum_{i=1}^{4}(-3)^i = -3+9-27+81 = 60$

11. Expanding the sum: $\displaystyle\sum_{i=3}^{6}(-2)^i = -8+16-32+64 = 40$

13. Expanding the sum: $\displaystyle\sum_{i=2}^{6}(-2)^i = 4-8+16-32+64 = 44$

15. Expanding the sum: $\displaystyle\sum_{i=1}^{5}\left(-\frac{1}{2}\right)^i = -\frac{1}{2}+\frac{1}{4}-\frac{1}{8}+\frac{1}{16}-\frac{1}{32} = -\frac{16}{32}+\frac{8}{32}-\frac{4}{32}+\frac{2}{32}-\frac{1}{32} = -\frac{11}{32}$

17. Expanding the sum: $\displaystyle\sum_{i=2}^{5}\frac{i-1}{i+1} = \frac{1}{3}+\frac{1}{2}+\frac{3}{5}+\frac{2}{3} = 1+\frac{5}{10}+\frac{6}{10} = \frac{21}{10}$

19. Expanding the sum: $\displaystyle\sum_{i=1}^{5}(x+i) = (x+1)+(x+2)+(x+3)+(x+4)+(x+5) = 5x+15$

21. Expanding the sum: $\displaystyle\sum_{i=1}^{4}(x-2)^i = (x-2)+(x-2)^2+(x-2)^3+(x-2)^4$

23. Expanding the sum: $\displaystyle\sum_{i=1}^{5}\frac{x+i}{x-1} = \frac{x+1}{x-1}+\frac{x+2}{x-1}+\frac{x+3}{x-1}+\frac{x+4}{x-1}+\frac{x+5}{x-1}$

25. Expanding the sum: $\displaystyle\sum_{i=3}^{8}(x+i)^i = (x+3)^3+(x+4)^4+(x+5)^5+(x+6)^6+(x+7)^7+(x+8)^8$

27. Expanding the sum: $\displaystyle\sum_{i=3}^{6}(x-2i)^{i+3} = (x-6)^6 + (x-8)^7 + (x-10)^8 + (x-12)^9$

29. Writing with summation notation: $2+4+8+16 = \displaystyle\sum_{i=1}^{4} 2^i$

31. Writing with summation notation: $4+8+16+32+64 = \displaystyle\sum_{i=2}^{6} 2^i$

33. Writing with summation notation: $5+9+13+17+21 = \displaystyle\sum_{i=1}^{5}(4i+1)$

35. Writing with summation notation: $-4+8-16+32 = \displaystyle\sum_{i=2}^{5} -(-2)^i$

37. Writing with summation notation: $\frac{3}{4}+\frac{4}{5}+\frac{5}{6}+\frac{6}{7}+\frac{7}{8} = \displaystyle\sum_{i=3}^{7}\frac{i}{i+1}$

39. Writing with summation notation: $\frac{1}{3}+\frac{2}{5}+\frac{3}{7}+\frac{4}{9} = \displaystyle\sum_{i=1}^{4}\frac{i}{2i+1}$

41. Writing with summation notation: $(x-2)^6+(x-2)^7+(x-2)^8+(x-2)^9 = \displaystyle\sum_{i=6}^{9}(x-2)^i$

43. Writing with summation notation: $\left(1+\frac{1}{x}\right)^2+\left(1+\frac{2}{x}\right)^3+\left(1+\frac{3}{x}\right)^4+\left(1+\frac{4}{x}\right)^5 = \displaystyle\sum_{i=1}^{4}\left(1+\frac{i}{x}\right)^{i+1}$

45. Writing with summation notation: $\dfrac{x}{x+3}+\dfrac{x}{x+4}+\dfrac{x}{x+5} = \displaystyle\sum_{i=3}^{5}\dfrac{x}{x+i}$

47. Writing with summation notation: $x^2(x+2)+x^3(x+3)+x^4(x+4) = \displaystyle\sum_{i=2}^{4}x^i(x+i)$

49. **a.** Writing as a series: $\frac{1}{3} = 0.3+0.03+0.003+0.0003+\dots$

　　 b. Writing as a series: $\frac{2}{9} = 0.2+0.02+0.002+0.0002+\dots$

　　 c. Writing as a series: $\frac{3}{11} = 0.27+0.0027+0.000027+\dots$

51. The sequence of values he falls is: 16, 48, 80, 112, 144, 176, 208
　　 During the seventh second he falls 208 feet, and the total he falls is 784 feet.

53. **a.** The series is $16+48+80+112+144$. 　　　 **b.** Writing in summation notation: $\displaystyle\sum_{i=1}^{5}(32i-16)$

55. The equation is: $x^2+(y-5)^2 = 36$

57. The center is $(-4,3)$ and the radius is $\sqrt{20}$.

59. Solving the equation for x:

$$\sum_{i=1}^{4}(x-i) = 16$$
$$(x-1)+(x-2)+(x-3)+(x-4) = 16$$
$$4x-10 = 16$$
$$4x = 26$$
$$x = \frac{13}{2}$$

11.3   Arithmetic Sequences

1. The sequence is arithmetic: $d = 1$

3. The sequence is not arithmetic.

5. The sequence is arithmetic: $d = -5$

7. The sequence is not arithmetic.

9. The sequence is arithmetic: $d = \frac{2}{3}$

11. Finding the general term: $a_n = 3 + (n-1) \cdot 4 = 3 + 4n - 4 = 4n - 1$. Therefore: $a_{24} = 4 \cdot 24 - 1 = 96 - 1 = 95$

13. Finding the required term: $a_{10} = 6 + (10-1) \cdot (-2) = 6 - 18 = -12$. Finding the sum: $S_{10} = \frac{10}{2}(6-12) = 5(-6) = -30$

15. Writing out the equations:
$$a_6 = a_1 + 5d \qquad a_{12} = a_1 + 11d$$
$$17 = a_1 + 5d \qquad 29 = a_1 + 11d$$
The system of equations is:
$$a_1 + 11d = 29$$
$$a_1 + 5d = 17$$
Subtracting yields:
$$6d = 12$$
$$d = 2$$
$$a_1 = 7$$
Finding the required term: $a_{30} = 7 + 29 \cdot 2 = 7 + 58 = 65$

17. Writing out the equations:
$$a_3 = a_1 + 2d \qquad a_8 = a_1 + 7d$$
$$16 = a_1 + 2d \qquad 26 = a_1 + 7d$$
The system of equations is:
$$a_1 + 7d = 26$$
$$a_1 + 2d = 16$$
Subtracting yields:
$$5d = 10$$
$$d = 2$$
$$a_1 = 12$$
Finding the required term: $a_{20} = 12 + 19 \cdot 2 = 12 + 38 = 50$. Finding the sum: $S_{20} = \frac{20}{2}(12 + 50) = 10 \cdot 62 = 620$

19. Finding the required term: $a_{20} = 3 + 19 \cdot 4 = 3 + 76 = 79$. Finding the sum: $S_{20} = \frac{20}{2}(3 + 79) = 10 \cdot 82 = 820$

21. Writing out the equations:
$$a_4 = a_1 + 3d \qquad a_{10} = a_1 + 9d$$
$$14 = a_1 + 3d \qquad 32 = a_1 + 9d$$
The system of equations is:
$$a_1 + 9d = 32$$
$$a_1 + 3d = 14$$
Subtracting yields:
$$6d = 18$$
$$d = 3$$
$$a_1 = 5$$
Finding the required term: $a_{40} = 5 + 39 \cdot 3 = 5 + 117 = 122$. Finding the sum: $S_{40} = \frac{40}{2}(5 + 122) = 20 \cdot 127 = 2540$

23. Using the summation formula:

$$S_6 = \tfrac{6}{2}\left(a_1 + a_6\right)$$
$$-12 = 3\left(a_1 - 17\right)$$
$$a_1 - 17 = -4$$
$$a_1 = 13$$

Now find d:

$$a_6 = 13 + 5 \cdot d$$
$$-17 = 13 + 5d$$
$$5d = -30$$
$$d = -6$$

25. Using $a_1 = 14$ and $d = -3$: $a_{85} = 14 + 84 \cdot (-3) = 14 - 252 = -238$

27. Using the summation formula:

$$S_{20} = \tfrac{20}{2}\left(a_1 + a_{20}\right)$$
$$80 = 10\left(-4 + a_{20}\right)$$
$$-4 + a_{20} = 8$$
$$a_{20} = 12$$

Now finding d:

$$a_{20} = a_1 + 19d$$
$$12 = -4 + 19d$$
$$16 = 19d$$
$$d = \tfrac{16}{19}$$

Finding the required term: $a_{39} = -4 + 38\left(\tfrac{16}{19}\right) = -4 + 32 = 28$

29. Using $a_1 = 5$ and $d = 4$: $a_{100} = a_1 + 99d = 5 + 99 \cdot 4 = 5 + 396 = 401$

Now finding the required sum: $S_{100} = \tfrac{100}{2}(5 + 401) = 50 \cdot 406 = 20,300$

31. Using $a_1 = 12$ and $d = -5$: $a_{35} = a_1 + 34d = 12 + 34 \cdot (-5) = 12 - 170 = -158$

33. Using $a_1 = \tfrac{1}{2}$ and $d = \tfrac{1}{2}$: $a_{10} = a_1 + 9d = \tfrac{1}{2} + 9 \cdot \tfrac{1}{2} = \tfrac{10}{2} = 5$. Finding the sum: $S_{10} = \tfrac{10}{2}\left(\tfrac{1}{2} + 5\right) = \tfrac{10}{2} \cdot \tfrac{11}{2} = \tfrac{55}{2}$

35.
a. The first five terms are: $18,000, \$14,700, \$11,400, \$8,100, \$4,800$
b. The common difference is $-\$3,300$.
c. Constructing a line graph:

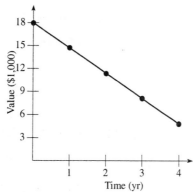

d. The value is approximately $9,750.
e. The recursive formula is: $a_0 = 18000; a_n = a_{n-1} - 3300$ for $n \geq 1$

37. **a.** The sequence of values is: 1500 ft, 1460 ft, 1420 ft, 1380 ft, 1340 ft, 1300 ft
     b. It is arithmetic because the same amount is subtracted from each succeeding term.
     c. The general term is: $a_n = 1500 + (n-1) \cdot (-40) = 1500 - 40n + 40 = 1540 - 40n$

39. **a.** The first 15 triangular numbers is: 1,3,6,10,15,21,28,36,45,55,66,78,91,105,120
     b. The recursive formula is: $a_1 = 1; a_n = n + a_{n-1}$ for $n \geq 2$
     c. It is not arithmetic because the same amount is not added to each term.

41. Finding the x-intercepts: Finding the y-intercepts:

$$\frac{x^2}{25} + \frac{0^2}{16} = 1 \qquad\qquad \frac{0^2}{25} + \frac{y^2}{16} = 1$$
$$x^2 = 25 \qquad\qquad\qquad y^2 = 16$$
$$x = \pm 5 \qquad\qquad\qquad y = \pm 4$$

43. Finding the x-intercepts: Finding the y-intercepts:

$$\frac{0^2}{9} - x^2 = 1 \qquad\qquad \frac{y^2}{9} - 0^2 = 1$$
$$x^2 = -1 \qquad\qquad\qquad y^2 = 9$$
$$x = \text{impossible} \qquad\qquad y = \pm 3$$

45. Graphing the equation:

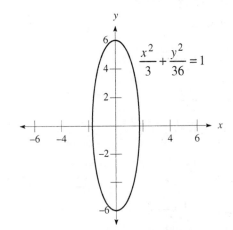

$$\frac{x^2}{3} + \frac{y^2}{36} = 1$$

47. Using the summation formulas:

$$S_7 = \tfrac{7}{2}\left(a_1 + a_7\right) \qquad\qquad\qquad S_{13} = \tfrac{13}{2}\left(a_1 + a_{13}\right)$$
$$147 = \tfrac{7}{2}\left(a_1 + a_7\right) \qquad\qquad\qquad 429 = \tfrac{13}{2}\left(a_1 + a_{13}\right)$$
$$a_1 + a_7 = 42 \qquad\qquad\qquad\qquad a_1 + a_{13} = 66$$
$$a_1 + \left(a_1 + 6d\right) = 42 \qquad\qquad\qquad a_1 + \left(a_1 + 12d\right) = 66$$
$$2a_1 + 6d = 42 \qquad\qquad\qquad\qquad 2a_1 + 12d = 66$$
$$a_1 + 3d = 21 \qquad\qquad\qquad\qquad a_1 + 6d = 33$$

So we have the system of equations:
$$a_1 + 6d = 33$$
$$a_1 + 3d = 21$$

Subtracting:
$$3d = 12$$
$$d = 4$$
$$a_1 = 33 - 6 \cdot 4 = 9$$

49. Solving the equation:
$$a_{15} - a_7 = -24$$
$$\left(a_1 + 14d\right) - \left(a_1 + 6d\right) = -24$$
$$a_1 + 14d - a_1 - 6d = -24$$
$$8d = -24$$
$$d = -3$$

51. Using $a_1 = 1$ and $d = \sqrt{2}$, find the term: $a_{50} = a_1 + 49d = 1 + 49\sqrt{2}$

Finding the sum: $S_{50} = \dfrac{50}{2}\left(a_1 + a_{50}\right) = 25\left(1 + 1 + 49\sqrt{2}\right) = 50 + 1225\sqrt{2}$

11.4   Geometric Sequences

1. The sequence is geometric: $r = 5$

3. The sequence is geometric: $r = \frac{1}{3}$

5. The sequence is not geometric.

7. The sequence is geometric: $r = -2$

9. The sequence is not geometric.

11. Finding the general term: $a_n = 4 \cdot 3^{n-1}$

13. Finding the term: $a_6 = -2\left(-\frac{1}{2}\right)^{6-1} = -2\left(-\frac{1}{2}\right)^5 = -2\left(-\frac{1}{32}\right) = \frac{1}{16}$

15. Finding the term: $a_{20} = 3(-1)^{20-1} = 3(-1)^{19} = -3$

17. Finding the sum: $S_{10} = \dfrac{10\left(2^{10}-1\right)}{2-1} = 10 \cdot 1023 = 10,230$

19. Finding the sum: $S_{20} = \dfrac{1\left((-1)^{20}-1\right)}{-1-1} = \dfrac{1 \cdot 0}{-2} = 0$

21. Using $a_1 = \frac{1}{5}$ and $r = \frac{1}{2}$, the term is: $a_8 = \frac{1}{5} \cdot \left(\frac{1}{2}\right)^{8-1} = \frac{1}{5} \cdot \left(\frac{1}{2}\right)^7 = \frac{1}{5} \cdot \frac{1}{128} = \frac{1}{640}$

23. Using $a_1 = -\frac{1}{2}$ and $r = \frac{1}{2}$, the sum is: $S_5 = \dfrac{-\frac{1}{2}\left(\left(\frac{1}{2}\right)^5 - 1\right)}{\frac{1}{2}-1} = \dfrac{-\frac{1}{2}\left(\frac{1}{32}-1\right)}{-\frac{1}{2}} = \frac{1}{32} - 1 = -\frac{31}{32}$

25. Using $a_1 = \sqrt{2}$ and $r = \sqrt{2}$, the term is: $a_{10} = \sqrt{2}\left(\sqrt{2}\right)^9 = \left(\sqrt{2}\right)^{10} = 2^5 = 32$

The sum is: $S_{10} = \dfrac{\sqrt{2}\left(\left(\sqrt{2}\right)^{10} - 1\right)}{\sqrt{2}-1} = \dfrac{\sqrt{2}(32-1)}{\sqrt{2}-1} = \dfrac{31\sqrt{2}}{\sqrt{2}-1} \cdot \dfrac{\sqrt{2}+1}{\sqrt{2}+1} = \dfrac{62+31\sqrt{2}}{2-1} = 62 + 31\sqrt{2}$

27. Using $a_1 = 100$ and $r = 0.1$, the term is: $a_6 = 100(0.1)^5 = 10^2\left(10^{-5}\right) = 10^{-3} = \frac{1}{1000}$

The sum is: $S_6 = \dfrac{100\left((0.1)^6 - 1\right)}{0.1-1} = \dfrac{100\left(10^{-6}-1\right)}{-0.9} = \dfrac{-99.9999}{-0.9} = 111.111$

29. Since $a_4 \cdot r \cdot r = a_6$, we have the equation:
$$a_4 r^2 = a_6$$
$$40r^2 = 160$$
$$r^2 = 4$$
$$r = \pm 2$$

31. Since $a_1 = -3$ and $r = -2$, the values are:
$$a_8 = -3(-2)^7 = -3(-128) = 384$$
$$S_8 = \dfrac{-3\left((-2)^8 - 1\right)}{-2-1} = \dfrac{-3(256-1)}{-3} = 255$$

33. Since $a_7 \cdot r \cdot r \cdot r = a_{10}$, we have the equation:

$$a_7 r^3 = a_{10}$$
$$13 r^3 = 104$$
$$r^3 = 8$$
$$r = 2$$

35. Using $a_1 = \frac{1}{2}$ and $r = \frac{1}{2}$ in the sum formula: $S = \dfrac{\frac{1}{2}}{1 - \frac{1}{2}} = \dfrac{\frac{1}{2}}{\frac{1}{2}} = 1$

37. Using $a_1 = 4$ and $r = \frac{1}{2}$ in the sum formula: $S = \dfrac{4}{1 - \frac{1}{2}} = \dfrac{4}{\frac{1}{2}} = 8$

39. Using $a_1 = 2$ and $r = \frac{1}{2}$ in the sum formula: $S = \dfrac{2}{1 - \frac{1}{2}} = \dfrac{2}{\frac{1}{2}} = 4$

41. Using $a_1 = \frac{4}{3}$ and $r = -\frac{1}{2}$ in the sum formula: $S = \dfrac{\frac{4}{3}}{1 + \frac{1}{2}} = \dfrac{\frac{4}{3}}{\frac{3}{2}} = \frac{4}{3} \cdot \frac{2}{3} = \frac{8}{9}$

43. Using $a_1 = \frac{2}{5}$ and $r = \frac{2}{5}$ in the sum formula: $S = \dfrac{\frac{2}{5}}{1 - \frac{2}{5}} = \dfrac{\frac{2}{5}}{\frac{3}{5}} = \frac{2}{5} \cdot \frac{5}{3} = \frac{2}{3}$

45. Using $a_1 = \frac{3}{4}$ and $r = \frac{1}{3}$ in the sum formula: $S = \dfrac{\frac{3}{4}}{1 - \frac{1}{3}} = \dfrac{\frac{3}{4}}{\frac{2}{3}} = \frac{3}{4} \cdot \frac{3}{2} = \frac{9}{8}$

47. Interpreting the decimal as an infinite sum with $a_1 = 0.4$ and $r = 0.1$: $S = \dfrac{0.4}{1 - 0.1} = \dfrac{0.4}{0.9} \cdot \dfrac{10}{10} = \frac{4}{9}$

49. Interpreting the decimal as an infinite sum with $a_1 = 0.27$ and $r = 0.01$: $S = \dfrac{0.27}{1 - 0.01} = \dfrac{0.27}{0.99} \cdot \dfrac{100}{100} = \frac{27}{99} = \frac{3}{11}$

51.
a. The first five terms are: $450,000, $315,000, $220,500, $154,350, $108,045
b. The common ratio is 0.7.
c. Constructing a line graph:

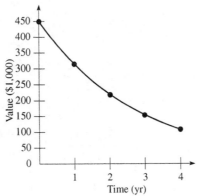

d. The value is approximately $90,000.
e. The recursive formula is: $a_0 = 450000; a_n = 0.7 a_{n-1}$ for $n \geq 1$

53.
a. Using $a_1 = \frac{1}{3}$ and $r = \frac{1}{3}$ in the sum formula: $S = \dfrac{\frac{1}{3}}{1 - \frac{1}{3}} = \dfrac{\frac{1}{3}}{\frac{2}{3}} = \frac{1}{2}$

b. Finding the sum: $S_6 = \dfrac{\frac{1}{3}\left(\left(\frac{1}{3}\right)^6 - 1\right)}{\frac{1}{3} - 1} = \dfrac{\frac{1}{3}\left(\frac{1}{729} - 1\right)}{-\frac{2}{3}} = \dfrac{\frac{1}{3}\left(-\frac{728}{729}\right)}{-\frac{2}{3}} = -\frac{1}{2}\left(-\frac{728}{729}\right) = \frac{364}{729}$

c. Finding the difference of these two answers: $S - S_6 = \frac{1}{2} - \frac{364}{729} = \frac{729}{1458} - \frac{728}{1458} = \frac{1}{1458}$

55. There are two infinite series for these heights. For the amount the ball falls, use $a_1 = 20$ and $r = \frac{7}{8}$:

$$S_{\text{fall}} = \frac{20}{1 - \frac{7}{8}} = \frac{20}{\frac{1}{8}} = 160$$

For the amount the ball rises, use $a_1 = \frac{7}{8}(20) = \frac{35}{2}$ and $r = \frac{7}{8}$: $S_{\text{rise}} = \frac{\frac{35}{2}}{1 - \frac{7}{8}} = \frac{\frac{35}{2}}{\frac{1}{8}} = \frac{35}{2} \cdot 8 = 140$

The total distance traveled is then 160 feet + 140 feet = 300 feet

57. **a.** The general term is: $a_n = 15\left(\frac{4}{5}\right)^{n-1}$ **b.** Finding the sum: $S = \frac{15}{1 - \frac{4}{5}} = \frac{15}{\frac{1}{5}} = 75$ feet

59. Expanding: $(x+5)^2 = (x+5)(x+5) = x^2 + 10x + 25$

61. Expanding: $(x+y)^3 = (x+y)(x+y)^2 = (x+y)\left(x^2 + 2xy + y^2\right) = x^3 + 3x^2y + 3xy^2 + y^3$

63. Expanding: $(x+y)^4 = (x+y)^2(x+y)^2 = \left(x^2 + 2xy + y^2\right)\left(x^2 + 2xy + y^2\right) = x^4 + 4x^3y + 6x^2y^2 + 4xy^3 + y^4$

65. Interpreting the decimal as an infinite sum with $a_1 = 0.63$ and $r = 0.01$: $S = \frac{0.63}{1 - 0.01} = \frac{0.63}{0.99} = \frac{63}{99} = \frac{7}{11}$

67. Finding the common ratio:

$$S = \frac{a_1}{1-r}$$
$$6 = \frac{4}{1-r}$$
$$6 - 6r = 4$$
$$-6r = -2$$
$$r = \frac{1}{3}$$

69. **a.** The areas are: stage 2: $\frac{3}{4}$;   stage 3: $\frac{9}{16}$;   stage 4: $\frac{27}{64}$

     b. They form a geometric sequence.

     c. The area is 0, since the sequence of areas is approaching 0.

     d. The perimeters form an increasing sequence.

11.5   The Binomial Expansion

1. Using the binomial formula:

$$(x+2)^4 = \binom{4}{0}x^4 + \binom{4}{1}x^3(2) + \binom{4}{2}x^2(2)^2 + \binom{4}{3}x(2)^3 + \binom{4}{4}(2)^4$$
$$= x^4 + 4 \cdot 2x^3 + 6 \cdot 4x^2 + 4 \cdot 8x + 16$$
$$= x^4 + 8x^3 + 24x^2 + 32x + 16$$

3. Using the binomial formula:

$$(x+y)^6 = \binom{6}{0}x^6 + \binom{6}{1}x^5y + \binom{6}{2}x^4y^2 + \binom{6}{3}x^3y^3 + \binom{6}{4}x^2y^4 + \binom{6}{5}xy^5 + \binom{6}{6}y^6$$
$$= x^6 + 6x^5y + 15x^4y^2 + 20x^3y^3 + 15x^2y^4 + 6xy^5 + y^6$$

5. Using the binomial formula:

$$(2x+1)^5 = \binom{5}{0}(2x)^5 + \binom{5}{1}(2x)^4(1) + \binom{5}{2}(2x)^3(1)^2 + \binom{5}{3}(2x)^2(1)^3 + \binom{5}{4}(2x)(1)^4 + \binom{5}{5}(1)^5$$
$$= 32x^5 + 5 \cdot 16x^4 + 10 \cdot 8x^3 + 10 \cdot 4x^2 + 5 \cdot 2x + 1$$
$$= 32x^5 + 80x^4 + 80x^3 + 40x^2 + 10x + 1$$

7.   Using the binomial formula:
$$(x-2y)^5 = \binom{5}{0}x^5 + \binom{5}{1}x^4(-2y) + \binom{5}{2}x^3(-2y)^2 + \binom{5}{3}x^2(-2y)^3 + \binom{5}{4}x(-2y)^4 + \binom{5}{5}(-2y)^5$$
$$= x^5 - 5 \cdot 2x^4y + 10 \cdot 4x^3y^2 - 10 \cdot 8x^2y^3 + 5 \cdot 16xy^4 - 32y^5$$
$$= x^5 - 10x^4y + 40x^3y^2 - 80x^2y^3 + 80xy^4 - 32y^5$$

9.   Using the binomial formula:
$$(3x-2)^4 = \binom{4}{0}(3x)^4 + \binom{4}{1}(3x)^3(-2) + \binom{4}{2}(3x)^2(-2)^2 + \binom{4}{3}(3x)(-2)^3 + \binom{4}{4}(-2)^4$$
$$= 81x^4 - 4 \cdot 54x^3 + 6 \cdot 36x^2 - 4 \cdot 24x + 16$$
$$= 81x^4 - 216x^3 + 216x^2 - 96x + 16$$

11.   Using the binomial formula:
$$(4x-3y)^3 = \binom{3}{0}(4x)^3 + \binom{3}{1}(4x)^2(-3y) + \binom{3}{2}(4x)(-3y)^2 + \binom{3}{3}(-3y)^3$$
$$= 64x^3 - 3 \cdot 48x^2y + 3 \cdot 36xy^2 - 27y^3$$
$$= 64x^3 - 144x^2y + 108xy^2 - 27y^3$$

13.   Using the binomial formula:
$$\left(x^2+2\right)^4 = \binom{4}{0}\left(x^2\right)^4 + \binom{4}{1}\left(x^2\right)^3(2) + \binom{4}{2}\left(x^2\right)^2(2)^2 + \binom{4}{3}\left(x^2\right)(2)^3 + \binom{4}{4}(2)^4$$
$$= x^8 + 4 \cdot 2x^6 + 6 \cdot 4x^4 + 4 \cdot 8x^2 + 16$$
$$= x^8 + 8x^6 + 24x^4 + 32x^2 + 16$$

15.   Using the binomial formula:
$$\left(x^2+y^2\right)^3 = \binom{3}{0}\left(x^2\right)^3 + \binom{3}{1}\left(x^2\right)^2\left(y^2\right) + \binom{3}{2}\left(x^2\right)\left(y^2\right)^2 + \binom{3}{3}\left(y^2\right)^3 = x^6 + 3x^4y^2 + 3x^2y^4 + y^6$$

17.   Using the binomial formula:
$$(2x+3y)^4 = \binom{4}{0}(2x)^4 + \binom{4}{1}(2x)^3(3y) + \binom{4}{2}(2x)^2(3y)^2 + \binom{4}{3}(2x)(3y)^3 + \binom{4}{4}(3y)^4$$
$$= 16x^4 + 4 \cdot 24x^3y + 6 \cdot 36x^2y^2 + 4 \cdot 54xy^3 + 81y^4$$
$$= 16x^4 + 96x^3y + 216x^2y^2 + 216xy^3 + 81y^4$$

19.   Using the binomial formula:
$$\left(\frac{x}{2}+\frac{y}{3}\right)^3 = \binom{3}{0}\left(\frac{x}{2}\right)^3 + \binom{3}{1}\left(\frac{x}{2}\right)^2\left(\frac{y}{3}\right) + \binom{3}{2}\left(\frac{x}{2}\right)\left(\frac{y}{3}\right)^2 + \binom{3}{3}\left(\frac{y}{3}\right)^3$$
$$= \frac{x^3}{8} + 3 \cdot \frac{x^2y}{12} + 3 \cdot \frac{xy^2}{18} + \frac{y^3}{27}$$
$$= \frac{x^3}{8} + \frac{x^2y}{4} + \frac{xy^2}{6} + \frac{y^3}{27}$$

21.   Using the binomial formula:
$$\left(\frac{x}{2}-4\right)^3 = \binom{3}{0}\left(\frac{x}{2}\right)^3 + \binom{3}{1}\left(\frac{x}{2}\right)^2(-4) + \binom{3}{2}\left(\frac{x}{2}\right)(-4)^2 + \binom{3}{3}(-4)^3$$
$$= \frac{x^3}{8} - 3 \cdot x^2 + 3 \cdot 8x - 64$$
$$= \frac{x^3}{8} - 3x^2 + 24x - 64$$

23. Using the binomial formula:

$$\left(\frac{x}{3}+\frac{y}{2}\right)^4 = \binom{4}{0}\left(\frac{x}{3}\right)^4 + \binom{4}{1}\left(\frac{x}{3}\right)^3\left(\frac{y}{2}\right) + \binom{4}{2}\left(\frac{x}{3}\right)^2\left(\frac{y}{2}\right)^2 + \binom{4}{3}\left(\frac{x}{3}\right)\left(\frac{y}{2}\right)^3 + \binom{4}{4}\left(\frac{y}{2}\right)^4$$

$$= \frac{x^4}{81} + 4\cdot\frac{x^3 y}{54} + 6\cdot\frac{x^2 y^2}{36} + 4\cdot\frac{xy^3}{24} + \frac{y^4}{16}$$

$$= \frac{x^4}{81} + \frac{2x^3 y}{27} + \frac{x^2 y^2}{6} + \frac{xy^3}{6} + \frac{y^4}{16}$$

25. Writing the first four terms:

$$\binom{9}{0}x^9 + \binom{9}{1}x^8(2) + \binom{9}{2}x^7(2)^2 + \binom{9}{3}x^6(2)^3$$

$$= x^9 + 9\cdot 2x^8 + 36\cdot 4x^7 + 84\cdot 8x^6$$

$$= x^9 + 18x^8 + 144x^7 + 672x^6$$

27. Writing the first four terms:

$$\binom{10}{0}x^{10} + \binom{10}{1}x^9(-y) + \binom{10}{2}x^8(-y)^2 + \binom{10}{3}x^7(-y)^3 = x^{10} - 10x^9 y + 45x^8 y^2 - 120x^7 y^3$$

29. Writing the first four terms:

$$\binom{25}{0}x^{25} + \binom{25}{1}x^{24}(3) + \binom{25}{2}x^{23}(3)^2 + \binom{25}{3}x^{22}(3)^3$$

$$= x^{25} + 25\cdot 3x^{24} + 300\cdot 9x^{23} + 2300\cdot 27x^{22}$$

$$= x^{25} + 75x^{24} + 2700x^{23} + 62100x^{22}$$

31. Writing the first four terms:

$$\binom{60}{0}x^{60} + \binom{60}{1}x^{59}(-2) + \binom{60}{2}x^{58}(-2)^2 + \binom{60}{3}x^{57}(-2)^3$$

$$= x^{60} - 60\cdot 2x^{59} + 1770\cdot 4x^{58} - 34220\cdot 8x^{57}$$

$$= x^{60} - 120x^{59} + 7080x^{58} - 273760x^{57}$$

33. Writing the first four terms:

$$\binom{18}{0}x^{18} + \binom{18}{1}x^{17}(-y) + \binom{18}{2}x^{16}(-y)^2 + \binom{18}{3}x^{15}(-y)^3 = x^{18} - 18x^{17}y + 153x^{16}y^2 - 816x^{15}y^3$$

35. Writing the first three terms: $\binom{15}{0}x^{15} + \binom{15}{1}x^{14}(1) + \binom{15}{2}x^{13}(1)^2 = x^{15} + 15x^{14} + 105x^{13}$

37. Writing the first three terms: $\binom{12}{0}x^{12} + \binom{12}{1}x^{11}(-y) + \binom{12}{2}x^{10}(-y)^2 = x^{12} - 12x^{11}y + 66x^{10}y^2$

39. Writing the first three terms:

$$\binom{20}{0}x^{20} + \binom{20}{1}x^{19}(2) + \binom{20}{2}x^{18}(2)^2 = x^{20} + 20\cdot 2x^{19} + 190\cdot 4x^{18} = x^{20} + 40x^{19} + 760x^{18}$$

41. Writing the first two terms: $\binom{100}{0}x^{100} + \binom{100}{1}x^{99}(2) = x^{100} + 100\cdot 2x^{99} = x^{100} + 200x^{99}$

43. Writing the first two terms: $\binom{50}{0}x^{50} + \binom{50}{1}x^{49}y = x^{50} + 50x^{49}y$

45. Finding the required term: $\binom{12}{8}(2x)^4(3y)^8 = 495\cdot 2^4\cdot 3^8 x^4 y^8 = 51,963,120x^4 y^8$

47. Finding the required term: $\binom{10}{4}x^6(-2)^4 = 210\cdot 16x^6 = 3360x^6$

49. Finding the required term: $\binom{12}{5}x^7(-2)^5 = -792\cdot 32x^7 = -25344x^7$

51. Finding the required term: $\binom{25}{2}x^{23}(-3y)^2 = 300 \cdot 9x^{23}y^2 = 2700x^{23}y^2$

53. Finding the required term: $\binom{20}{11}(2x)^9(5y)^{11} = \dfrac{20!}{11!9!}(2x)^9(5y)^{11}$

55. Writing the first three terms:
$$\binom{10}{0}(x^2y)^{10} + \binom{10}{1}(x^2y)^9(-3) + \binom{10}{2}(x^2y)^8(-3)^2$$
$$= x^{20}y^{10} - 10 \cdot 3x^{18}y^9 + 45 \cdot 9x^{16}y^8$$
$$= x^{20}y^{10} - 30x^{18}y^9 + 405x^{16}y^8$$

57. Finding the third term: $\binom{7}{2}\left(\frac{1}{2}\right)^5\left(\frac{1}{2}\right)^2 = 21 \cdot \frac{1}{128} = \frac{21}{128}$

59. Graphing the inequality:

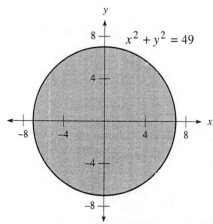

61. Solving the second equation for x yields $x = y + 2$. Substituting into the first equation:
$$(y+2)^2 + y^2 = 4$$
$$y^2 + 4y + 4 + y^2 = 4$$
$$2y^2 + 4y = 0$$
$$2y(y+2) = 0$$
$$y = 0, -2$$
$$x = 2, 0$$
The solutions are $(2,0)$ and $(0,-2)$.

63. They both equal 56.   **65.** They both equal 125,970.

67. Smaller triangles (Serpinski triangles) begin to emerge.

Chapter 11 Review

1. Writing the first four terms: 7,9,11,13   **3.** Writing the first four terms: 0,3,8,15

5. Writing the first four terms: 4,16,64,256   **7.** The general term is: $a_n = 3n - 1$

9. The general term is: $a_n = n^4$ **11.** The general term is: $a_n = \left(\frac{1}{2}\right)^n = 2^{-n}$

13. Expanding the sum: $\displaystyle\sum_{i=1}^{4}(2i+3) = 5+7+9+11 = 32$ **15.** Expanding the sum: $\displaystyle\sum_{i=2}^{3}\dfrac{i^2}{i+2} = 1 + \frac{9}{5} = \frac{14}{5}$

17. Expanding the sum: $\displaystyle\sum_{i=3}^{5}\left(4i+i^2\right) = 21 + 32 + 45 = 98$

19. Writing in summation notation: $\displaystyle\sum_{i=1}^{4} 3i$

21. Writing in summation notation: $\displaystyle\sum_{i=1}^{5} (2i+3)$

23. Writing in summation notation: $\displaystyle\sum_{i=1}^{4} \frac{1}{i+2}$

25. Writing in summation notation: $\displaystyle\sum_{i=1}^{3} (x-2i)$

27. The sequence is geometric.

29. The sequence is arithmetic.

31. The sequence is geometric.

33. The sequence is arithmetic.

35. Finding the general term: $a_n = 2+(n-1)3 = 2+3n-3 = 3n-1$. Now finding a_{20}: $a_{20} = 3(20)-1 = 59$

37. Finding a_{10}: $a_{10} = -2+9\cdot4 = 34$. Now finding S_{10}: $S_{10} = \frac{10}{2}(-2+34) = 5\cdot32 = 160$

39. First write the equations:
$$a_5 = a_1 + 4d \qquad a_8 = a_1 + 7d$$
$$21 = a_1 + 4d \qquad 33 = a_1 + 7d$$
We have the system of equations:
$$a_1 + 7d = 33$$
$$a_1 + 4d = 21$$
Subtracting yields:
$$3d = 12$$
$$d = 4$$
$$a_1 = 33 - 28 = 5$$
Now finding a_{10}: $a_{10} = 5+9\cdot4 = 41$

41. First write the equations:
$$a_4 = a_1 + 3d \qquad a_8 = a_1 + 7d$$
$$-10 = a_1 + 3d \qquad -18 = a_1 + 7d$$
We have the system of equations:
$$a_1 + 7d = -18$$
$$a_1 + 3d = -10$$
Subtracting yields:
$$4d = -8$$
$$d = -2$$
$$a_1 = -18 + 14 = -4$$
Now finding a_{20}: $a_{20} = -4+19\cdot(-2) = -42$. Finding S_{20}: $S_{20} = \frac{20}{2}(-4-42) = 10\cdot(-46) = -460$

43. Using $a_1 = 100$ and $d = -5$: $a_{40} = 100 + 39\cdot(-5) = 100 - 195 = -95$

45. The general term is: $a_n = 5(-2)^{n-1}$. Now finding a_{16}: $a_{16} = 5(-2)^{15} = -163,840$

47. Finding the sum: $S = \dfrac{-2}{1-\frac{1}{3}} = \dfrac{-2}{\frac{2}{3}} = -3$

49. Since $a_3 \cdot r = a_4$, we have:
$$12r = 24$$
$$r = 2$$
Finding the first term:
$$a_3 = a_1 r^2$$
$$12 = a_1 \cdot 2^2$$
$$12 = 4a_1$$
$$a_1 = 3$$
Now finding a_6: $a_6 = a_1 r^5 = 3\cdot2^5 = 96$

51. Evaluating: $\dbinom{8}{2} = \dfrac{8!}{6!2!} = \dfrac{8\cdot7}{2} = 28$

53. Evaluating: $\dbinom{6}{3} = \dfrac{6!}{3!3!} = \dfrac{6\cdot5\cdot4}{3\cdot2} = 20$

55. Evaluating: $\binom{10}{8} = \frac{10!}{8!2!} = \frac{10 \cdot 9}{2} = 45$

57. Using the binomial formula:

$$(x-2)^4 = \binom{4}{0}x^4 + \binom{4}{1}x^3(-2) + \binom{4}{2}x^2(-2)^2 + \binom{4}{3}x(-2)^3 + \binom{4}{4}(-2)^4$$

$$= x^4 - 4 \cdot 2x^3 + 6 \cdot 4x^2 - 4 \cdot 8x + 16$$

$$= x^4 - 8x^3 + 24x^2 - 32x + 16$$

59. Using the binomial formula:

$$(3x+2y)^3 = \binom{3}{0}(3x)^3 + \binom{3}{1}(3x)^2(2y) + \binom{3}{2}(3x)(2y)^2 + \binom{3}{3}(2y)^3$$

$$= 27x^3 + 3 \cdot 18x^2 y + 3 \cdot 12xy^2 + 8y^3$$

$$= 27x^3 + 54x^2 y + 36xy^2 + 8y^3$$

61. Using the binomial formula:

$$\left(\frac{x}{2}+3\right)^4 = \binom{4}{0}\left(\frac{x}{2}\right)^4 + \binom{4}{1}\left(\frac{x}{2}\right)^3(3) + \binom{4}{2}\left(\frac{x}{2}\right)^2(3)^2 + \binom{4}{3}\left(\frac{x}{2}\right)(3)^3 + \binom{4}{4}(3)^4$$

$$= \tfrac{1}{16}x^4 + 4 \cdot \tfrac{3}{8}x^3 + 6 \cdot \tfrac{9}{4}x^2 + 4 \cdot \tfrac{27}{2}x + 81$$

$$= \tfrac{1}{16}x^4 + \tfrac{3}{2}x^3 + \tfrac{27}{2}x^2 + 54x + 81$$

63. Writing the first three terms:

$$\binom{10}{0}x^{10} + \binom{10}{1}x^9(3y) + \binom{10}{2}x^8(3y)^2 = x^{10} + 10 \cdot 3x^9 y + 45 \cdot 9x^8 y^2 = x^{10} + 30x^9 y + 405x^8 y^2$$

65. Writing the first three terms:

$$\binom{11}{0}x^{11} + \binom{11}{1}x^{10}y + \binom{11}{2}x^9 y^2 = x^{11} + 11x^{10}y + 55x^9 y^2$$

67. Writing the first two terms: $\binom{16}{0}x^{16} + \binom{16}{1}x^{15}(-2y) = x^{16} - 16 \cdot 2x^{15}y = x^{16} - 32x^{15}y$

69. Writing the first two terms: $\binom{50}{0}x^{50} + \binom{50}{1}x^{49}(-1) = x^{50} - 50x^{49}$

71. Finding the sixth term: $\binom{10}{5}x^5(-3)^5 = 252 \cdot (-243)x^5 = -61,236x^5$

Chapters 1-11 Cumulative Review

1. Simplifying: $\dfrac{5(-6)+3(-2)}{4(-3)+3} = \dfrac{-30-6}{-12+3} = \dfrac{-36}{-9} = 4$

3. Simplifying: $\dfrac{18a^7 b^{-4}}{36a^2 b^{-8}} = \tfrac{1}{2}a^{7-2}b^{-4+8} = \tfrac{1}{2}a^5 b^4 = \dfrac{a^5 b^4}{2}$

5. Simplifying: $8^{-2/3} = \left(8^{1/3}\right)^{-2} = 2^{-2} = \tfrac{1}{4}$

7. Factoring completely: $ab^3 + b^3 + 6a + 6 = b^3(a+1) + 6(a+1) = (a+1)(b^3 + 6)$

9. Solving the equation:

$$6 - 2(5x-1) + 4x = 20$$
$$6 - 10x + 2 + 4x = 20$$
$$-6x + 8 = 20$$
$$-6x = 12$$
$$x = -2$$

11. Solving the equation:

$$(x+1)(x+2) = 12$$
$$x^2 + 3x + 2 = 12$$
$$x^2 + 3x - 10 = 0$$
$$(x+5)(x-2) = 0$$
$$x = -5, 2$$

13. Solving the equation:
$$t - 6 = \sqrt{t - 4}$$
$$(t - 6)^2 = t - 4$$
$$t^2 - 12t + 36 = t - 4$$
$$t^2 - 13t + 40 = 0$$
$$(t - 5)(t - 8) = 0$$
$$t = 8 \qquad (t = 5 \text{ does not check})$$

15. Solving the equation:
$$8t^3 - 27 = 0$$
$$(2t - 3)\left(4t^2 + 6t + 9\right) = 0$$
$$t = \frac{3}{2}, \frac{-6 \pm \sqrt{36 - 144}}{8} = \frac{-6 \pm 6i\sqrt{3}}{8} = \frac{-3 \pm 3i\sqrt{3}}{4}$$

17. Solving the inequality:
$$-3y - 2 < 7$$
$$-3y < 9$$
$$y > -3$$
Graphing the solution set:

19. Graphing the line:

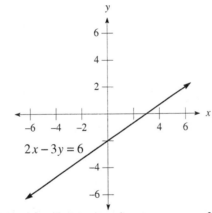

$$2x - 3y = 6$$

21. First complete the square: $y = x^2 - 2x - 3 = \left(x^2 - 2x + 1\right) - 1 - 3 = (x - 1)^2 - 4$
Graphing the curve:

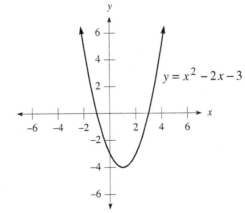

$$y = x^2 - 2x - 3$$

23. Multiply the first equation by 2 and the second equation by 3:
$$10x - 6y = -8$$
$$3x + 6y = 21$$
Adding yields:
$$13x = 13$$
$$x = 1$$
Substituting to find y:
$$1 + 2y = 7$$
$$2y = 6$$
$$y = 3$$
The solution is $(1,3)$.

25. Multiplying: $\dfrac{3y^2 - 3y}{3y - 12} \cdot \dfrac{y^2 - 2y - 8}{y^2 + 3y + 2} = \dfrac{3y(y-1)}{3(y-4)} \cdot \dfrac{(y-4)(y+2)}{(y+2)(y+1)} = \dfrac{y(y-1)}{y+1}$

27. Multiplying: $(2 + 3i)(1 - 4i) = 2 + 3i - 8i - 12i^2 = 2 - 5i + 12 = 14 - 5i$

29. Combining radicals: $4\sqrt{50} + 3\sqrt{8} = 4 \cdot 5\sqrt{2} + 3 \cdot 2\sqrt{2} = 20\sqrt{2} + 6\sqrt{2} = 26\sqrt{2}$

31. Finding the inverse:
$$4y + 1 = x$$
$$4y = x - 1$$
$$y = \frac{x-1}{4}$$
$$f^{-1}(x) = \frac{x-1}{4}$$

33. Finding the logarithm: $\log_6 14 = \dfrac{\log 14}{\log 6} \approx 1.47$

35. The general term is: $a_n = 16\left(\frac{1}{2}\right)^{n-1}$

37. Expanding the sum: $\displaystyle\sum_{i=1}^{5}(2i - 1) = 1 + 3 + 5 + 7 + 9 = 25$

39. First find the slope: $m = \dfrac{-3-5}{6-2} = \dfrac{-8}{4} = -2$. Using the point-slope formula:
$$y - 5 = -2(x - 2)$$
$$y - 5 = -2x + 4$$
$$y = -2x + 9$$

41. Completing the square:
$$x^2 - 10x + y^2 + 4y = -25$$
$$\left(x^2 - 10x + 25\right) + \left(y^2 + 4y + 4\right) = -25 + 25 + 4$$
$$(x - 5)^2 + (y + 2)^2 = 4$$
The center is $(5,-2)$ and the radius is 2.

43. First find the determinants:
$$D = \begin{vmatrix} 3 & -5 \\ 2 & 4 \end{vmatrix} = 12 + 10 = 22$$
$$D_x = \begin{vmatrix} 2 & -5 \\ 1 & 4 \end{vmatrix} = 8 + 5 = 13$$
$$D_y = \begin{vmatrix} 3 & 2 \\ 2 & 1 \end{vmatrix} = 3 - 4 = -1$$
Using Cramer's rule:
$$x = \frac{D_x}{D} = \frac{13}{22} \qquad\qquad y = \frac{D_y}{D} = -\frac{1}{22}$$
The solution is $\left(\frac{13}{22}, -\frac{1}{22}\right)$.

45. Finding the value: $(-7-4)+(-8)=-11-8=-19$

47. Let x and $5x$ represent the two angles. The equation is:
$$x+5x=180$$
$$6x=180$$
$$x=30$$
The angles are $30°$ and $150°$.

49. Let x and $3x$ represent the two numbers. The equation is:
$$\frac{1}{x}+\frac{1}{3x}=\frac{4}{3}$$
$$3x\left(\frac{1}{x}+\frac{1}{3x}\right)=3x\left(\frac{4}{3}\right)$$
$$3+1=4x$$
$$4=4x$$
$$x=1$$
The numbers are 1 and 3.

Chapter 11 Test

1. The first five terms are: $-2,1,4,7,10$

2. The first five terms are: $3,7,11,15,19$

3. The first five terms are: $2,5,10,17,26$

4. The first five terms are: $2,16,54,128,250$

5. The first five terms are: $2,\frac{3}{4},\frac{4}{9},\frac{5}{16},\frac{6}{25}$

6. The first five terms are: $4,-8,16,-32,64$

7. Writing the general term: $a_n=4n+2$

8. Writing the general term: $a_n=2^{n-1}$

9. Writing the general term: $a_n=\left(\frac{1}{2}\right)^n=2^{-n}$

10. Writing the general term: $a_n=(-3)^n$

11. **a.** Expanding the sum: $\displaystyle\sum_{i=1}^{5}(5i+3)=8+13+18+23+28=90$

 b. Expanding the sum: $\displaystyle\sum_{i=3}^{5}(2^i-1)=7+15+31=53$

 c. Expanding the sum: $\displaystyle\sum_{i=2}^{6}(i^2+2i)=8+15+24+35+48=130$

12. First write the equations:
$$a_5=a_1+4d \qquad a_9=a_1+8d$$
$$11=a_1+4d \qquad 19=a_1+8d$$
We have the system of equations:
$$a_1+8d=19$$
$$a_1+4d=11$$
Subtracting yields:
$$4d=8$$
$$d=2$$
$$a_1=19-8\cdot2=3$$

13. Since $a_3\cdot r\cdot r=a_5$, we have:
$$18\cdot r^2=162$$
$$r^2=9$$
$$r=\pm3$$
Since $a_2\cdot r=a_3$, $a_2=\pm6$.

14. Using $a_1=5$ and $d=6$: $a_{10}=5+9\cdot6=59$. Now finding the sum: $S_{10}=\frac{10}{2}(5+59)=5\cdot64=320$

15. Using $a_1=25$ and $d=-5$: $a_{10}=25+9\cdot(-5)=-20$. Now finding the sum: $S_{10}=\frac{10}{2}(25-20)=5\cdot5=25$

16. Using $a_1 = 3$ and $r = 2$: $S_{50} = \dfrac{3\left(2^{50} - 1\right)}{2 - 1} = 3\left(2^{50} - 1\right)$

17. Using $a_1 = \frac{1}{2}$ and $r = \frac{1}{3}$: $S = \dfrac{\frac{1}{2}}{1 - \frac{1}{3}} = \dfrac{\frac{1}{2}}{\frac{2}{3}} = \frac{1}{2} \cdot \frac{3}{2} = \frac{3}{4}$

18. Using the binomial formula:

$$(x - 3)^4 = \binom{4}{0}x^4 + \binom{4}{1}x^3(-3) + \binom{4}{2}x^2(-3)^2 + \binom{4}{3}x(-3)^3 + \binom{4}{4}(-3)^4$$
$$= x^4 - 4 \cdot 3x^3 + 6 \cdot 9x^2 - 4 \cdot 27x + 81$$
$$= x^4 - 12x^3 + 54x^2 - 108x + 81$$

19. Using the binomial formula:

$$(2x - 1)^5 = \binom{5}{0}(2x)^5 + \binom{5}{1}(2x)^4(-1) + \binom{5}{2}(2x)^3(-1)^2 + \binom{5}{3}(2x)^2(-1)^3 + \binom{5}{4}(2x)(-1)^4 + \binom{5}{5}(-1)^5$$
$$= 32x^5 - 5 \cdot 16x^4 + 10 \cdot 8x^3 - 10 \cdot 4x^2 + 5 \cdot 2x - 1$$
$$= 32x^5 - 80x^4 + 80x^3 - 40x^2 + 10x - 1$$

20. Finding the first three terms: $\binom{20}{0}x^{20} + \binom{20}{1}x^{19}(-1) + \binom{20}{2}x^{18}(-1)^2 = x^{20} - 20x^{19} + 190x^{18}$

21. Finding the sixth term: $\binom{8}{5}(2x)^3(-3y)^5 = -56 \cdot 1944x^3y^5 = -108,864x^3y^5$

Appendices

Appendix A   Synthetic Division

1. Using synthetic division:

$$\begin{array}{r|rrr} -2 & 1 & -5 & 6 \\ & & -2 & 14 \\ \hline & 1 & -7 & 20 \end{array}$$

The quotient is $x - 7 + \dfrac{20}{x+2}$.

3. Using synthetic division:

$$\begin{array}{r|rrr} 1 & 3 & -4 & 1 \\ & & 3 & -1 \\ \hline & 3 & -1 & 0 \end{array}$$

The quotient is $3x - 1$.

5. Using synthetic division:

$$\begin{array}{r|rrrr} 2 & 1 & 2 & 3 & 4 \\ & & 2 & 8 & 22 \\ \hline & 1 & 4 & 11 & 26 \end{array}$$

The quotient is $x^2 + 4x + 11 + \dfrac{26}{x-2}$.

7. Using synthetic division:

$$\begin{array}{r|rrrr} 3 & 3 & -1 & 2 & 5 \\ & & 9 & 24 & 78 \\ \hline & 3 & 8 & 26 & 83 \end{array}$$

The quotient is $3x^2 + 8x + 26 + \dfrac{83}{x-3}$.

9. Using synthetic division:

$$\begin{array}{r|rrrr} 1 & 2 & 0 & 1 & -3 \\ & & 2 & 2 & 3 \\ \hline & 2 & 2 & 3 & 0 \end{array}$$

The quotient is $2x^2 + 2x + 3$.

11. Using synthetic division:

$$\begin{array}{r|rrrrr} -4 & 1 & 0 & 2 & 0 & 1 \\ & & -4 & 16 & -72 & 288 \\ \hline & 1 & -4 & 18 & -72 & 289 \end{array}$$

The quotient is $x^3 - 4x^2 + 18x - 72 + \dfrac{289}{x+4}$.

13. Using synthetic division:

$$\begin{array}{r|rrrrrr} 2 & 1 & -2 & 1 & -3 & -1 & 1 \\ & & 2 & 0 & 2 & -2 & -6 \\ \hline & 1 & 0 & 1 & -1 & -3 & -5 \end{array}$$

The quotient is $x^4 + x^2 - x - 3 - \dfrac{5}{x-2}$.

15. Using synthetic division:

$$\begin{array}{r|rrr} 1 & 1 & 1 & 1 \\ & & 1 & 2 \\ \hline & 1 & 2 & 3 \end{array}$$

The quotient is $x + 2 + \dfrac{3}{x-1}$.

17. Using synthetic division:

$$\begin{array}{r|rrrrr} -1 & 1 & 0 & 0 & 0 & -1 \\ & & -1 & 1 & -1 & 1 \\ \hline & 1 & -1 & 1 & -1 & 0 \end{array}$$

The quotient is $x^3 - x^2 + x - 1$.

19. Using synthetic division:

$$\begin{array}{c|cccc} 1 & 1 & 0 & 0 & -1 \\ & & 1 & 1 & 1 \\ \hline & 1 & 1 & 1 & 0 \end{array}$$

The quotient is $x^2 + x + 1$.

Appendix B   Matrix Solutions to Linear Systems

1. Form the augmented matrix: $\begin{bmatrix} 1 & 1 & | & 5 \\ 3 & -1 & | & 3 \end{bmatrix}$

Add –3 times row 1 to row 2: $\begin{bmatrix} 1 & 1 & | & 5 \\ 0 & -4 & | & -12 \end{bmatrix}$

Dividing row 2 by –4: $\begin{bmatrix} 1 & 1 & | & 5 \\ 0 & 1 & | & 3 \end{bmatrix}$

So $y = 3$. Substituting to find x:

$$x + 3 = 5$$
$$x = 2$$

The solution is $(2,3)$.

3. Using the second equation as row 1, form the augmented matrix: $\begin{bmatrix} -1 & 1 & | & -1 \\ 3 & -5 & | & 7 \end{bmatrix}$

Add 3 times row 1 to row 2: $\begin{bmatrix} -1 & 1 & | & -1 \\ 0 & -2 & | & 4 \end{bmatrix}$

Dividing row 2 by –2 and row 1 by –1: $\begin{bmatrix} 1 & -1 & | & 1 \\ 0 & 1 & | & -2 \end{bmatrix}$

So $y = -2$. Substituting to find x:

$$x + 2 = 1$$
$$x = -1$$

The solution is $(-1,-2)$.

5. Form the augmented matrix: $\begin{bmatrix} 2 & -8 & | & 6 \\ 3 & -8 & | & 13 \end{bmatrix}$

Dividing row 1 by 2: $\begin{bmatrix} 1 & -4 & | & 3 \\ 3 & -8 & | & 13 \end{bmatrix}$

Add –3 times row 1 to row 2: $\begin{bmatrix} 1 & -4 & | & 3 \\ 0 & 4 & | & 4 \end{bmatrix}$

Dividing row 2 by 4: $\begin{bmatrix} 1 & -4 & | & 3 \\ 0 & 1 & | & 1 \end{bmatrix}$

So $y = 1$. Substituting to find x:

$$x - 4 = 3$$
$$x = 7$$

The solution is $(7,1)$.

7. Form the augmented matrix: $\begin{bmatrix} 1 & 1 & 1 & | & 4 \\ 1 & -1 & 2 & | & 1 \\ 1 & -1 & -1 & | & -2 \end{bmatrix}$

Add –1 times row 1 to both row 2 and row 3: $\begin{bmatrix} 1 & 1 & 1 & | & 4 \\ 0 & -2 & 1 & | & -3 \\ 0 & -2 & -2 & | & -6 \end{bmatrix}$

Multiply row 2 by –1 and add it to row 3: $\begin{bmatrix} 1 & 1 & 1 & | & 4 \\ 0 & 2 & -1 & | & 3 \\ 0 & 0 & -3 & | & -3 \end{bmatrix}$

Divide row 3 by –3: $\begin{bmatrix} 1 & 1 & 1 & | & 4 \\ 0 & 2 & -1 & | & 3 \\ 0 & 0 & 1 & | & 1 \end{bmatrix}$

So $z = 1$. Substituting to find y:
$$2y - 1 = 3$$
$$2y = 4$$
$$y = 2$$

Substituting to find x:
$$x + 2 + 1 = 4$$
$$x = 1$$
The solution is (1,2,1).

9. Form the augmented matrix: $\begin{bmatrix} 1 & 2 & 1 & | & 3 \\ 2 & -1 & 2 & | & 6 \\ 3 & 1 & -1 & | & 5 \end{bmatrix}$

Add –2 times row 1 to row 2 and –3 times row 1 to row 3: $\begin{bmatrix} 1 & 2 & 1 & | & 3 \\ 0 & -5 & 0 & | & 0 \\ 0 & -5 & -4 & | & -4 \end{bmatrix}$

Dividing row 2 by –5: $\begin{bmatrix} 1 & 2 & 1 & | & 3 \\ 0 & 1 & 0 & | & 0 \\ 0 & -5 & -4 & | & -4 \end{bmatrix}$

So $y = 0$. Substituting to find z:
$$-4z = -4$$
$$z = 1$$
Substituting to find x:
$$x + 0 + 1 = 3$$
$$x = 2$$
The solution is (2,0,1).

11. Form the augmented matrix: $\begin{bmatrix} 1 & 2 & 0 & | & 3 \\ 0 & 1 & 1 & | & 3 \\ 4 & 0 & -1 & | & 2 \end{bmatrix}$

Add –4 times row 1 to row 3: $\begin{bmatrix} 1 & 2 & 0 & | & 3 \\ 0 & 1 & 1 & | & 3 \\ 0 & -8 & -1 & | & -10 \end{bmatrix}$

Add 8 times row 2 to row 3: $\begin{bmatrix} 1 & 2 & 0 & | & 3 \\ 0 & 1 & 1 & | & 3 \\ 0 & 0 & 7 & | & 14 \end{bmatrix}$

Dividing row 3 by 7: $\begin{bmatrix} 1 & 2 & 0 & | & 3 \\ 0 & 1 & 1 & | & 3 \\ 0 & 0 & 1 & | & 2 \end{bmatrix}$

So $z = 2$. Substituting to find y:

$y + 2 = 3$

$y = 1$

Substituting to find x:

$x + 2 = 3$

$x = 1$

The solution is $(1,1,2)$.

13. Form the augmented matrix: $\begin{bmatrix} 1 & 3 & 0 & | & 7 \\ 3 & 0 & -4 & | & -8 \\ 0 & 5 & -2 & | & -5 \end{bmatrix}$

Add –3 times row 1 to row 2: $\begin{bmatrix} 1 & 3 & 0 & | & 7 \\ 0 & -9 & -4 & | & -29 \\ 0 & 5 & -2 & | & -5 \end{bmatrix}$

Add 2 times row 3 to row 2: $\begin{bmatrix} 1 & 3 & 0 & | & 7 \\ 0 & 1 & -8 & | & -39 \\ 0 & 5 & -2 & | & -5 \end{bmatrix}$

Add –5 times row 2 to row 3: $\begin{bmatrix} 1 & 3 & 0 & | & 7 \\ 0 & 1 & -8 & | & -39 \\ 0 & 0 & 38 & | & 190 \end{bmatrix}$

Dividing row 3 by 38: $\begin{bmatrix} 1 & 3 & 0 & | & 7 \\ 0 & 1 & -8 & | & -39 \\ 0 & 0 & 1 & | & 5 \end{bmatrix}$

So $z = 5$. Substituting to find y:

$y - 40 = -39$

$y = 1$

Substituting to find x:

$x + 3 = 7$

$x = 4$

The solution is $(4,1,5)$.

15.   Form the augmented matrix: $\begin{bmatrix} \frac{1}{3} & \frac{1}{5} & 2 \\ \frac{1}{3} & -\frac{1}{2} & -\frac{1}{3} \end{bmatrix}$

Multiplying row 1 by 15 and row 2 by 6: $\begin{bmatrix} 5 & 3 & 30 \\ 2 & -3 & -2 \end{bmatrix}$

Dividing row 2 by 2: $\begin{bmatrix} 5 & 3 & 30 \\ 1 & -\frac{3}{2} & -1 \end{bmatrix}$

Add -5 times row 2 to row 1: $\begin{bmatrix} 0 & \frac{21}{2} & 35 \\ 1 & -\frac{3}{2} & -1 \end{bmatrix}$

Multiplying row 1 by $\frac{2}{21}$: $\begin{bmatrix} 0 & 1 & \frac{10}{3} \\ 1 & -\frac{3}{2} & -1 \end{bmatrix}$

So $y = \frac{10}{3}$. Substituting to find x:

$$x - \frac{3}{2}\left(\frac{10}{3}\right) = -1$$
$$x - 5 = -1$$
$$x = 4$$

The solution is $\left(4, \frac{10}{3}\right)$.

17.   Form the augmented matrix: $\begin{bmatrix} 2 & -3 & 4 \\ 4 & -6 & 4 \end{bmatrix}$

Add -2 times row 1 to row 2: $\begin{bmatrix} 2 & -3 & 4 \\ 0 & 0 & -4 \end{bmatrix}$

The second row states that $0 = -4$, which is false. There is no solution.

19.   Form the augmented matrix: $\begin{bmatrix} -6 & 4 & 8 \\ -3 & 2 & 4 \end{bmatrix}$

Divide row 1 by -2: $\begin{bmatrix} 3 & -2 & -4 \\ -3 & 2 & 4 \end{bmatrix}$

Adding row 1 to row 2: $\begin{bmatrix} 3 & -2 & -4 \\ 0 & 0 & 0 \end{bmatrix}$

The second row states that $0 = 0$, which is true. The system is dependent.

Appendix C   Conditional Statements

1.   The hypothesis is "you argue for your limitations", the conclusion is "they are yours".
3.   The hypothesis is "x is an even number", the conclusion is "x is divisible by 2".
5.   The hypothesis is "a triangle is equilateral", the conclusion is "all of its angles are equal".
7.   The hypothesis is "$x + 5 = -2$", the conclusion is "$x = -7$".
9.   The converse is "If $a^2 = 64$, then $a = 8$", the inverse is "If $a \neq 8$, then $a^2 \neq 64$", and the contrapositive is "If $a^2 \neq 64$, then $a \neq 8$".
11.   The converse is "If $a = b$, then $\frac{a}{b} = 1$", the inverse is "If $\frac{a}{b} \neq 1$, then $a \neq b$", and the contrapositive is "If $a \neq b$, then $\frac{a}{b} \neq 1$".
13.   The converse is "If it is a rectangle, then it is a square", the inverse is "If it is not a square, then it is not a rectangle", and the contrapositive is "If it is not a rectangle, then it is not a square".
15.   The converse is "If good is not enough, then better is possible", the inverse is "If better is not possible, then good is enough", and the contrapositive is "If good is enough, then better is not possible".

17. If E, then F.

19. If it is misery, then it loves company.

21. If the wheel is squeaky, then it gets the grease.

23. (c)

25. (a)

27. (c)

29. (b)